On the Moon
The *Apollo* Journals

Grant Heiken and Eric Jones

On the Moon

The *Apollo* Journals

 **Springer**

Published in association with
Praxis Publishing
Chichester, UK

Dr Grant Heiken
Independent Consultant
Freeland
Washington
USA

Dr Eric Jones
Wodonga
Victoria
Australia

SPRINGER–PRAXIS BOOKS IN SPACE EXPLORATION
SUBJECT *ADVISORY EDITOR*: John Mason, M.Sc., B.Sc., Ph.D.

ISBN 978-0-387-48939-1 Springer Berlin Heidelberg New York

Springer is part of Springer-Science + Business Media (springer.com)

Library of Congress Control Number: 2007920176

Cover design: Jim Wilkie
Copy editing: Jody Heiken
Typesetting: Originator Publishing Services, Gt Yarmouth, Norfolk, UK

Printed on acid-free paper

Contents

Contents

Foreword

"A ridge of mountains of prodigious height which appear to consist of nothing but barren rocks, covered in many places with large patches of snow which appear to have lain there since creation. No country upon earth can appear with a more rugged and barren aspect than this doth."

It has been customary over the centuries during voyages of exploration for the Captain of the ship to record daily events of significance in the "Captain's Log." The above entry was made by Captain James Cook in the log of His Majesty's Bark, *Endeavour*, during February 1770. It reflects his amazement at seeing "mountains of strange, volcanic appearance totally unlike anything in Europe ..." Captain Cook, during his 3-year voyage of scientific discovery in the South Pacific, was the first known explorer to view the Southern Alps in New Zealand (Syme, Ronald, *The Travels of Captain Cook*, McGraw-Hill, New York, 1971).

Just over two hundred years later, I too had the opportunity to view an amazing and unfamiliar scene of "mountains of strange, volcanic appearance totally unlike anything in Earth ..." As our spaceship *Endeavor* circled the Moon, I stared at our spectacular landing site in the vast Apennine Mountains. But unlike Capt Cook, I had no time to record my observations in a Captain's Log. Every minute of our 12-day voyage was packed with tasks, objectives, experiments, observations, investigations, and a whole manner of activities necessary to complete the first extended scientific exploration of the Moon. As an illustration of contrasts, every 15-minute period during our 1971 voyage of Apollo 15 was comparable to a full day (24 hours) during Cook's 3-year voyage of 1768–1771.

Before Captain James Cook embarked on his historic first voyage to the South Pacific and Australia, he invested 3 months in its preparation, using his crew of 70 to prepare the bark *Endeavour*; gather equipment, provisions, and instruments; and generally plan the expedition. After departing from Plymouth on 26 August 1768, he and his crew of "94 persons" spent almost 3 years conducting the first truly scientific expedition by sea, arriving back in England on 15 July 1771. His original "Journal" was not published until 13 years after he returned (London, W. and A. Strahan, 1784). However, it was not until 171 years later that a comprehensive analysis of his entries, including a detailed description of his ship, equipment, and instruments, scientific results, and anecdotes of interest, was published (Beaglehole, *The Journals of Captain James Cook on His Voyages of Discovery*, Cambridge, University Press, Vol. I, 1955).

In comparison, NASA-of-Apollo invested 20 months in the preparation of the Apollo 15 mission – using more than 100,000 people to prepare the launch vehicles and spacecraft; gather equipment, provisions, and instruments; and generally plan the expedition. After departing Cape Kennedy on 26 July 1971, our crew of three journeyed to the Moon, spent 3 days in lunar orbit, three days on the surface, and returned to Earth on 7 August 1917. However, because of the enormous amount of intellectual capital invested in preparing the mission (100,000 times 20 months) as well as the added dimensions of "modern technology," the exploratory information gathered during the mission could not be comprehensively recounted in brief daily journal entries (as per Captain Cook). Only now, with the time available to compile and analyze these highly compressed minute-by-minute events and activities, do we see the emergence of comprehensive records of the mission to include personal observations, technical explanations, scientific results, and anecdotes of interest.

Fortunately, Eric Jones and Grant Heiken have taken it upon themselves to integrate all of these features in the preparation of *On The Moon: The Apollo Journals*. And thank goodness they have ... !!

In 1989, inspired by the work of 20th century New Zealand historian J.C. Beaglehole, Eric Jones began to review and correct the recorded transcripts of surface activities during the six Apollo lunar landing missions. Using the transcripts as a guide, he subsequently conducted over 450 hours (56 days) of personal interviews with 10 of the 12 astronauts who explored the surface of the Moon. He then intertwined these personal observations, explanations, descriptions and memorable anecdotes with the edited transcripts. And finally, he integrated with the narrative the 6,998 photographs taken by the astronauts to document the spectacular scenes and the *in-situ* locations of lunar samples returned to Earth. The result is the precise and highly detailed *Apollo Lunar Surface Journals* (available on the Internet only: www.ha.nasa.gov/alsj), the equivalent of four 600-page books (2,400 pages). Because of worldwide interest, the "*ALSJ*" has become a "living" record; updated frequently with new analyses and new findings and continuous probes of "what did you mean when you said that at (some particular) GET (Ground Elapsed Time)."

Eric has now joined with Grant Heiken to write *On the Moon: The Apollo Journals*, a combination of the major sections in the *ALSJ* with summaries of the scientific data and technical results. Grant, whom I have also known for many years, is a renowned geologist and co-author of the *Lunar Sourcebook – A User's Guide to the Moon*, a comprehensive compendium of lunar science and engineering data. During Apollo, Grant was a member of the Lunar Sample Examination Team and a geology instructor in the astronaut-training program. The result of their collaboration, *On the Moon: The Apollo Journals*, is essentially a condensed combination of the *ALSJ* and the *Lunar Sourcebook*, an invaluable record of the most complex and carefully planned expeditions in the history of exploration. Other than the 381 kilograms of lunar rocks and soil returned by the Apollo crews, this unique treasure is a major source of accurate information and careful analysis of the 80 hours of actual human physical exploration of the lunar surface. Basically, *On The Moon* is intended to complement the *ALSJ* and the *Lunar Sourcebook* by providing the reader with a flavor of the activities and events on the lunar surface, what was involved in getting the work done, and some fundamentals regarding the resulting science.

In preparing the *Journals of Captain Cook*, J.C. Beaglehole became the foremost authority on European exploration of the Pacific, and especially the three voyages of Capt Cook. It can now be argued that Eric Jones and Grant Heiken have become the foremost authorities on the six Apollo expeditions to the surface of the Moon. Without their tireless, determined, and dedicated efforts, the recorded legacy of human lunar exploration would be left in random piles of decaying paper stored in some remote warehouse long since abandoned by NASA.

Most importantly, this superlative book and the supporting *ALSJ* and *Lunar Sourcebook* record and elucidate the results of the bold decisions to send men to the Moon – six times ... !! – with, by today's standards, crude technology, limited resources, and the many unknowns and uncertainties of, for the first time, humans living and working on a hostile "planet" 240,000 miles from our comfortable home on Earth. However, the real significance of this narrative may very well be its value to future planetary explorers, including robotic explorers and even "virtual" explorers. When these future explorers look back, they will surely ask: what did we learn from Apollo? How does it apply to the future? When will such an adventure happen again?

In any event, it should be pointed out, for whatever meaning it may have today, that in historical equivalents the age of planetary exploration is barely post-Colombian in its development. Columbus has returned to Spain and perhaps the Earth is not flat after all. But Cabot and Magellan have yet to make their voyages – and Capt. James Cook of the *Endeavour* has not even been born.

And until we humans once again embark on a venture of planetary exploration, historians, writers, journalists, and the human inquisitive will continue to probe and learn from the six remarkable Apollo missions that explored the surface of the Moon. And their root source will surely be *On The Moon: The Apollo Journals* and its treasured companions, *The Apollo Lunar Surface Journals* and the *Lunar Sourcebook*, which are exceptional legacies of the first human expeditions to another world.

David R. Scott
Commander, Apollo 15
Los Angeles
December 2006

Preface

On the Moon – The Apollo Journals describes both the challenges and the exhilaration that the Apollo astronauts experienced on the Moon. The pages provide a closeup view of the harsh lunar landscape, which has a stark beauty all its own. The chapters also reveal the six teams' courageous, sometimes creative, and occasionally humorous adaptation to the field conditions on another planet. *On the Moon* is based on selected transcripts from Apollo crew communications both with Earth and between themselves while on the lunar surface. The astronauts' recorded commentary from the missions is interspersed with commentary about the landing sites and the work, including discussions about the importance of their specific tasks and how the results from their experiences changed the way we look at space exploration. Extracts from post-mission and more recent interviews conducted with the astronauts in the early 1990s highlight their thoughts not only during their missions but also after more than 25 years of reflection.

Many people who are interested in space exploration are too young to remember much about the events that led to the Apollo Program and the global excitement that accompanied the missions. That first lunar landing transcended all political, economic, and social borders. For a brief moment, the world was united by a grandness of vision, the courage of the voyagers, and the wonder of their accomplishments. *On the Moon* is a call to memory for many, but it also tells a new generation about these events, how they affected our understanding of the universe, and the value of exploration to the human psyche.

Explorers from the Earth will return to the Moon – three nations have already announced their plans to make another journey. But before those next trips, it is vitally important to revisit the roots of mankind's boldest exploration, examining the astronauts' observations, tallying the accuracy of our assumptions, and gaining new perspective for this century's missions.

Author Eric Jones' interest in exploration of the Earth's Moon began on July 20, 1969, when the lunar module *Eagle* landed on the Sea of Tranquillity. A graduate student at the University of Wisconsin, he invited a few friends to watch the lunar landing in his small apartment. More than 45 people showed up with beer and food; conversations grew louder until the TV signal from the Moon started. From then on, the moonwalk had everyone's undivided attention. Co-author Grant Heiken was already working in the Lunar Receiving Laboratory at the NASA Manned Spacecraft Center as a junior member of the Lunar Sample Preliminary Examination Team and part of the astronaut-training team. He stayed with NASA through Apollo and Skylab and later moved on to the Los Alamos Scientific Laboratory where the two authors met. Heiken later co-edited *Lunar Sourcebook – A User's Guide to the Moon* (Cambridge University Press, 1991) with David Vaniman and Bevan French. Jones co-edited *Interstellar Migration and the Human Experience* (University of California Press, 1985) with Ben Finney. Jones is the originator and leader of the international team that created the *Apollo Lunar Surface Journal* (ALSJ), a comprehensive online, evolving history of the Apollo program that is based on transcripts from the Apollo missions and interviews with the astronauts (http://www.hq.nasa.gov/office/pao/History/alsj/frame.html). *On the Moon – The Apollo Journals* is based on material in that Journal.

The ALSJ would not exist without the participation of ten of the twelve moon-walkers. Their active support of the effort to accurately document the lunar surface operations conducted during Apollo has been invaluable. Apollo 15 Commander Dave Scott provided very useful comments on portions of the draft of *On the Moon*.

The authors profusely thank Jody Heiken, who agreed, on very short notice, to edit this book and Dave Scott for contributing the humbling foreword. The many collaborators who work with Eric Jones on the Apollo Lunar Surface Journal are to be thanked, especially ALSJ Co-Editor Ken Glover, Ulli Lotzmann, who created the marvelous drawings featured in Chapter 2 and provided telescopic photographs of the Moon, and Kipp Teague, editor of the online Project Apollo Archive (http://www.apolloarchive.com/apollo_archive.html). Clive Horwood and his team at Praxis have been both marvelously quick and thoroughly professional.

And, finally, our thanks to Jody and Di for their ongoing patience and under-standing of their husbands' lunar mania.

Grant Heiken
Freeland, Washington, USA

Eric Jones
Wodonga, Victoria, Australia

(*left*) Moon photographed by Ulli Lotzmann (Marburg, Germany) using an 8-inch Newtonian telescope with a 1.2-m focal length. The box indicates area shown on the right. (*right*) Closeup of Tranquillity Base taken by Lotzmann with a 7-inch refractor and an effective 12-m focal length.

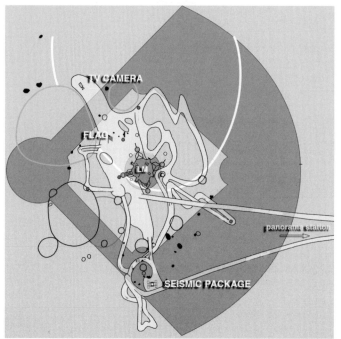

Site Map. This schematic Apollo 11 traverse map was adapted from the *Apollo 11 Preliminary Science Report* by Thomas Schwagmeier. For scale, the site has been superimposed on a baseball diamond, with the LM on the pitcher's mound. The light gray paths are routes taken by the crew as they moved about during the EVAs. The path at lower right was the route taken by Armstrong to take a photographic panorama of the site. Ovals are edges of subdued craters, and dark blotches represent large rocks.

1

Adapting to a New World

Apollo 11's lunar landing was undoubtedly one of the most memorable events of the 20th Century. The Apollo Program, created to send a man to the Moon, was the culmination of a promise made by President John Kennedy during the heat of the Cold War, but ultimately, it became the most important scientific expedition of the century.

The purpose of the first lunar landing was to see if it could be done. The mission was a success because of the ability and bravery of the crew and the dedication and determination of the 400,000 people who worked on Apollo. It was accomplished with clever engineering coupled with the flexibility made possible by having a pilot in the spacecraft. The landings were successful in part because of the simplicity of the technology: astoundingly, the computer on board the spacecraft was equivalent to a calculator readily available for pocket change in a grocery store only 40 years later. With a tight deadline, the men and women of Apollo had to get it right and didn't have time to tinker and improve. Apollo was truly a team effort, led by individuals who knew every subsystem and how they all were linked.

After the lunar module (LM) had successfully landed on the lunar surface, the crew of Neil Armstrong and Buzz Aldrin faced the reality of exploring another planet – one with no air or water, serious radiation hazards, and only one-sixth of the Earth's gravity.

In all directions, the land was West Texas flat. The horizon was broken here and there by the subtle rims of distant craters. In the middle distance, Armstrong and Aldrin could see boulders and ridges, some of the latter perhaps 20 or 30 feet high. Close at hand, a hodgepodge of craters pockmarked the surface; small rocks and pebbles were scattered everywhere. It was a flat, level site but, like Australia's Nullarbor (Latin for "Treeless") Plain, the coastal plains of North Africa or – as Neil Armstrong himself suggested – the high desert of the western United States, its small variations gave the landscape a subtle beauty of its own. Of course, because this was the first landing on the Moon, everything outside the LM was novel and enormously interesting. However, before Armstrong and Aldrin could pay much attention to the view or explore the surface, they had to be sure that they had a healthy spacecraft and that the navigation computer was properly loaded with the information needed to get them back to orbit for a rendezvous with Collins. Finally, 2 hours after the landing, they and the NASA engineers were satisfied that the LM was prepared for a return home and that it was safe to stay for a while.

According to the flight plan, Armstrong and Aldrin were scheduled to take a 5-hour rest break before going outside. However, it came as no surprise when they suggested to Houston Mission Control, that after a scheduled hour-long meal, they prepare for what was called, in NASA jargon, an EVA – a period of Extra Vehicular Activity. Normally, EVA preparations were expected to take about 2 hours; but because this was to be the shortest of all Apollo EVAs and largely symbolic, no one – except, perhaps, the worldwide TV audience waiting impatiently – was concerned when the EVA preparations actually took 3 and one-half hours. With no room for error, meticulous preparation was worth the extra hour and one-half.

APOLLO 11 FACTLIST

Crew	Commander, Neil Armstrong; CMP, Mike Collins; LMP, Buzz Aldrin
EVA CapCom	Bruce McCandless
Command Module	Columbia
Lunar Module	Eagle
Launch from Earth	13:32 GMT 16 July 1969
Lunar landing	20:18 GMT 20 July 1969
Popular site designation	Tranquillity Base
Lunar stay	21 hours 36 minutes
Splashdown	16:51 GMT 24 July 1969
Command Module lunar orbits	30
Time outside the LM	1 EVA, 2 hours 24 minutes
Maximum distance from the LM	60 meters (Armstrong only)
Distance walked or driven	120 meters (Armstrong only)
Lunar samples	21 kilograms
Lunar surface photographs	339
Surface equipment	102 kilograms
Highlights	First moonwalk

Finally, about 6 and a half hours after the landing, the astronauts had the hatch open and Armstrong crawled out onto the porch on his hands and knees. Moments later he was on the top rung of the ladder and pulled a lanyard to release a workbench/stowage area that was attached to the side of the LM. This Modular Equipment Storage Assembly (MESA) pivoted at the bottom so that, when Armstrong pulled the lanyard, it swung down into a horizontal position. The most important piece of gear on the platform was undoubtedly the black-and-white TV camera. It was mounted in such a way that, when the MESA swung down, the camera was pointed directly at the foot of the ladder, a carefully planned EVA detail. For the astronauts, the landing had been the big moment of the mission, but for the waiting world, the supreme moment was still to come – that first footstep.

From the bottom rung of the ladder, Armstrong had to make a 3-foot jump down to the LM's footpad. Extra length was added in the event that a less-than-gentle landing might have compressed and shortened the landing strut. From the footpad, he had only several inches to step down to the surface itself. He stood on the pad for a moment or two, testing the soil with the tip of his boot before he made the epochal "small step."

The soil was very fine grained, with a powdery appearance; as he stepped down, his boot sank perhaps a couple of inches, making a sharply defined print. Because of the Moon's relatively weak gravity field (one-sixth as strong as the Earth's), Armstrong's total weight – proportionately one-half astronaut and one-half suit and backpack – was only about 60 pounds. Movement in this environment wasn't particularly tiring, but because of the dramatic upward shift in his center of mass caused by the backpack, he had to lean forward to keep his balance, and it took a few minutes before he could walk comfortably. In case he had to end the EVA suddenly, it was important to take a quick sample of the lunar material so Armstrong used a long-handed tool called the Contingency Sampler to scoop up a bit of rock and soil into a Teflon bag. He then removed the bag from the Sampler, folded it, and stored it in his shin pocket.

We pick up their conversation here after both astronauts had stepped onto the lunar surface and began to go about their assigned chores. [The lunar mission transcript is on a tinted background; the later interviews are indented and set in Helvetica.] The dialogue is between Neil Armstrong (Mission Commander), Buzz Aldrin (Lunar Module Pilot, or LMP), Michael Collins (Command Module Pilot), and Houston's Mission Control Capsule Communicator Captain Bruce McCandless, who later flew on two space shuttle flights. The interviewer is author Eric Jones.

Aldrin: *(to Houston)* *[As per checklist,]* I'd like to evaluate the various paces that a person can (garbled) traveling on the lunar surface. I believe I'm out of your field-of-view. Is that right, now, Houston?

Buzz ran toward the west about 10 to 15 meters, then turned and came back toward the MESA

McCandless: That's affirmative, Buzz. *(pause)* You're in our field-of-view now.

Aldrin: Okay. You do have to be rather careful to keep track of where your center of mass is. Sometimes, it takes about two or three paces to make sure you've got your feet underneath you. *(pause)* About two to three or maybe four easy paces can bring you to a fairly smooth stop. *(pause as he turns at the TV camera and then heads toward the LM)* *(Garbled)* change directions, like a football player, you just have to put a foot out to the side and cut a little bit. *(pause as he turns toward the TV camera and starts to do a two-footed hop)* The so-called kangaroo hop does work, but it seems as though your forward mobility is not quite as good as it is in the more conventional one foot after another. *(pause as he turns at the TV camera and heads back to the LM)* As far as saying what a sustained pace might be, I think that one that I'm using now would get rather tiring after several hundred *(garbled)*. But this may be a function of this suit, as well as the lack of gravity forces. *(pause)*

Figure 1.1. Neil Armstrong and Buzz Aldrin didn't expect to have any particular problems walking and working on the Moon; however, as a precaution, for the first 15 minutes or so that Armstrong spent outside the Lunar Module, Aldrin remained inside and watched him out the LM window. When Buzz crawled out backwards through the hatch and made his way down the ladder, Neil took a planned series of photos to document his progress. Dave Byrne (member of the *Apollo Lunar Surface Journal* team) created this photomosaic later by combining one of those pictures with some frames from Neil's earlier panoramic sequence. Because the Moon has no atmosphere – and therefore no atmospheric haze – objects in the distance look as clear and sharp as those nearby. With no familiar objects to give a sense of scale, it is difficult to determine if a boulder on the horizon is small and nearby, or large and far away. For technical reasons, all the panorama frames were shot at a 74-foot focus, which means that, as spectacular as high-resolution versions of the photos are, distant objects do not appear as sharp and clear as the astronauts actually saw them.

Aldrin – "You can't get very specific about [planning] what you're going to do in the way of evaluating mobility. Excepting kind of think about it. You can't script and choreograph left turns and right turns and all of this. But I had felt, really, that there would be a lot more evaluation made of what I was doing and maybe my comments. So when I came back and went through debriefing, I was really kind of disappointed that there wasn't some person or group that had done some evaluations. But, they did or they didn't. And that's it. They took our subjective comments. And we said to later crews, 'take the first couple of minutes and familiarize; and that's all you need.' But I was still kind of disappointed until, 20 years after the flight, I ran into a guy from Japan, named Seki, and he said that [reading from the letter], 'at the time, July 20, 1969, I gazed your jumping on television, holding tightly a stop watch in my hand. The television was the first color television of my home. We bought the instrument for the purpose to see the color of the Moon and to observe the activity – walking, moving – of you on the surface of the Moon. And I was relieved to certify that your jumping steps were just the same order of our calculated values on the

Moon, based on our experimental data on Earth.' It [Seki's letter] renewed my faith that people cared about those things. And I got to say that I don't think it's a good testimony to our American way of thoroughness and looking at things that, if we're going to have competition in the world, we're going to have it from people who *do* do those things."

McCandless: Tranquility Base, this is Houston. Could we get both of you on the camera for a minute, please?

Armstrong – "We were behind."

Aldrin – "Possibly, the original schedule hadn't included the plaque and the flag."

Armstrong – "That's true." [the absence can be seen in both Neil's and Buzz's checklists]

Aldrin – "And the Oval Office is just about to put us even further behind."

Armstrong – "Something like the flag was going to be done, but there was an uncertainty about what the flag would be and how it would be handled."

A 1993 NASA Contractor Report, *Where No Flag Has Gone Before* by Anne M. Platoff, details the history of the decision to deploy the U.S. flag during the Apollo 11 EVA. Although design of the flag assembly began about 3 months before the launch and training versions were delivered to the Kennedy Space Center on 25 June 1969, the final decision was made late enough that the flag and plaque were installed on the Apollo 11 LM on launch day.

Interviewer – "Had there been many walkthroughs?"

Armstrong – "Quite a few. And they used those to determine the timeline and made measurements of different tasks. And we went through almost start to finish, finally, maybe a couple of times."

Aldrin – "But not pressurized."

Armstrong – "In the sandbox in Houston. The sandbox was a test and practice area of about 100 feet square, located in one of the buildings at MSC [the Manned Spacecraft Center, now the Johnson Space Center]. We practiced there a number of times."

Aldrin – "But, of course, when you do something in the sandbox you have to make sure all the photographers are kept clear of you."

Armstrong – "The final one was a real PR event. It was a photo opportunity; but we did go through the whole thing on that occasion, I think."

Aldrin – "But it's not the kind of opportunity where you want to note difficulties and then sit there and discuss them in front of the media cameras. And if you don't do it then – we're not taking notes – then not all of it's going to be debriefed afterwards."

Aldrin: Say again, Houston.

McCandless: Roger ...

Armstrong: He wants us *(garbled)* camera.

McCandless: ... We'd like to get both of you in the field-of-view of the camera for a minute. *(pause)* Neil and Buzz, the President of the United States is in his office now and would like to say a few words to you. Over.

Armstrong: That would be an honor.

McCandless: All right. Go ahead, Mr. President. This is Houston. Out.

Nixon: Hello, Neil and Buzz. I'm talking to you by telephone from the Oval Room at the White House, and this certainly has to be the most historic telephone call ever made. I just can't tell you how proud we all are of what you *(garbled)*. For every American, this has to be the proudest day of our lives. And for people all over the world, I am sure they, too, join with Americans in recognizing what an immense feat this is. Because of what you have done, the heavens have become a part of man's world. And as you talk to us from the Sea of Tranquility, it inspires us to redouble our efforts to bring peace and tranquility to Earth. For one priceless moment in the whole history of man, all the people on this Earth are truly one; one in their pride in what you have done, and one in our prayers that you will return safely to Earth. *(pause)*

Armstrong: Thank you, Mr. President. It's a great honor and privilege for us to be here representing not only the United States but men of peace of all nations, and with interests and the curiosity and with the vision for the future. It's an honor for us to be able to participate here today.

Nixon: And thank you very much and I look forward ... All of us look forward to seeing you on the Hornet on Thursday.

Aldrin: I look forward to that very much, sir. *(pause)*

Buzz salutes a second time, as does Neil a few seconds later. [The aircraft carrier U.S.S. Hornet was the Apollo 11 recovery ship that would pick up the astronauts after splashdown in the Pacific Ocean about 900 miles southwest of Hawaii.]

McCandless: Columbia. Columbia. This is Houston. Over.

Collins: Loud and clear, Houston.

After being out of sight for just a few seconds, Buzz walks part way toward the flag, turns to face the LM, looks down and kicks at the soil, sending a spray in a southerly direction. This is the "Scuff/Cohesion/Adhesion" activity on his checklist. In the TV view, he is at the extreme right and it is more difficult to see what he is doing than in the 16-mm-film shot from a camera in Buzz's LM window.

Buzz now turns to face more or less west and kicks the surface again, probably wanting to see how the spray looks under different lighting conditions. Next Buzz walks beyond the flag to a relatively undisturbed area and kicks a spray of dirt toward the north. From this spot, he launches another half dozen or so sprays. Although Buzz is out of the TV field-of-view, the sprays of dirt can be seen coming from the right and landing in view of the TV.

Finally, like a kid at the beach, Buzz switches to a left-footed kick for one last spray before moving farther from the LM.

Aldrin: Houston, it's very interesting to note that when I kick my foot *(garbled)* material, with no atmosphere here, and this gravity *(garbled)* they seem to leave, and most of them have about the same angle of departure and velocity. From where I stand, a large portion of them will impact at a certain distance out. Several *(garbled)* percentage is, of course, that will impact *(garbled)* different regions out *(garbled)* it's highly dependent upon *(garbled)* the initial trajectory upwards *(garbled)* determine where the majority of the particles come down, *(garbled)* terrain.

McCandless: Roger, Buzz.

Aldrin – "It catches your attention."

Armstrong – "When you kick the surface, it makes a little fan which, to me, is in the shape of a rose petal or something. The particles all go out, and there's nothing behind 'em. There's just a little ring of particles – nothing behind 'em – no dust, no swirl, no nothing. They just go out in little, like, fans; and they kind of all hit together."

Aldrin – "And the ones at the side don't go quite as far, so they hit first ..."

Armstrong – "It's really unique."

Aldrin – "And I would guess that, for some reason, there must be an angle of departure when you kick the soil that way, that causes it to all leave at the same angle and velocity. Because, otherwise, it would leave at different angles and velocities and it wouldn't all land in the same ring. You don't see that here (on Earth), and maybe it's because the atmosphere immediately breaks things up into dust (sorts things by size), and there, it doesn't do that."

Aldrin: *(back at the LM)* I've noticed several times in going from the sunlight into the shadow, that just as I go in, I catch an additional reflection off the LM that, along with the reflection off my face onto the visor, makes visibility very poor just at

Figure 1.2. Contrary to commonly held belief, this photo is not a picture of the first footprint on the Moon. Buzz Aldrin took this photograph much later in the EVA to show a boot print he'd just made as part of the soil mechanics team's investigation of the lunar surface's physical properties. This boot print – and all the others – laid to rest a wildly dramatic theory that the lunar surface was fluffy and that the astronauts would meet their end by sinking into some sort of dusty quicksand. At the immediate surface, the soil was fluffy enough that Buzz's boot sank in a fraction of a centimeter or so, but at greater depths it was compact enough to easily support his weight. Most important was the fact that man had left his mark on a surface that, in any one spot, might not have been disturbed in millions – or perhaps even billions – of years. The astronauts were aware that in this truly pristine environment, an incautious step could ruin a remarkable sampling opportunity.

the transition *(garbled)* sunlight into the shadow. I essentially have so much glare coming onto my visor *(garbled)* shadow until *(garbled)* helmet actually gets in the shadow. Then it takes a short while for my eyes to adapt to the different lighting conditions. *(garbled)* inside the shadow area, visibility, as we said before, is not too great, but with both visors up *(garbled)* you can certainly *(garbled)* what sort of footprints we have and the general condition of the soil. Then, after being out in the sunlight a while, it takes … Watch it, Neil! Neil, you're on the *[TV]* cable.

Armstrong: Okay.

As can be seen in the shadows in the 16-mm film, Neil has been at the MESA, getting the scoop with which he will collect the bulk sample (a nonselected, or random, sample of rock and soil). He has gone a short distance north of the LM to get his first sample. Buzz comes over to help him get untangled.

Aldrin: Yeah. Lift up your right foot. Right foot. It's still … Your toe is still hooked in it.

Armstrong: That one?

Aldrin: Yeah. It's still hooked in it. Wait a minute. Okay. You're clear now.

Armstrong: Thank you.

The TV cable had been coiled up inside the LM for several weeks and, after being pulled out during the TV deployment, retained memory of the loops it had in storage.

Aldrin: *(approaching the MESA)* Now, let's move that over this way. *(pause as Buzz gets the cable out of the way)* Okay. I've got it. *(long pause)*

As Buzz comes over to the MESA, Neil turns to face him and, when Buzz suggests that they get the cable out of the way, Neil lifts the cable with the scoop he is carrying and extends the cable so that Buzz can take hold of it.

Aldrin: *(with the cable in hand)* Okay. I've got it. *(long pause)*

A full frame shows Buzz moving left with the TV cable. A second full frame shows the TV cable more clearly. In a third frame, Buzz appears to be pulling the cable into a more-or-less straight line past the flag before getting the excess out of the way under the spacecraft.

Aldrin – "This is one of the pitfalls of testing things in 1 g that you're going to use in reduced-g, because I'm sure that when we uncoiled the cable in 1 g it would lay flat.

But not so there [on the Moon]. There may be other reasons why it took a set, but it did not want to lay flat. It was just a trap when you hook a foot into it, because you can't see it too well."

Armstrong – "You really can't see your feet."

Aldrin – "The same sort of thing came up later on, with the leveling BB [on the seismometer]. The experiment [originally] had a fluid bubble, but they came up with a new and different way of doing it. It had a depression with a BB in it, because they didn't want to deal with the fluid for some reason. Well, when I looked down at it, the damn thing was just going around and around and around at the top. It wouldn't do that in 1 g because it would be pulled down from there. [So, both the TV cable and the leveling BB demonstrate that], if you build something and it's a little different and you test it in 1 g, that isn't the way it's going to perform, necessarily, at one-sixth."

The following interview comments are taken from the crew's technical debriefing at the Manned Spacecraft Center (MSC) in Houston on July 31, 1969.

Armstrong – "The TV was operated as planned with no particular difficulties. The one thing that gave us more trouble than we expected was the TV cable; I kept getting my feet tangled up in it. It's a white cable and was easily observable for a while. But it soon picked up the black dust, which blended it in with the terrain, and it seemed that I was forever getting my foot caught in it. Fortunately, Buzz was able to notice this and keep me untangled. Here was good justification for the two men helping each other. There was no question about that, either; he was able to tell me which way to move my foot to keep out of trouble. We knew this might be a problem from our simulations, but there just was no way that we could avoid crossing back and forth across that cable. There was no camera location that could prevent a certain amount of traverse of this kind."

Aldrin – "Neil initially pulled out about 20 feet of cable, and then I pulled out the rest of it. It seemed to reach a stop; it seemed to have a certain amount of resistance, and I thought that was the end of the cable. However, when I pulled normal to the opening, I found that I could then extract the cable to the point where I saw the black and white marks on it. The cable, being wound around the mounting inside the MESA, developed a set to it so that, when it was lying on the surface in one-sixth g, it continued to have a spiral set to it, which would leave it sticking up from the surface 3 or 4 inches. It would be advantageous if we could get rid of that some way."

Armstrong – "Your foot is continually going underneath it as you walk, rather than over the top of it."

Aldrin – "One time when Neil did get the cable wrapped around his foot, the cable very nearly wrapped itself over the top of the tab on the back of the boot. That created a problem in disentanglement. I don't know whether it's worth moving that tab or not."

Like Neil and Buzz, the later Apollo 12 and 14 crews had a continual problem with the TV cable and with the cable for the large umbrella-shaped, S-band antenna that they erected on each of the missions. For the last three, Rover-equipped missions (also known as the *J missions*), the TV camera and high-gain antenna were mounted on the Rover and there were, of course, no cables connecting the Rover communication system to the LM. The changeover to the wireless, Rover-based communication eliminated cable problems around the LM. However, later crews had to contend with connecting multiple cables on the much more elaborate arrays of scientific gear that they deployed and, on one occasion when the Apollo 16 was setting up some surface experiments, a cable was actually torn loose, disabling the experiment.

Aldrin: The blue color of my boots has completely disappeared now into this ... Still don't know exactly what color to describe this other than [a] grayish-cocoa color. It seems to be covering most of the lighter part of the boot *(garbled)* color that *(garbled)* very fine particles. *(Garbled)*.

McCandless: Buzz, this is Houston. You're cutting out on the end of your transmissions. Can you speak a little more closely into your microphone? Over.

Aldrin: Roger. I'll try that.

McCandless: Beautiful.

Aldrin: Well, I had that one inside my mouth that time.

McCandless: It sounded a little wet. *(long pause)*

Aldrin *(at the MESA)*: In general, time spent in the shadow doesn't seem to have any *(garbled)* thermal effects. *(garbled)* feel inside the suit. There is a difference, of course, in the *(garbled)* radiation and the helmet. So I think there's a tendency to feel a little cooler in the shadow than we feel out in the Sun. *(pause)*

Neil is going out some distance from the MESA to get material for the bulk sample, mostly in the area of the Solar Wind Collector, but also off-camera to the left.

Armstrong – "We wanted to not take too much time to collect that [bulk] sample but, at the same time, we wanted to minimize contamination from engine exhaust and tried to go out into areas that were, one, untrampled and, second, a little bit farther out."

Interviewer – "Did you make any attempt to get a representative set of samples?"

Armstrong – "In the bulk we were just supposed to get ..."

Aldrin – "Where was that close-up camera [Apollo Lunar Surface Close-up Camera, or ALSCC]? Has that been used at this point?"

Armstrong – "No. We get to it at the end. *(chuckling)* with great reluctance. They get nervous about that one at the end, if you remember."

Interviewer – "Why were you reluctant to use the ALSCC?"

Armstrong – "Professor Gold got his camera placed on the manifest very late and over crew objections. He hoped to support his erroneous theory of a 'cotton candy' surface. We had little enthusiasm for the intruder."

The ALSCC was also known as the Gold camera, after the Principal Investigator, the late Dr. Tommy Gold. Gold was a long-standing proponent of a theory – based on radio astronomical observations of the Moon – in which the surface was covered with a deep layer of fine dust into which spacecraft and astronauts might sink into oblivion. Neither the fact that pictures of the Moon taken by the Ranger spacecraft showed small craters that would not have survived in Gold's "dust sea" nor the fact the Surveyor spacecraft all landed safely on very firm surfaces made Gold's theory sink into oblivion; the close-up camera was his seeming retaliation to the Apollo program and to the crews. The close-up camera was designed to take very high resolution pictures of very small surface areas and, indeed, they showed that the top millimeter or two usually had a "fairy castle structure" that would explain the radio returns.

Armstrong (from the 1969 Technical Debrief) – "The bulk sample took longer than in the simulations because the area where the bulk sample was collected was significantly farther from the MESA table than the way we had done it in training. The MESA table was in deep shadow and collecting samples in that area was far less desirable than collecting them out there in the sunlight where we could see what we were doing. In addition, [by going farther from the MESA] we were farther from the exhaust plume and the contamination of propellants. So I made a number of trips back and forth in the sunlight, and then carried the sample back over to the scale where the sample bag was mounted. I probably made 20 trips back and forth from sunlight to shade. I took a lot longer, but by doing it that way, I was able to pick up both a hard rock and ground mass [soil] in almost every scoopful. I tried to choose various types of hard rocks out there so that, if we never got to the documented sample, at least we would have a variety of types of hard rock in the bulk sample. This was at the cost of probably double the amount of time that we normally would take for the bulk sample."

Apollo 17 astronaut Jack Schmitt, Jim Gooding (NASA's Lunar Sample Curator), and others have said that Neil did a superb job of gathering a large, representative collection of samples in a relatively short period of time. Indeed, the fact that they are so representative of the site and of mare (lunar 'sea') sites in general has meant that, over the years when researchers have wanted mare samples for use in procedures that would result in sample destruction, they frequently were given Apollo 11 samples.

Figure 1.3. To protect lunar soil and rocks from contamination by the Earth's atmosphere, the samples were placed in machined-aluminum boxes, which were then sealed to keep their contents in a vacuum for the trip home to Houston. After arriving at the Lunar Receiving Laboratory (LRL), the boxes were taken into an ultra-clean vacuum chamber before they were opened. In this photograph, you can see the lower half of the rock box (48 by 30 by 20 centimeters) partly filled with rock samples. Early concern about possible biological contamination from the samples played a large role in the way the samples were isolated and workers were protected in the LRL's quarantine environment. Thick rubber gloves protected the samples and personnel while allowing an operator to move equipment inside the vacuum chamber. Technicians were rotated frequently because the stiff gloves made movement difficult and tiring. After the samples were cataloged by description and individual photographs, they were assigned to individual scientists for both preliminary examination and tests for organic material, as required by the quarantine. Maintaining a large vacuum chamber was difficult and expensive; by the time of Apollo 14, all sample processing was moved into clean stainless steel glove boxes filled with nitrogen.

For example, T.D. Lin of Construction Technology Laboratories was looking into the feasibility of using lunar soil in the manufacture of concrete on the Moon – a process that could greatly reduce construction costs at a future lunar base. After doing a great deal of preliminary work on simulated lunar soil, created by breaking up terrestrial basalt, he requested a small quantity of lunar soil for a final test. Because Neil had collected such a large quantity of representative soil, NASA decided that it would be appropriate to give Lin a small amount from the Apollo 11 sample suite for his experiments, knowing that there was a plentiful supply left for future work.

Figure 1.4. The black shadow of the lunar module is silhouetted against the Moon's surface in this photograph taken from inside the spacecraft. The lunar soil, created over billions of years by impacting bodies of all sizes, has been disturbed by astronauts' activities as they collected samples and set up the geophysical station.

Aldrin: As I look around the area, the contrast, in general, is … comes about completely by virtue of the shadows. Almost *(garbled)* looking down-Sun at zero-phase very light-colored gray, light gray color *(garbled)* a halo around my own shadow, around the shadow of my helmet.

Aldrin: Then, as I look off cross-Sun, the contrast becomes strongest in that the surrounding color is still fairly light. As you look down into the Sun *(garbled)* a larger amount of *(garbled)* shadowed area is looking toward us. The general color of the *(garbled)* surrounding *(garbled)* darker than cross-Sun. The contrast is not as great. Surveying all the dusty area that we've kicked up *(garbled)* considerably darker in texture. Now, I've kicked up one, and I imagine that this is *(garbled)*. The same is true when I survey cross-Sun along the area that we're walking. In general, in addition to the fact that there are footprints there, the general terrain where I've been kicking up a lot of this surface material is generally of a darker contrast *(garbled)* color. *(long pause)*

On the later missions when the actual landing site was known soon after touchdown, photographs taken from the Command Module in orbit show that the surface around the LM appears to be significantly brighter than the surrounding soils. In addition, each of the last three crews drove up onto hillsides several kilometers from their LM and, with long focal length lens, took pictures of the spacecraft. In these pictures, too, the ground near the LM is light in color. In his commentary on Apollo 17, Jack Schmitt suggests that, during the landing, the engine plume sweeps the surface of small particles, increasing the proportion of larger particles and, hence, the amount of reflected sunlight. In discrete places where the soil is subsequently disturbed by footprints or Rover tracks, the original state of the soil is more or less restored and those disturbed areas appear darker against the lightened soil. In support of this hypothesis, he notes that areas of disturbed soil well away from the LM do not appear darker.

Aldrin: *(garbled)* panorama I'll be taking is about 30 to 40 feet out the plus *(garbled)* ...

McCandless: Say again which strut, Buzz?

Aldrin: The plus-Z strut.

McCandless: Roger.

Buzz is taking the first of two planned panoramic sequences of photos and, when Bruce asks, "Say again which strut, Buzz," he is asking Buzz which of the pans he is taking.

Aldrin (from the 1969 Technical Debrief) – "I took the first panorama out in front without having the camera mounted on the RCU, and it did not appear to be unnatural to do so. It's much easier to operate with it mounted; however, I didn't find that the weight of the camera was as much a hindrance to operation as pre-flight simulations indicated it would be. There is no doubt that having the mount frees you to operate with both hands on other tasks. The handle is adequate to perform the job of pointing the camera. I don't think we took as many inadvert pictures as some pre-flight simulations would have indicated. It seems as though, in all the simulations where we picked up the camera, we always managed to take [unintended] pictures. I don't think that was the case in this mission as much as we thought it was going to be. We'll know if a number of pictures taken are pointed at odd angles."

Aldrin: And right in this area, there are two craters. The one that's right in front of me now as I look off in about the eleven o'clock position from the spacecraft, [is] about 30 to 35 feet across. There's several rocks to boulders [that is, pieces that could be called rocks or boulders] 6 to 8 inches across *(garbled)* sizes.

* * *

Aldrin: How's the bulk sample coming, Neil?

Buzz enters the LM shadow from the south and takes up a position west of the ladder.

Armstrong: Bulk sample is just being sealed *(as per checklist)*.

Interviewer – "Did the sample go directly into the SRC (Sample Return Container) or into a bag first?"

Armstrong – "We had individual bags for documented samples which we didn't use very much. I didn't take the boxes away from the MESA with me out when I was making collections."

Judy Allton of Lockheed Engineering at NASA JSC (Johnson Spacecraft Center, renamed from MSC in 1973) compiled a catalog of geology tools. The interviewer and astronauts were examining pictures of weigh bags in that reference document during the following discussion.

Interviewer – "Based on your comment from the 1969 Technical Debrief, Neil, I had the impression that one of these hung from the MESA to receive the bulk sample."

Armstrong – "Yeah, I think this kind of thing is the kind of thing we used. That's not me in the picture, but it was something like that."

Interviewer – "By the end of the bulk sample collection, Neil seemed to be moving with some ease. Were there any problems carrying soil in the scoop?"

Armstrong – "It was a process in which one had to use some care, or you would lose the sample in the transition from the surface up to the collection device [bag]. It was a fixed head scoop, quite small."

Buzz has moved in closer to the ladder and remains there during the following discussion between Bruce McCandless and Mike Collins. Buzz is probably still inspecting the LM.

* * *

Aldrin: The exhaust deflector that's mounted on quad 1 seems to be a good bit more wrinkled *(garbled)* on quad 4.

McCandless: You're breaking up again, Buzz.

If communications between the crew and Houston had been unsatisfactory during the EVA, the astronauts had the option of deploying a large, umbrella-shaped S-band antenna. On Apollos 12 and 14, the S-band antenna was deployed, primarily to provide good signals from the color TV camera. However, because the deployment would have taken 20 minutes and communication was acceptable, the Apollo 11 antenna was never deployed.

Armstrong – "It takes a while to put that thing up."

Interviewer – "On both 12 and 14, it was basically a two-man operation getting it set up and aligned."

Armstrong – "My recollection is that I always practiced doing that alone. Is that right, Buzz? You were always doing something else at the time I was fiddling with that."

Aldrin – "The only time I remember screwing around with that was at the factory."

Armstrong – "Unless we had a problem, I was expected to do it myself."

Interviewer – "What the 12 and 14 crews found was that, as you turned the crank, the thing tended to tip over and so you needed to have the other guy hold it down."

Armstrong – "Was that a characteristic of the lunar gravity that made it [behave] different?"

Interviewer – "I would think so. Thank you; I didn't know they were planning on doing it as a one-man operation."

Armstrong – *(Laughing)* "It would probably have toppled over."

Aldrin – "Wise choice [not to deploy it]!"

Aldrin: I say the jets deflector that's mounted on quad 4 seems to be ... the surface of it seems to be more wrinkled than the one that's on quad 1. Generally, underneath part of the LM seems to have stood up quite well to the *(garbled)* pictures in the aft part of the LM that will illuminate the thermal effects much better than we could get them up here in the front.

McCandless: Roger. Out.

The following interview comments again are taken from the transcript of the crew's technical debriefing.

Aldrin – "I don't think we noticed anything that was abnormal. I guess the only thing that I noticed that I made a note of was the jet plume deflectors. The one of the right side as I was looking at the LM – which would make it Quad 1 [outside Neil's window] – appeared to be a bit more wrinkled than the one on Quad 4 [outside Buzz's window]. Of course, there's nothing to compare it with, because I'd never seen them before. As a matter of fact, the first time we really saw them was when we looked out of the Command Module and got a pretty good idea of their structure."

Armstrong – "The only abnormality I noticed – and it wasn't an abnormality – was that the insulation had been thermally damaged and broken on the secondary [horizontal] struts of the forward leg."

Aldrin – "This is true in the rear, also."

Armstrong – "We didn't carefully check every secondary strut, but the primary struts didn't seem to be damaged."

Aldrin – "There was a sooting or darkening or carboning; I don't know what you call it. At least, I feel it was a deposit rather than just a baking or singeing of the material."

Armstrong – "We have some pictures of the struts."

Aldrin – "The part that had been melted, separated, and rolled back or peeled back on the secondary strut appeared as though it was a much more flimsy design than any other thermal coverings on there. I don't think there is anything significant in the fact that part of the thermal coating that was higher up had separated, whereas the material lower down [and farther from the thrusters] had not. I didn't notice anything peculiar about the [descent stage fuel and oxidizer] vents. There didn't seem to be anything at all deposited on the surface from any of the vents underneath or from the oxidizer fuel vent above."

Armstrong – "The most pronounced insulation damage was on the front, plus-Z strut. Its being in deep shadow obviated the possibility of getting a good close-up picture in that dark environment."

Aldrin – "I think the best pictures we got were of the minus-Z strut."

Armstrong – "There was less damage than on the examples we looked at pre-flight. Just the very outer layers were penetrated."

Aldrin – "From what I could see of the probes, they had just bent or broken at the upper attach point. I didn't observe that they had any other fractures in them. One of them – on the minus-Y strut – was sticking almost straight up."

Aldrin: Want to get some particular photographs of the bulk sample area, Neil?

McCandless: Okay. *(long pause)*

Aldrin: And, Houston, Buzz here. I'm showing 3.78 psi, 63 percent (oxygen), no flags, *(garbled)* adequate, slight warming *(garbled)* fingered. *(long pause)*

Buzz moves out of the TV field-of-view to the left.

Armstrong: Roger. And Neil has 66 percent O_2, no flags, minimum cooling, and the suit pressure is 3.82.

McCandless: Houston. Roger. Out. *(long pause)*

McCandless: Buzz, this is Houston. Have you removed the close-up camera from the MESA yet? Over.

Buzz comes into view and joins Neil at the MESA.

Aldrin: Negative. Thank you. *(long pause)*

> **Armstrong** – "The Gold camera got added very late. It was an irritation. It was not that the crew was inflexible or anything. But, over the previous years and other flights, our history had told us that, when we added experiments or procedures very late – without the chance of having to find out how they interacted with other events on the timeline – we inevitably found ourselves in difficulty in flight as a result of those additions. So, we grew a natural reluctance to accept additional items late in the game."
>
> **Aldrin** – "They're saying they want this thing out now, before we deployed the EASEP?" [Early Apollo Scientific Experiments Package, later known as the ALSEP, Apollo Lunar Scientific Experiments Package]
>
> **Armstrong** – "Yeah. I'm sure Tommy was urging them through the Backroom."

Aldrin: *(Garbled)* get the panorama *(garbled)*

Here, Buzz may be suggesting that Neil stay at the MESA and offload the close-up camera while he, Buzz, takes the Hasselblad and starts a pan from a spot north of the spacecraft, as per the checklist.

Armstrong: Okay. *(pause)* Got it?

"It" may have been the Hasselblad camera.

Aldrin: Okay, got it. *(long pause)*

When Buzz got the Hasselblad camera – either directly from Neil or by picking it up off the MESA where Neil may have put it as per his cuff checklist – he took an accidental frame.

During the 1991 mission review, we did not realize that this camera exchange took place and therefore were confused.

> **Aldrin** – "For all the panorama that I was supposed to be doing and I maybe was

doing in here, there are damn few pictures that show up that indicate that that many were taken."

Aldrin: Houston, how does our timeline appear to be going?

McCandless: Roger. It looks like you're about a half hour slow on it. We're working on consumables. Over.

Aldrin: All right. *(long pause)*

Neil removes the Gold camera from the MESA and takes his first picture between the MESA and the north (plus-Y) footpad. He then moves off to the northeast. This sequence represents a slight departure from the checklists in that Buzz was to have off-loaded the ALSCC before taking his second panorama.

The interval between the time that Neil starts his first Gold-camera photo and the time he starts the second is about 25 seconds. In all, Neil took 17 pictures – each of them actually a stereopair.

The stereo pairs were later analyzed by NASA geologists and engineers. The results were a stunning view of undisturbed (and sometimes purposely disturbed) lunar regolith surfaces. Images included views of small particles coated with dark glass blobs. Gold presented a theory that they were bits of regolith melted by solar radiation focused by the semi-spherical craters. In reality, they were spatters formed by melting of the regolith by high-velocity impacts of meteors into the lunar surface. These heterogeneous "dirty" melts are found at all scales throughout the lunar regolith and form the major component of lunar soils.

The close-up photographs also demonstrated that there is a delicate crust on the lunar surface, which is only a few millimeters thick. When broken by a footprint or wheel, the crust is broken into thin plates. Similar thin plates of consolidated regolith are also present in cores where older surfaces were broken and stirred into the lunar soil. What binds particles together to form these plates continues to be a mystery.

McCandless: Neil and Buzz, this is Houston. To clarify my last [transmission], your consumables are in good shape at this time. The 30-minute reference was with respect to the nominal timeline. Over.

Aldrin: Roger. I understand that.

Armstrong – "The close-up camera was a stereo, 3-D camera. They really wanted three-dimensional pictures of small particles. He [Gold] had no interest in the exhaust plume or anything like that. He was interested in the character of the lunar soil. I think he was really looking for a defense of his cotton-candy-surface theory. I think I was just looking for different kinds of surface contours. If that's what I was doing there. My recollection was that I did it rather later in the EVA. But, it could be [here]."

The following interview comments again are taken from the transcript of the crew's technical debriefing.

> **Armstrong** – "The stereo camera worked fine. We had no problems with it; however, it was hard to operate. I found that the angle that I had to put my hand on the handle to pull [the trigger] and the force that it took was excessive... I found my hand getting tired very soon while taking pictures with that camera. It was wearing out my grip."

> **Aldrin** – "Would you say that the angle was too horizontal?"

> **Armstrong** – "Yes."

> **Aldrin** – "You would like to have had it sloped down more towards you."

> **Armstrong** – "Yes. It was requiring the wrist to be cocked down."

> **Aldrin** – "The initial opening up or deploying of it went quite smoothly. The extension of the handle and the opening up of the case was quite well engineered. Separating the cover, taking it off, cutting the film, and removing the cassette also went quite smoothly. I think that the big area for re-engineering might be just a change in the angle that the handle comes out. We might have to add a hinge or something like that to it. What about the height of the handle? That would probably not be too bad."

> **Armstrong** – "I think that probably was reasonable. The other problem we had with the camera is that it was falling over all the time. I think this was the result of a little bit of difficulty in figuring out the local vertical."

> **Aldrin** – "Yes."

> **Armstrong** – "You'd set it down and think it was level; but apparently it wasn't because, the next time you looked, it would be laying over on its side. Or you would bump it inadvertently while you were looking somewhere else and knock it over. I picked it up three different times off the surface and it's a major effort to get down to the surface to pick the thing up."

> **Aldrin** – "How'd you do that? By going down on the knee?"

> **Armstrong** – "On one occasion I got it with the knee, one time I got it with the tongs, and the last time I had something else in my hand like a scoop or something that I could lean on and go down and get it. In general, there were a lot of times that I wanted to get down closer to the surface for one reason or another. I wanted to get my hand down to the surface to pick up something. This was one thing that restricted us more than we'd like. We really didn't have complete clearance to go put our knees on the surface any time we wanted. We thought the suit was qualified to do that in an emergency, but it wasn't planned as a normal operation."

The difficulty the astronauts had in bending the leg enough to be able to kneel was due to the internal pressure of the suit, which resisted flexing. Bellows-like joints called *convolutes* were built into the suits at the knees and elbows to make it easier for the astronauts to bend those joints. It wasn't difficult to get on hands and knees, but even with the convolutes, it was difficult to bend the knee enough to actually kneel. For some reason, John Young and Gene Cernan – the Apollo 16 and 17 commanders, respectively – could get on their knees with relative ease, albeit not all the time. Others, such as Charlie Duke and Jack Schmitt, the 16 and 17 LMPs, respectively, couldn't get to their knees unless their feet were in a small crater and they were kneeling on the rim. On all of the missions, picking things up off the ground – whether rocks or dropped tools – was a continuing problem. Neil's solution of leaning on a scoop to get within arm's reach of the surface was used successfully by several of the astronauts, including Cernan.

The following interview comments again are taken from the transcript of the crew's technical debriefing.

> **Armstrong** – "We didn't let ourselves settle on our knees a lot of the times to get our hand on the surface. Now I think that is one thing that should be done more on future flights. We should clear that suit so that you could go down to your knees, and we should work more on being able to do things on the surface with your hands. That will make our time a lot more productive, and we will be less concerned about little inadvertent things that happen."

> **Aldrin** – "Now we can say that we have the confidence to know that we could get back up from the surface. You might have to put your hand down into all this. The thing that discouraged me was the powdery nature of the surface and the way that it adhered to everything. I didn't see any real need in getting down. I had no concern about doing it. But I agree. I think if we need something on the suit to qualify it to do this, then we ought to go ahead and do that. If it doesn't, if it just requires looking at the suits that we brought back and saying that they're qualified for kneeling, we ought to do that."

> **Armstrong** – "If you have a grip on something like a scoop, or a stick to hold on, then there's no problem at all in getting back up. You can go right down and just push on your hand and push yourself right back up. It was easy the one time I did it with the scoop in my hand. That's one thing that we hadn't done a lot in our simulations, and it would be a help, I think."

Astronauts on later missions found that, if they fell, they could get up relatively easily by getting on their hands and knees and then pushing back with their hands until their center of mass had rotated over their feet, at which point they could stand with the help of the internal pressure of the suit.

Armstrong: *(still northeast of the LM, taking a break from the close-up photos and performing the "LM Inspect" item in his checklist)* I don't note any abnormalities in the LM. The quads seem to be in good shape. The primary and secondary struts are in good shape. Antennas are all in place. There's no evidence of problem under-neath the LM due to either engine exhaust or drainage of any kind.

McCandless: Roger. Out.

Aldrin: It's very surprising, the very surprising lack of penetration of all four of the foot pads. I'd say if we were to try and determine just how far below the surface they would have penetrated, you'd measure [depths of] 2 or 3 inches, wouldn't you say, Neil?

Armstrong: At the most, yes. That Y-strut there is probably even less than that. *(long pause)*

Buzz comes back to the LM near the plus-Y (north) strut and takes both close-ups of the plus-Y footpad.

Collins: *(garbled)* *(pause)*

Aldrin: *(going east of the strut)* I'll get a picture of the plus-Y strut taken from near the descent stage, and I think we'll be able to see a little bit better what the thermal effects are. Seem to be quite minimal. *(long pause, then turning to look under the spacecraft)* There's one picture taken *(pause)* in the right rear of the spacecraft looking at the skirt of the descent stage, shows a slight darkening of the surface color, a rather minimal amount of radiating or etching away or erosion of the surface. On descent, both of us remarked that we could see a large amount of very fine dust particles moving out. *(pause)* It was reported beforehand that we would probably see an outgassing from the surface after actual engine shutdown, but as I recall, I was unable to verify that. *(long pause)*

Aldrin: Just too big an angle, Neil.

Armstrong: Yeah. I think you are right. *(long pause)*

The idea was to point the camera up at the Earth and Buzz isn't sure that it is possible to do so. Certainly, with the Earth nearly overhead, Buzz had to take the camera off the RCU (remote control unit) bracket and then either guess at aiming the camera or lean far enough back that he could hold the camera over his head and sight along it. Because of the stiffness of the suit, it was very difficult to lean far enough back. One puzzling aspect of these pictures is the fact that, during the 1991 mission review, Neil said that it was he that took the Earth pictures. One possible explanation of this memory is that Buzz took the camera off, decided that he couldn't get the pictures, and then Neil took the camera for a moment, took the frames, and then gave the camera back to Buzz. Another possibility is that Neil is mistaken about having taken Earth pictures during the EVA. Certainly, after the EVA he took pictures of Earth through the overhead rendezvous window and, 20 years after the fact, he may have gotten the memories confused. After reading a draft of the discussion in 1995, Neil said he had nothing more to add to the solution of this minor mystery.

Just prior to Buzz's next transmission, an astronaut comes into view near the east footpad. The picture is very grainy, but it appears to be Neil, still using the close-up camera. This astronaut goes out of the TV field-of-view after a few seconds and then, after a few more seconds, an astronaut – this time, probably Buzz – comes into the field-of-view.

Aldrin: We're back at the minus-Z strut now. [The] stereopair we're taking [of the] footpad will *(garbled)* very little force of impact that we actually had. *(long pause)*

Aldrin: And, Neil, if you'll take the camera [as indicated in Neil's checklist], I'll get to work on the SEQ [Scientific Equipment] bay.

Armstrong: Okay.

McCandless: Columbia. Columbia. This is Houston.

Aldrin: Houston, I notice that … *(listens)*

McCandless: Go ahead, Buzz.

At this point in the TV record, we see Buzz hand the camera to Neil, who has just come back into the TV picture. Buzz's next task is to off-load the EASEP (Early Apollo Scientific Experiment Package) while Neil takes documentation photos of the off-load.

Aldrin: *(to Neil)* Try to get some close-up pictures [meaning Gold-camera photos] of that rock. *(to McCandless)* I was saying that, Houston, *(garbled)* stop and take a photograph or something and then want to start moving again sideways, there's quite a tendency to start doing it with just gradual sideways hops until you start getting momentum.

McCandless: Roger.

The largest sample of lunar regolith, which is now used for many analyses and experiments, was collected as a very clever afterthought by Armstrong. Many of the rocks were partly coated with glass that was splashed onto their surfaces by nearby impact melt. Rocks that were partially buried provided information about the flux of micrometeorites striking the lunar surface. The buried parts were fresh and angular, whereas the exposed portions were rounded, pitted, and partly coated with glass spatter. Realizing that these glass surfaces would be damaged by movement within the stainless steel rock box during re-entry into the Earth's atmosphere, Armstrong quickly filled all of the space between the rocks with regolith. He shoveled in about 38 kilograms (84 pounds, measured at 1 g) – not only ensuring padding for these unique rock surfaces, but also providing scientists the largest of all lunar soil samples.

Neil Armstrong (left) and Buzz Aldrin finished off a Hasselblad magazine with some pictures of themselves in the LM after mankind's first moonwalk. Compared with the difficulty and stress of performing the first lunar landing, the EVA was a stroll in the park. Although largely ceremonial and symbolic, the time Neil and Buzz spent on the surface had demonstrated that humans could adapt fairly easily to the lunar environment. Still ahead of them: lift off, return to orbit to rejoin Mike Collins in the CM, the trip home, and a safe landing on Earth – before the conditions of President Kennedy's challenge would be completely met. But with that first, risky landing behind them and with the first 21 kilograms of priceless sample in the cabin with them, they had every reason for confidence.

The splendid isolation and desolation of the lunar surface is evident in this view from the southwestern rim of Surveyor Crater. Note the rearward tilt of the spacecraft. On-board measurements indicated that the LM was pitched up (backwards) by 3 degrees and was rolled left (toward the camera in this picture) by 3.8 degrees. This tilt was noticeable from within the spacecraft but was not alarming.

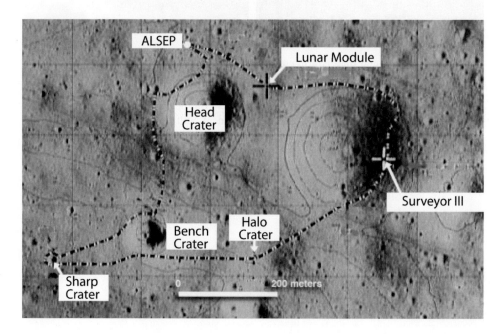

Site Map. The dotted line shows the route of the geology traverse that Conrad and Bean made during the second EVA. They started out going west from the LM toward the northwest rim of Head Crater, then south and west to Sharp Crater, finishing the traverse at Surveyor Crater. In addition to sampling soil and rocks and collecting cores from the lunar regolith, they visited the Surveyor III spacecraft that landed on the Moon April 1, 1967. From their vantage point in the Lunar module and from orbital photography, the crew was worried that the slopes of Surveyor Crater would be too steep and they would not be able to collect parts from the unmanned spacecraft. It wasn t a problem; the steep slopes were an illusion related to sun angle and perceptions of distance on a planet with nothing familiar to serve as a scale.

2

Pinpoint Landing, Great Science, and a Lot of Fun

On April 20, 1967, more than 2 years before Apollo 11, the unmanned spacecraft Surveyor III landed in a crater on the plains of Oceanus Procellarum (Ocean of Storms). The craft had a stormy landing, bouncing twice because the descent rockets had not shut off. Unlike Surveyor II, which crashed into the lunar surface and was never heard from again, the robust Surveyor III had a short – but productive – life digging a trench, conducting soil mechanics tests, and sending more than 6,000 photographs back to Earth. The mission proved without a doubt that the lunar surface was firm and that the astronauts would not disappear into "quicksand," thus clearing the way for the landing of the first Apollo lunar module.

Apollo 11 demonstrated that a Moon landing was possible for manned missions, and Apollo 12 demonstrated that astronauts could make a pinpoint landing. Testing the planners' navigational skills, Apollo 12 would land within a few hundred meters of Surveyor III on a flat, relatively safe mare plain. Accurate landings would be needed for future missions, which were sited in complex, rugged terrain. With Apollo 12, mission planners successfully demonstrated that they could put the lunar module down within a few hundred meters of a target site and ensure that the craft wouldn't slam into a mountainside or a end up in a rugged crater.

In a 2001 interview with historians Stephen Ambrose and Douglas Brinkely, Neil Armstrong was asked which part of the mission had been of greatest concern. He replied that the final descent to landing "was far and away the most complex part of the flight . . . walking around on the surface, you know, on a ten scale, was a one, and I thought that the lunar descent on a ten scale was probably a thirteen." He got a good laugh.

Although the goal for Apollo 12 was to demonstrate a precision landing, a mare site was chosen because it was relatively flat and free of mountainsides and large craters. It was a safe site but geologically interesting – not only because of the mare lavas and regolith, but also because the area had been showered by ejecta from the large impact that formed the crater Copernicus. (One of the youngest large-scale impacts on the Moon, the impact crater and its rays of debris are so large that they can be seen from Earth with a pair of common binoculars.) Knowing the age of the Copernicus impact is important when studying lunar history because its ejecta provide a "marker horizon" used by geologists to separate the younger from the older events that shaped the Moon.

In addition to demonstrating precision landing and collecting a variety of samples, the Apollo 12 crew was assigned the task of establishing the first geophysical observatory on the Moon. Apollo 11 had set up a solar-powered seismometer, but the crew of Apollo 12 would install a very complex and

APOLLO 12 FACTLIST

Crew	Commander, Pete Conrad; CMP, Dick Gordon; LMP, Alan Bean
EVA CapCom	Ed Gibson
Command Module	Yankee Clipper
Lunar Module	Intrepid
Launch from Earth	16:22 GMT 14 November 1969
Lunar landing	06:55 GMT 19 November 1969
Popular site designation	Oceanus Procellarum
Lunar stay	31 hours 31 minutes
Splashdown	20:58 GMT 24 November 1969
Command Module lunar orbits	45
Time outside the LM	2 EVAs, 7 hours 29 minutes
Maximum distance from the LM	450 meters
Distance walked or driven	1.3 kilometers
Lunar samples	34 kilograms
Lunar surface photographs	583
Surface equipment	166 kilograms
Highlights	first precision landing; first multiple-EVA mission

sophisticated collection of instruments on the surface of the Ocean of Storms. This Apollo Lunar Surface Experiments Package was known as the *ALSEP*. Astronauts and NASA ground personnel alike used a vast list of acronyms to refer to myriad engineering components, actions, and scientific instruments. These shorter, often single-syllable words were easier to say and understand during both training and the missions. Acronyms began to form the unique language of the space program [see the Acronym List in the Appendix for a selected list of those used in this book]. The partial list of instruments from the first ALSEP makes the point.

- Passive Seismic Experiment (PSE), a seismometer to monitor moonquakes and meteor impacts.
- Suprathermal Ion Detector Experiment (SIDE), to monitor the energy and composition of positive ions at the lunar surface. The data were useful for evaluating the solar wind hazard and the shielding that would be required to protect humans working on the lunar surface for long periods of time.
- Solar Wind Spectrometer Experiment (SWS) to detect whatever solar plasmas strike the Moon's surface.
- Lunar Surface Magnetometer Experiment (LSM) to measure the magnetic field on the lunar surface and evaluate the Moon's deep electrical properties.
- Cold Cathode Ion Gauge (CCIG), to analyze any lunar atmosphere, no matter how tenuous.

Figure 2.1. Appearing so small and lonesome when viewed from a distance, the LM is quite impressive at close quarters. In this photo, Al Bean stands on the top rung of the ladder with his hands on the porch rails. He has closed the cabin door behind him, knowing that they can open it from the outside – but fully aware of the consequences if they couldn't. He would climb down the ladder to become the fourth person to walk on the lunar surface. The experience served him well after his career as an astronaut: now a well-known, talented painter, he has provided the world a view of the Moon through his eyes.

These instruments were integrated into or linked to a central station, which was powered by a radioisotope thermoelectric generator (RTG) that was in turn fueled by plutonium-238. Needless to say, the launch of a plutonium fuel element in a spacecraft was controversial, but the material was stored in a fuel cask that was designed to withstand Earth re-entry in the event of an accident.

By 1969, Surveyor III had been on the lunar surface for 2.5 years, exposed to the hostile lunar environment and bombarded by micrometeorites and radiation. To evaluate the effects of this exposure, the Apollo 12 crew brought back Surveyor's scoop (the one used to dig its trench) and Surveyor's TV camera. Retrieving the instruments not only gave the mission planners a boost for planning the accurate landing, but also gave the engineers and scientists who designed the unmanned Surveyor lander a chance to evaluate the robustness of the older spacecraft.

Pete Conrad and Alan Bean, the crew of the Apollo 12 LM, were efficient and upbeat; they also had a lot of fun. It's important to remember, however, that the hazards of exploring on the lunar surface were not trivial. The astronauts discussed the difficulties and risks of simply getting out of the tiny lunar module cabin in their "Michelin Man" suits during the Apollo 12 Crew Debriefing held December 1, 1969, and during interviews with author Jones. As soon as they stepped onto the lunar surface, they encountered another problem that dogged all of the Apollo explorers: dust. It clung to everything – getting into seals and dirtying space suits and equipment, which then absorbed more sunlight and got hotter. Armstrong and Aldrin had noted the dust problem earlier, but it really became evident during the EVAs of Conrad and Bean, who spent much more time loping and striding across the lunar surface.

THROUGH THE HATCH

Although the work on the lunar surface (particularly on Apollo 11) wasn't risky in the same way the landing was, it could be challenging. Especially the first part of the EVA: getting out through the hatch. Complete cooperation was vital as the EVA-suited crew members were maneuvering within the tight spaces of the LM and performing their delicate exit.

The LM was by necessity a small, lightweight spacecraft and that meant that there wasn't much spare room in the cabin. Grumman Aerospace LM project manager Tom Kelly described it as being "the size of a modest walk-in closet." The front part of the cabin, where the crew stood, was about 40 inches deep, from the hatch back to the bulkhead that marked the transition to the back of the cabin, where the ascent engine cover was in the center, the Environmental Control System was on the right-hand side behind the LMP, and various stowage compartments were behind the Commander. The decking in the rear of the cabin was about 18 inches above the flooring in the front. The rise between the two areas was sometimes called the *midstep*.

The floor in the front part of the cabin was about 48 inches wide, but because of the cabin's cylindrical shape, the maximum usable width halfway up was about 72 inches. That gave the astronauts some extra space, which they desperately needed after they had the suits pressurized and were getting ready to open the hatch. After he and Charlie Duke returned from the Moon, John Young commented, "As everybody [on the other LM crews] has remarked, once you get the gear on, the only thing you can do is get out of the spacecraft, because you've run out of room to do anything else." By turning to face each other and backing as far as they could against the circuit breaker panels on the side walls, they could create enough open space between them to get the hatch open – but just barely.

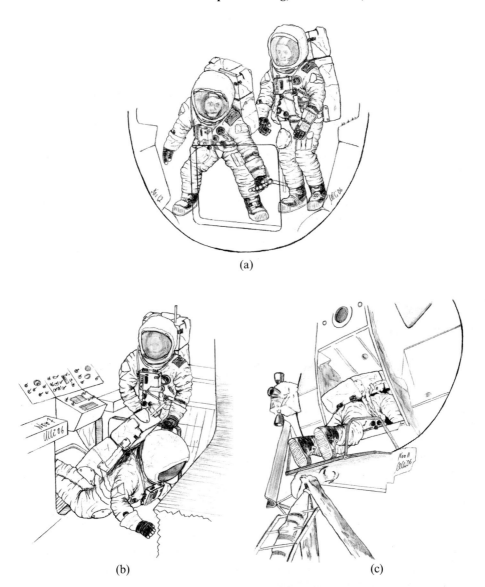

(a)

(b)   (c)

Figure 2.2. Egress from the LM was an exacting process. The astronauts exited from the cramped cabin space through a small door while wearing large spacesuits; the exercise required patience, strength, and cooperation. (*These illustrations from Apollo 12 and Apollo 14 were created by Ulli Lotzmann, member of the* Apollo Lunar Surface Journal *team*) (a) Ed Mitchell (Apollo 14) is opening the hatch with Al Shepard backed into his corner. After the door is opened, Al Shepard prepares to back out of the cabin. (b) Pete Conrad (Apollo 12) must turn, face the rear of the cabin, get down on his knees, and inch backwards. Even by arching his back, Conrad and his backpack barely clear the hatch and his helmet just clears the midstep. (c) Almost through the hatch, the astronaut must feel for the top of the ladder with his boots before he can descend to the lunar surface.

Here's what Ed Mitchell had to say about the first time he and Alan Shepard opened the hatch during Apollo 14.

Mitchell – "I had to get out of the way so that he could swing the hatch open. It was tight. I had to turn around and back right up against the instrument panel – the circuit breaker panel – so that the door could come open and Al could turn around and back his way out through it."

Interviewer – "While the door is actually opening, is it mostly over on your side, or is it centered?"

Mitchell – "The handle was on his side and the hinge was on my side. So it opened from his side toward me. He had to back in the corner while it swung open."

Interviewer – "Right. But, once it was swung open and you were backed into your corner, then he had two-thirds of the cabin or something like that."

Mitchell – "Right."

The established procedure was that when the hatch was open as far it could go, with the LPM behind it, the Commander got himself turned to face the right rear of the cabin. After he was in position, he needed to get down to his knees with his feet pointing out the hatch, move to his left, get lined up with the hatch, and then work his way backwards until he was clear of the midstep and could get down on all fours. With the suit fully pressurized, that wasn't easy. To complicate matters, the rigid backpack alone was 36 inches tall – the combined height of suit and backpack left only 4 inches between the midstep and the hatch.

In 1964–5, astronauts Pete Conrad, Ed White, and Roger Chaffee had worked with the Grumman team to fine-tune the layout of the cabin and, in particular, the size and shape of the hatch, to make sure that it would be possible to get out. The square hatch was 32 inches on a side, just enough to accommodate a suited astronaut wearing a Portable Life Support System (PLSS or backpack). At the front of the cabin, just above the hatch, the LM's computer display panel and keyboard (DSKY) stuck out about 12 inches into the cabin, with just enough clearance under it for the hatch. The DSKY's position meant that, as the Commander was working his feet backwards toward the open hatch, he had to arch his back to get his hips close enough to the floor so that he and the PLSS could slide under the DSKY.

As Jim Irwin (Apollo 15) described it later, "you had to drop down and then [get] almost on your belly to kind of squirm out backwards. It was always a difficult maneuver."

Until they had the PLSS part way out through the hatch, their heads were still over the midstep, and they had to move cautiously until there was room to get their hands on the floor and their heads down. Ed Mitchell said, "Getting your head down with the LEVA (the visor assembly) on and trying to get down [to the floor], that rear step came right in front of your face. There was barely room for you to get your butt and the bottom of the PLSS out the door. You brushed the LEVA right across the

[mid]step. I mean, it was tight." They certainly didn't want to scratch the visor, so they took it slowly and carefully.

The astronauts had all done simulations on the ground and then in an airplane flying parabolic trajectories that gave them the experience of 30 seconds of 1/6 g at a time; however, exiting the LM on the Moon was always a challenge – especially the first time.

Getting back in was pretty much a reverse of the process. Buzz Aldrin said, "I had very little difficulty [getting in] ... About halfway in, [you have to] make a concerted effort to arch your back to keep the PLSS down by keeping your belly down against the floor. This affords you the least profile going in. There didn't seem to be any exertion at all associated with raising yourself up and transitioning to a point where you can bring your knees on inside the cockpit, and then moving from a kneeling [position] to an upright position. It all seemed to work quite smoothly. When there is a large bulk [that is, the PLSS] attached to you, you have to be careful. Once you get inside, before you start to turn around, you must make adequate allowance for all this material behind you."

It was a challenge but it had to be met. Gene Cernan's perspective after Apollo 17: "By this time, I was really tired and it was a lot of trouble to get back in the cabin. It's such a small space; and you've got to twist and turn to get in there. You've got to crawl in with that big PLSS on your back until you hit the back bulkhead with your helmet. Then you've got to start arching your back so that you can get your head up so that you can get further in. But you've got to keep the PLSS clear of the DSKY, so that means really arching your back. Then you push with your feet until you're in and you can stand up and then turn around. And it didn't help that a lot of space, particularly on the engine cover, was taken up by the boxes and bags we'd brought in from the surface. It was really crowded."

DOWN THE LADDER

We pick up the transcript as the Apollo 12 crew is in the process of crawling through the hatch and descending to the lunar surface. The dialogue is between Pete Conrad (Mission Commander, CDR), Alan Bean (LMP), Michael Collins (Command Module Pilot, CMP), and Houston's Mission Control Capsule Communicator (MCCC or Cap Com) Edward Gibson. Eric Jones is the interviewer who conducted mission reviews with the Apollo crews from 1989 through 1992, about 20 years after the flights.

Conrad: Okay, it's in Intermediate. I'm ready to go over the sill.

Bean: Just a second.

Interviewer – "Al's standing in his corner, with the hatch open at his knees. How do you go about getting out?"

Conrad – "I got to turn to the face the wall, my back to Al. Then I got to get down and go back feet first."

Interviewer – "You couldn't get down on your hands and knees."

Conrad – "It was more me sliding down the back corner and getting my feet out the hatch."

Bean – "And then you could get down on your knees."

Conrad: Consult my [cuff] checklist.

Pete and Al brought their cuff checklists to Santa Fe for the 1991 mission review.

Conrad – "Don't blow the lunar dust off it!"

Interviewer – "I won't."

The checklist is mounted on an aluminum arc of about 120 degrees, which fits over the sleeve of the bulky suit. A cloth strap loops through some hooks on the underside and has Velcro material along its entire length so that the metal arc can be strapped almost anywhere on the forearm. The checklist pages are made of a double laminate of fairly heavy photographic paper and have text and/or drawings and/or pictures on both

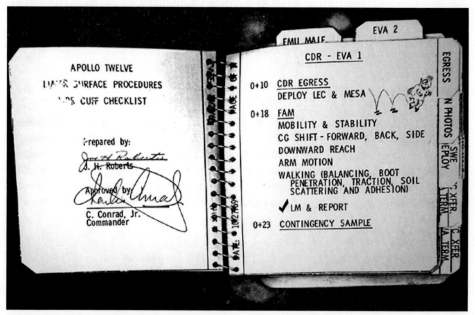

Figure 2.3. This cuff checklist, worn by Pete Conrad, outlines details of the EVA activities against a timeline. An identical checklist was used by the Capcom in Houston.

sides. The square pages are about 3.5 inches on a side with index tabs on the outer edge on the top. This book's metal spiral spine is attached to the metal arc of the wristband; the pages are easy to read if you hold your arm as if to read a watch. The really clever aspect of the design is that, because the metal spiral conforms to the shape of the wristband, an open page will stay open – it actually takes a little force to turn a page.

In addition to the expected list of tasks and sketches illustrating aspects of the tasks, the checklist contains some "extras." Both checklists contain pictures of Playboy Playmates – provided by the backup crew of Scott, Worden, and Irwin – and at the back of Pete's checklist there are two pages of handy geology phrases.

> **Conrad** – "I always called everything [especially rocks and other material of geologic interest] 'stuff.' And there was a lot of money on how often I was going to call stuff 'stuff.' And so, Jack Schmitt put these pages in for me. And I was going to dazzle the geologists if I so much as mentioned any of these."

> **Bean** – *(quoting)* " 'Shows localized subtle evidence of graded and possibly … ' *(laughing)* Read the others. Oh, you got more!"

> **Conrad** – "Oh, yeah. 'Bedrock exposed in the bottom of the craters. The bedrock surface resembles Moshia breccia *(laughing)* with a consistent organization to its internal fabric.' And they would have known that I was reading from notes."

> **Bean** – "No they wouldn't. They would have never guessed."

Many of the pages also feature small cartoons drawn by Ernie Reyes, a member of the crew operations team and responsible for stowing various items in both the Command Module and the Lunar Module.

> **Conrad** – "The backup crew had that all done for me. We never saw that until we got to the surface. Al didn't see his. I didn't see mine."

> **Interviewer** – "So this isn't your signature here on the front page?"

> **Conrad** – "Yeah, it's my signature. That was written on approval of the checklist [content], but, [with regard to] the actual flight checklist, the backup crew verified that this all got put together. And they actually loaded this on the LM. We never saw it until we got to the lunar surface. We practiced with another one that was identical [in official content] but you never practiced with the flight gear. Very little did you ever see the flight hardware. As a matter of fact, we [only] saw the flight hardware once prior to flight. Like, we inspected the hammer and all that stuff that was laid out [in a lab prior to final packing]."

Bean: Find the LEC; hey, come on, babe.

The Lunar Equipment Conveyor (LEC) is a clothesline-like device with which Pete and Al will transfer equipment and samples to and from the surface. The LEC is threaded over a yellow bar in the overhead inside the cabin and either astronaut – Pete on the surface or Al in the cabin – can move equipment up and down by pulling on it like a "Brooklyn clothesline."

Conrad: *(reading)* "Deploy the LEC and the MESA [Modularized Equipment Stowage Assembly]; Mobility; c.g. [center-of-gravity] shift, downward reach, arm motion, walking balance," and all that good stuff. *(pause)* All right.

Once he gets down to the surface, Pete will take a few minutes to check his ability to control his c.g. which, because the weight of the PLSS, is displaced up and back from its normal position. He will check his ability to reach downward, and so on. Al will repeat the process when he gets out.

Bean: Just a second.

Conrad: Pull the door all the way back.

Bean: Let me hand you something. *(pause)* When you go out, and then I'll hand it to you. *(pause)*

Conrad: Yeah.

Bean: Bye-bye; see you in a minute. *(pause)*

Conrad: How am I doing? Am I hanging on something here? I get the feeling I'm stuck under something here.

Bean: You're bumping into the purse there. [Move] forward.

Pete is facing the rear of the cabin, trying to get his feet out the hatch and Al wants him to move toward the back of the cabin.

Conrad: Purse?

Bean: McDivitt purse.

Conrad: Okay.

Bean – "The purse was a white, cloth bag that was up there under the computer DSKY that we just threw loose items in. In fact, I've still got it at home. It had a snap [closure] and you opened it up and threw stuff in and closed it. That's why we called it a purse. It was also shaped like a purse. It was made out of beta cloth. That's what Pete was hung up on, and I probably just lifted it up so that he could get out."

Bean: All right; now you're in good shape.

Conrad: Okay.

Bean: Go straight down from where you are.

Conrad: Okay.

Bean: Good; that's good.

Conrad: Okay.

Bean: Doing good. You're headed right square out the hatch. You'll have to bend over more, though. *(pause)* Wait. Wait. Wait. Oops. Come forward a little. Move to your right, you're ... There you are. Now go. You were getting that little ... *(pause)* You got to kneel down a little more. *(pause)* Well, I'll push you [down] if you don't mind. *(pause)*

According to Pete and Al, the original LM design had a round hatch.

> **Bean** – "We found out, with the PLSS and everything, you couldn't get out through the hatch they had. So they made a major change and made that a square hatch, which allowed you then – without changing the dimensions of the LM – to still scoot down and get your square PLSS out. But it was the minimum size you could get away with for the biggest astronaut. You talk about the LM [cabin] being small, I think this was the critical point right here."

Conrad: What am I hung on? Nothing.

Bean: You're okay.

Conrad: Got this garbage bag in my way.

> **Conrad** – "It's probably the jettison bag."
>
> **Bean** – "I bet it's over there against the wall ..."
>
> **Conrad** – "Yeah, hanging under the console ..."
>
> **Bean** – "And you pushed it out of the way, to your right as you go out backwards."

The other possibility is that the jettison bag was sitting on the engine cover, where it would have also been in the way. In either case, the only place to get it out of the way is on the left side of the cabin, the space that Pete has just vacated.

Bean: *(moving the jettison bag)* Okay. *(pause)* You're headed out the door. *(pause)* Looks real good. *(pause)*

Conrad: Okay. *(pause)* Okay; I'm out on the porch. *(pause)*

Bean: Just a second.

Conrad (from the 1969 Technical Debriefing) – "It took me a moment to get oriented and Al gave me a GCA [Ground Controlled Approach] and I apparently was rolled to my right slightly on my way out so that the left lower corner of my PLSS tore about a 6-inch rip in the hatch insulation. That was the only problem. I didn't notice it going out, except I did fray the insulation a little bit."

Conrad: Okay; let me pull a pip pin *(holding the D-ring on the MESA release)*. Deploy the MESA. *(pause)*

Bean: Okay.

Conrad: Man, that's a heck of a tug with that handle. [I'll go] down another step [on the ladder].

Bean: Okay. While you're doing that, let me get the LEC ready for you. *(pause)*

Conrad: Good Godfrey. That handle's in there [in its mounting bracket] like something I never saw before.

As Al will comment later, in normal conversation Pete uses a great deal of profanity. His use of terms like "Good Godfrey" while he is on the Moon indicates that he is aware that the world is listening and that a little self-censorship is in order.

In 1997, Andy Chaikin (author of *A Man on the Moon*) called author Jones' attention to the following comment by Brian Duff of NASA's Public Affairs Office (PAO) during a post-Apollo interview.

"Was the open program already in place when you came on board?"

DUFF: "No. That's why I mentioned this thing about Al Sehlstad and how it evolved. I was trying to describe how it actually came incrementally. I'll give an example of the other side, the temptation to not do it. Pete Conrad – someone else I'm very fond of – came in before his flight and said that we had to reinstate the ten-second delay [in providing audio to the media], which is one of the devices that protect television people from mistakes on the air. I said, 'Pete, I'm not going to do it. We worked too hard to get this. It's part of the tradition of the space program now. We do not have a delay, we have live air-to-ground with no delay. They hear it absolutely as it comes down, they hear it at the same time, and they see it at the same time you do.' I said, 'The scientists out at JPL [Jet Propulsion Laboratory in Pasadena, California] don't like it. They want to see this first. We're not going to do that. It's going to come straight down from the spacecraft; they're going to see it the same time you see it.' He said, 'I'm going to embarrass you. We can't clean up our language, and we're going to embarrass our families and NASA and the whole space program.' He said, "You've got to do something to clean it up and cut out the swear words." I said, "Pete, we're not going to do it." He said, "Well, it's your problem. It's going to be your fault." I said, "No, it's going to be your fault." The result was – if you've ever listened to a tape of that flight – it is the most Boy Scout, full of expressions like 'gee whiz, golly whiskers, holy smoke, gosh, isn't that a great big rock there?' and things like that. Of course there was no swearing."

Bean: How's the lock doing? Was it easy to get … Can you get the lock out?

Conrad: Yeah. *(pause)* Now that's better. I couldn't even get the handle out of the deal. I just pulled the cable; the MESA's down.

Bean: Okay.

Conrad: There you go. *(pause)*

Conrad (from the 1969 Technical Debriefing) – "I got out on the platform okay, and I released the lock-lock mechanism on the MESA and the MESA handle was free in its holder. I tried it and it wouldn't come out. I pulled on that thing as hard as I could pull, two or three times; [and I] jerked it and everything else and I couldn't get it out. I got tired of wrestling with it, so I just reached over, pulled the cable, released the MESA, and down it went."

Bean: You haven't got anything to do, so you can take this *(possibly the jettison bag)* with you.

Conrad: Hey, I'll tell you what we're parked next to.

Bean: What?

Conrad: We're about 25 feet in front of the Surveyor Crater. *(guffaws)*

Bean: That's good. That's where we wanted to be.

Conrad: I bet you when I get down to the bottom of the ladder, I can see the Surveyor.

Bean: Rog. Hey, guy; you want to take this [free end of the LEC] with you, Pete?

Gibson: Sounds good, Pete. Just like you wanted.

Conrad: Just swing her out here. *(to Gibson)* That's right. *(pause)*

Bean: Okay; now hold her there a second, Pete.

Conrad: Okay.

Bean: One second. *(pause)*

Conrad: Do you have any TV, Houston?

Gibson: Roger. We've got a TV. *(pause)* No Pete Conrad as yet.

Conrad: No; I'm at the top of the ladder. *(to Al)* Okay. Now look, this thing is all the way out of the bag. How do you want me to do it? This way?

Bean: Just keep doing.

Conrad: Huh?

Bean: Adios.

Conrad: No, but this thing isn't all the way out of the ...

Bean: That's okay.

Conrad: Here; let me have this end of it. Let me come back up the ladder a notch.

Bean: Okay.

Conrad: That a boy! Which end [of the LEC] is that? Which end do I want? This is the end I want.

Bean: There you go.

Conrad: There we go. *(giggling)*

Bean: Look at that stuff go. Sure flies at 1/6 [g], doesn't it?

Conrad: Yeah. Wait a minute. *(pause)* Looks like we got 900 feet of this stuff!

Bean: Okay; just a second. Don't go down yet. I've got to get my camera on you, babe.

Bean – "I couldn't get down there, so I just turned the camera upside down and flashed a few. And they all turned out pretty good."

Because the hatch is so small, the astronauts cannot wear their cameras as they get out of the spacecraft.

Conrad: I can't go down yet, anyhow. I got to – Whoop! – get the LEC all the way down.

Bean: Okay.

Conrad: There we go. *(long pause)*

The end of the LEC is visible in the TV picture. This marks the first use of an Apollo color TV camera on the lunar surface. This particular camera had been used in the Apollo 10 CM; after that mission, it was modified for use on the lunar surface during Apollo 12.

Conrad: All right. Still can't figure out what kind of snarl I've got here. *(long pause)*

On Apollos 14 and 15, the crews carried some of the gear up and down the ladder by hand. For Apollos 16 and 17, the crews decided that the LEC was more trouble than it was worth and got rid of it entirely. The only relic was a lanyard they used to raise and lower the ETB (Equipment Transfer Bag) with the cameras in it.

Interviewer – "Was there any discussion of getting rid of the LEC at this point in Apollo?"

Conrad – "No, I don't think so. It worked fine, once we got it out."

Conrad: Hey, Al.

Bean: Yeah?

Conrad: Can you look out your window?

Bean: Sure. *(pause)*

Conrad: All right; I think I see what's wrong.

Bean: What's the problem?

Conrad: Oh; that LEC came out of the bag in three pieces and, as you would might well imagine, I picked the wrong piece.

Bean: Do you want me to pull it back in and throw you the end?

Conrad: No. That's not the problem.

Bean: It's no trouble.

Conrad: I got it right now. Man, they aren't kidding when they say things get dusty.

The end of the LEC has dragged on the ground and has picked up a lot of dirt.

Conrad: Whew! I'm headed down the ladder.

Bean: Okay; wait. Let me get the old [70-mm Hasselblad] camera on you, babe.

Conrad: Okay. *(long pause)*

On the TV picture, Pete's feet appear at the top of the picture as he makes his way cautiously down the ladder, stepping down from one step to the next with his left foot and following with his right. As the astronauts gained confidence, they were able to take advantage of lunar gravity and hop up and down the ladder, using their hands on the outer rail to guide themselves.

> **Conrad:** Man, is that a pretty looking sight, that LM.
>
> **Gibson:** You're coming into the picture now, Pete.
>
> **Conrad:** Okay.
>
> **Bean:** Okay: got the old [16-mm sequence] camera running.
>
> **Conrad:** Okay. *(pause)* Down to the pad.
>
> **Bean:** Okay.

After pausing on the next-to-last rung, Pete steps down to the last one, gets his hands in position and jumps down, sliding his hands along the outside rails as he drops. After he gets down to the footpad, the bottom rung is about level with his waist.

> **Conrad:** *(as he lands)* Whoopie! Man, that may have been a small one for Neil, but that's a long one for me. *(pause)*

> **Interviewer** – "I understand that there was a bet on your saying that."
>
> **Bean** – "Who'd you bet?"
>
> **Conrad** – "You know who I bet."
>
> **Bean** – "Nope. I forget."
>
> **Conrad** – "A reporter, who thought the government put words in our mouths."
>
> **Bean** – "Oh!"
>
> **Conrad** – *(laughing)* "I also had $500 riding on it, but I never got paid."
>
> **Bean** – *(laughing)* "I didn't know that! Is that right? I kind of remember it, a little. Oh, well."
>
> **Interviewer** – "Do you want that story as part of the record?"
>
> **Bean** – "Put it in. It will be good for the myth. We're trying to create a Conrad Myth. Big Bucks on this. Can't have too many human interest things."
>
> **Conrad** – "I tell the story, but I don't tell who I bet."

Actually, Pete does occasionally reveal that the reporter was the late Oriana Fallaci. A more detailed version of the story can be found in Andrew Chaikin's *A Man on the Moon*.

Pete Conrad was probably the most playful of the moonwalkers – challenged only by Charlie Duke. Ulli Lotzmann (*Apollo Lunar Surface Journal*, ALSJ, contributor) tells us that Pete's motto was always, "If you can't be good, be colorful" – and Pete was both.

Conrad: I'm going to step off the pad.

Conrad: *(as his foot touches the surface)* Mark. Off the ... Oooh, is that soft and queasy. *(pause)* *(holding on to the ladder as he tests the footing)* Hey, that's neat. *(pause)* I don't sink in too far. *(pause)* I'll try a little ... *(letting go of the ladder and stepping out of the LM shadow)* Boy, that Sun is bright. That's just like somebody shining a spotlight in your hand. *(pause)* Well, I can walk pretty well, Al, but I've got to take it easy and watch what I'm doing.

Conrad: *(gleeful)* Boy, you'll never believe it. Guess what I see sitting on the side of the crater!

Bean: The old Surveyor, right?

Conrad: The old Surveyor. Yes, sir. *(laughing)* Does that look neat! It can't be any further than 600 feet from here. How about that?

Gibson: Well planned, Pete.

Generally, all the astronauts had a difficult time estimating sizes and distances on the Moon. For example, during the site description, Pete and Al badly underestimated the distance to the large crater on their western horizon. However, in the case of the Surveyor, Pete is dealing with a familiar object. Because he knows the actual size of the Surveyor, Pete's distance estimate of 600 feet is almost right on.

Conrad – "You're right. It was a familiar object. We had been around it a lot. And, as I remember, the Surveyor stands about that tall *(gesturing)* ... And, as it turned out, we were, what, 500 feet?"

Interviewer – "535 feet, it says in the mission report. Not bad."

Bean – "I'd call it okay."

Conrad – "We knew what size it was, because we worked with an exact mock-up – exact, at least, in dimension – maybe three or four times in training. We would practice with it and even cut the appropriate tubing; and then they would replace the tubing and we'd do it over again. I don't think we cut the scoop off, but we practiced with those big bolt cutters. And the big thing was that we had to know what was the right thing to cut because that thing [the actual Surveyor] still had propellant in it and, for all they knew, it was still pressurized and everything. And, again, [as was used in the LM] that was hypergols [fuel components that ignite as soon as they are mixed]."

Bean – "But there's one thing we don't say here but we do later. I can remember the first time I looked at it and I thought it was on a slope of about 40 degrees [instead of the actual slope of about 10 degrees]. And I remember us talking about it in the cabin, about having to use ropes. How are we going to get down there? How come they screwed up so badly [on the slope estimate]? And I think I was fooled because, on Earth, if something is sunny on one side and very dark on the other, it has to be a tremendous slope. We weren't getting [scattered] light in there like you do on Earth. So when light finally did strike, it was real ..."

Conrad – "It turned out it was real flat."

Bean – "Yeah. But I can remember us talking about ropes and how were we going to get there and what can we do. And there it was, sitting there at 11 degrees like it should be."

Conrad – "I guess it was still in the shadow [here at the start of the first EVA]. But, by the time we got to the going down into the crater [toward the end of the second EVA], it was no longer in the shadow. Almost 20 hours had gone by."

Interviewer – "For Dick Gordon to have seen it, there had to have been sun on the spacecraft, but not on the ground it was sitting on."

Bean – "That's what it was. He saw it shining in there, I'm sure."

Conrad – "Something is shining on it."

Bean – "But you can see here how steep it looks like it could be."

Conrad – "Yeah, that does look real steep. And it turned out we just walked right straight down to it."

Bean – "Yeah, it was nothing. It was just like they said. Shows how the lighting can screw you up, because of those shadows being much darker because there isn't that light [scattered by] the atmosphere filling in."

Conrad: Okay. Let me see. I've got a little chore [the contingency sample] to do here, right? *(to Gibson)* Say again.

Gibson: I say that was well planned, Pete.

Conrad: Yeah. Just a couple of months [of planning and training] with a lot of people. Let's see. *(reading)* "Deploy the LEC and the MESA." That's done. I'm looking at my mobility; c.g. shift. *(moving off camera to the right)* I have the decided impression I don't want to move too rapidly, but I can walk quite well. The Surveyor really is sitting on the side of a steep slope, I'll tell you that. Okay. Now I'll work on my contingency sample. *(pause)* Got to walk real careful, Al.

Bean: Okay.

Interviewer – "After Apollo 15, somebody estimated that Dave and Jim increased their average walking/running speed from 1 meter per second on the first day to 1.5 on the second day and up to 2.0 on the third. Do you have any recollections about how quickly walking and running got easier?"

Bean – "Don't forget, he didn't really have a lot of acclimation time, here. I don't know how many minutes has passed since he got on the dirt, but it gradually gets better and better and, of course, like any training, it gets better faster initially."

Interviewer – "It's 2 minutes since he stepped off the LM footpad."

Bean – "That's what I mean. He hasn't hardly even been out there for any time at all and, yet, he's walked over and looked at the Surveyor and he's also run around and he's getting ready to do some work and it's only been 2 minutes. He really didn't have to get acclimated all that much."

Pete had been out of the room, and we told him we'd been talking about acclimation.

Conrad – "It still took a while. I said in there somewhere, 'You got to take it easy.'"

Bean – "The fact that you can even do this much [this quickly] shows how fast humans adapt."

Conrad – "Of course, we also knew that it was going to work, because Neil and Buzz got around okay."

Interviewer – "Did you have much interaction with them after they got back? They went off on their around-the-world PR tour pretty quick."

Conrad – "We had one debrief, I think, and that was about it."

Bean – "It seems to me you got an initial debriefing because, one day, you came back with a whole list of things that we hadn't talked about, and you said, 'These are things I've talked with Neil about that we needed to put in our plan. So you had some kind of a private conversation with Neil and then, about a day or two later, we all had the crew debriefing. That lasted about 4 or 5 hours and we got some of those things [that Pete had talked about with Armstrong], plus some others. So you must have gone over to where they were in quarantine. And you got all the things you wanted."

Conrad – "That's right. I never felt that we missed anything by them being run off."

Conrad (from the 1969 Technical Debriefing) – "I went down the ladder and the lighting was excellent. I had no trouble seeing where I was. It took me about 5 to 10 minutes to acclimate to what was going on. I didn't have any trouble moving around, but I felt a little rocky. It just took me a while to get organized. This feeling was not bad the second time I got out. As soon as I got out the second time, away I went. So, like anything else, there's a slight learning curve which took all of 5 minutes and away we went."

Conrad: *(to Al)* Can you see me all right?

Bean: Not yet. *(garbled)* back of the window for ...

Conrad: Okay.

Bean: ... just a second.

Conrad: *(garbled)* in a hurry. *(long pause)* *(to Houston)* As you might suspect from some of the pictures Neil brought back, gang, I have several small rocks sitting out in front of me that have a neat amount of dirt built up around them. I'm not sure my descent engine didn't blow 'em there. But then again, it may not have.

Gibson: Roger, Pete. Copy that. Is the dirt built up on the side closest to the LM?

Conrad: Well, let me ... I'm going over to get my contingency sample, and I'll get one of the rocks in the sample. And yeah, as a matter of fact, it is built up on the side that *(pause)* the LM landed on. Let me get a ... *(pause)* Well, there's one scoop. And another with some more rocks in it. *(pause)* Whoo! This dirt's just like the 1/6-g airplane, Al ...

Bean: Flies up in the air.

Conrad: ... *(chuckling)* and you get to chase it around. It's wild. Now, I'll tell you ... You know, this Sun ... It really is ... It's just like somebody's got a super-bright spotlight. Here's another good-looking rock – whoops – in the sample. *(pause)* Here's another rock I want to get in it. *(pause)* I think that's about enough, don't you? Except there's one big rock that's too pretty to pass up. No; I may not be a hog. It won't fit. I'll go over here and get this other one, though.

Conrad – "If my memory serves me right, the contingency sample got sent up right away on the LEC. I think we wanted to get it in the spacecraft so that we had it [in case they had to leave immediately]."

Bean: Boy, you sure lean forward, Pete.

Conrad: Hey, "lean forward": I feel like I'm going to fall over in any direction.

Bean: You're leaning about ...

Conrad: Say, Houston; one of the first things that I can see, by golly, is little glass beads. I got a piece about a quarter of an inch in sight, and I'm going to put it in the contingency sample bag, if I can get it. I got it. Am I really leaning over, Al?

Bean: You sure are. On Earth, you'd fall over, I believe.

Conrad: Huh?

Bean: On Earth, you'd fall over leaning that far forward.

Conrad: It seems a little weird, I'll tell you. Don't think you're going to steam around here quite as fast as you thought you were.

Interviewer – "Did you think, pre-mission, that you'd be able to move pretty quickly?"

Conrad – "Well, I said in the technical debrief that [when] we finally started to move around fast [it was] because it was harder to move around slow. So, maybe I'm trying to move around slow [at this early stage of the EVA] and I'm thinking I'm not going to move around fast, because slow is hard. But once we got the hang of it, we really moved out."

The following exchange is from the 1969 Technical Debriefing.

Conrad – "We listened to Neil's and Buzz's comments [on walking traction, balance, distance and direction, pace and stability] and ours are exactly the same. There's no need to go over them, other than just to remind you to lead [that is, anticipate] your direction changes slightly; but you acclimate very rapidly and it's no problem."

Bean – "I never noticed any slippery surfaces such as Neil and Buzz pointed out. The ground never felt slippery at all to me. The c.g. problems they had were the same. I was very careful not to walk backward; because, I noticed a couple of times when I did, I usually stepped in a crater or on uneven ground and it put me off balance. *(to Pete)* What do you think about the slipperiness?"

Conrad – "I didn't notice any slipperiness, but I think the other comment about moving backwards is the fact that you have such a mass on your back. Al commented to me and I noticed, watching him, that you look like you're standing with quite a forward tilt; but all you're doing is putting your c.g. over your feet. Your c.g. is quite far aft with the PLSS, so you have the tendency to lean at what, at first glance, looks quite far forward; it's not."

Bean – "Pete mentioned that it only takes you about 5 minutes to learn how to move around; the second time you go out, you really don't need the 5 minutes. Neil pointed out that this was the best thing to allow for acclimation. I concur 100 percent. Another good thing is both those Pogos. The mobile Pogo that FCSD has is good except that it needs a Z-axis freedom [meaning up-down] that it doesn't have now."

The Flight Control Systems Division provided simulators for 1/6-g environment – such as the Pogos, a set of harnesses and supports that gave a partial feel for the 1/6 g but not as faithful as the atmosphere in airplanes flying 1/6-g parabolas.

Bean – "The one on the centrifuge is excellent. I found that running around on the lunar surface, moving from side to side, hopping, and so on were almost precisely like using the one in the centrifuge. I'd recommend having a couple of exercises over there before you go and recommend changing the terrain over there so that the simulation includes a few more big craters, little hills and dales. I think it would be very good training."

Conrad – "And there's another thing. There's no such thing as 'walking' on the lunar surface. Wherever you go, you just want to go at a lope. If you walk, it takes more energy to move slowly and take a normal step than it does to lope."

Bean – "It's interesting, and I know we commented about it when we were doing it. If you look at somebody's footprints on the Moon, it's almost exactly the opposite of the way they are on the Earth. On the Moon, you can see a flat footprint as the guy lands and then he pushes off with his toe so it ends up being sort of dug in at the toe and flat in the rest of the print. On Earth, a fellow steps forward, lands on his heel, which digs in, and he kind of drives off on his toe. This sort of bouncing along [that we did on the Moon], using your toes for springing and moving from side to side so that the c.g. is always over the foot that's landing, allows you to move at a pretty good pace and to move a good distance. I had the feeling that, if our TV had been working and if the TV hadn't been pointed in exactly the right place when we went out to 450 feet to lay out the ALSEP, it wouldn't have taken us over 2 minutes to run back, position the TV exactly right, and return to the ALSEP. It would have been no trouble and would have been the thing to do."

During Apollos 15, 16, and 17, several of the astronauts ran distances of up to 150 meters (500 feet). Their typical running speeds were about 5 kilometers/hour and they would have made the 900-foot run out and back in about 3 minutes.

Bean: I'll tell you, your boots are digging in the soil quite a bit. If you don't pick up your feet, you really kick a load of dirt ahead of you. Your left foot's got a big mound ahead of it right now that it's just pushing along.

Evidently, Pete is not yet lifting his feet as he walks.

Conrad: Uh-oh, did I hear a tone?

Bean: Yeah: I've got an H_2O A [flag].

This flag appears when the feedwater pressure drops below 1.2 to 1.7 psi.

Conrad: You do?

Bean: Yeah. I wonder why? *(pause)* Hey, Houston?

Gibson: Al, verify feedwater's On.

Bean: It's been on. It's still on. *(pause)*

Conrad: Boy, do I sink in. Wow!

Bean: Feedwater's on and it's still real cool in here.

Gibson: Al, Diverter Valve to Minimum.

Bean: Okay. It's Minimum now. What do you think I may have done? Broken through the sublimator or something?

Gibson: That's affirmative, Al. *(long pause)*

They are speculating that part of the porous plate in the sublimator has lost its ice layer.

Gibson: We have a good shot of you there [on the TV], Pete.

Conrad: Okay. Well I'm starting to take this baby [the contingency sampler] apart. While I'm doing that ... *(pause)*

Conrad: Houston.

Gibson: Go ahead.

Conrad: The descent engine, it's just like Neil's. I didn't dig any crater at all! *(pause)* Al, you've really got to watch your step down here.

Bean: Okay. *(long pause while Pete removes the thermal blankets from the MESA)*

Conrad: Look at all those good things in this MESA. Things that I've seen before. Didn't hardly stroke the gear [the struts] at all, and it looks like I landed just about vertical. *(pause)* Whee! *(chuckling)* Just like Neil and Buzz said, Al. You get on a little slope and you tend to keep going. *(long pause)* Almost fell over. *(pause)*

> **Bean** – "You don't have to hold yourself up much [because of the weak lunar gravity], but your momentum is the same as it is on Earth. And, so, when you get moving, you can't just stop. You kind of have to ... just like Pete said. It's like when you're going downhill and get going, on Earth, you have a hard time stopping. If you watch the TV, particularly of Apollo 11, you can see them run by the TV camera. At one point, Buzz says 'I'll go out and get the camera.' And he runs right by the sucker! 'Cause he hadn't learned to slow down in advance."

Conrad: Would you believe it? The MESA's too low for once.

When Pete released the MESA, it swung down 120 degrees and, had the surface been level and the LM perfectly upright, it would have been at a comfortable working height. Because of the uneven ground and the slight tilt of the spacecraft, the MESA was a little too low. Pete's "almost fell over," came as he started to raise the MESA to the proper height.

> **Gibson:** Al, how's the feedwater look now?
>
> **Bean:** Well, it's still got an 'A' in there [in the warning flag window on the top of the RCU], but I'm plenty cool. I went back to Minimum, and I'm sitting here in Minimum. What do you think happened? *(pause)*
>
> **Gibson:** Al, we would like to watch it a little bit. It could be instrumentation [that is, a sensor malfunction]. Standby.
>
> **Bean:** Okay. *(long pause)*
>
> **Conrad:** How long have I been out, Houston?
>
> **Gibson:** Pete, you've been out 25 minutes, and you're about 4 minutes ahead [of the planned timeline].

Pete and Al are each wearing an *Omega Speedmaster Professional* wristwatch on their suit sleeves and could use them, probably in stopwatch mode, to time the EVA duration. In fact, page 31 in the Apollo 12 LM checklist – the one they used in the cabin – includes "Start Wrist Watches" among the final tasks to be done just before they opened the hatch. Evidently they skipped that "start watch" step.

Conrad – "We were wearing watches. It's probably covered up by the checklist at this point. That's why I asked the question to the ground."

Interviewer – "Did you set the watches to 12 o'clock or something like that, when you started out?"

Conrad – "I don't think the watch is running in accordance with our EVA time. I think it's just got Houston time on it."

Bean – "You sure? I thought we did. Maybe we didn't. I thought we used [the watches to keep track of the EVA elapsed time]."

Conrad – "We may have. I know we had the watches on all the time, but I don't ever remember starting or stopping them. Because they're the world's hardest thing to start or stop with your glove on. See, I'm asking the question, 'How long have I been outside?' I wouldn't know how long I'd been outside (with the watch on Houston time), unless I was doing a bunch of mental calculations, which I didn't do."

Bean – "Well, how'd we know we were on the timeline, then?"

Conrad – "We had to ask the ground."

Bean – "Well, isn't that strange. Maybe we just said, 'It's too much trouble [to start the watches], so we'll get Ed to keep up with it'."

Conrad – "Yeah."

Conrad: Okay. I got the table out, adjusted the MESA [to a comfortable working height], and I'm setting up ETB at this time.

Conrad (from the 1969 Technical Debriefing) – "The nicest part of the exercise was that everything went according to the checklist as best I could see. It went exactly the way we practiced it. And we had no trouble with the equipment. I had excellent mobility in 1/6 g. I missed the fact that I couldn't bend over. That's something I knew I was going to face the whole time and it didn't bother me too much."

Conrad: Let's see. How is this packed? Very nice. Very nice. *(pause)* Hey, Al, you can work out here all day. Just take your time. *(pause)* Almost too cold on Intermediate [cooling]. I'm thinking seriously about going to Min. *(pause)*

Pete is standing in the LM shadow and isn't getting any solar heating. He still needs to get rid of excess body heat, but at a very low rate.

Bean – "It was just amazing how quickly you could get cold with that thing, depending on the setting and how hard you'd been working. If you were working hard, you'd change [the cooling setting] and, in 30 seconds, you were comfy again. But, boy, you could really get too cold too fast, if you quit putting out heat or got out of the Sun. It was an amazing machine, that water cooling. It was like you were in water. You weren't wet, but it cooled you that fast."

Conrad – "The other thing I sort of remember is being amazed at the narrow range of the change in water temperature ... You know, it's only about like a few degrees between freezing your ass off and chilling your kidney to being warm. I also remember it ran around 55 or 58 degrees or something like that, where they measured it. Right where it came in, which was down around your kidney, it used to freeze your right side off."

Conrad: *(to himself)* There's some color charts. *(long pause)* Dum dee dum dum dum. Dum dee dum dum dum. No. Which is right side out? The other way. No, that's not right. No. *(pause)* I think our next big surprise, Al, is getting this thing [the ETB] up [to the cabin with the LEC].

Bean: Getting what up?

Conrad: The ETB.

Bean: Oh, is that right?

Conrad: *(laughing)* We'll see what happens.

Bean: Okay.

Conrad: How's your water?

Bean: Oh, it still shows an A, but it's cool. It may be instrumentation.

Conrad: Let's hope so. Just beginning to warm up to this task. *(long pause)* LCG [Liquid Cooled Garment] water pump sounds like a diesel truck running out here. Comforting to know that it's running. I'm off to get the [PLSS replacement] battery[s].

Bean: Okay. I think I know what happened, Houston. I think I know what happened. *(pause)*

Gibson: *(making a rare misidentification)* Pete, go ahead.

Bean: Ahhhh!

Conrad: What did you just do, Al?

Bean: Man, I just figured it out.

Conrad: You sure did. You just blew water out the front of the cabin. *(correcting himself)* Ice crystals.

Al has just noticed that he had accidentally knocked the hatch closed and, now, has re-opened it. With the hatch closed, the output of the sublimator slightly raised the cabin pressure. The build up of water vapor pressure was quite small, rising only until the output of the sublimator was balanced by outflow through the dump valve. The build up wasn't enough to keep Al from re-opening the hatch, but was enough to interfere with operation of the sublimator.

> **Bean** (from the 1969 Technical Debriefing) – "Just as soon as Pete got out, I had to move over to the right window to take some motion pictures; [and] when I did, I pushed the door partially closed and went to work. About this time, I got a low feedwater pressure [warning]. We stood around and tried to figure that out for a while and, finally, I happened to glance down and noticed the door was closed. I realized what had happened. The outgassing of my sublimator had closed the door, with the result that I didn't have a good vacuum inside the cabin anymore. I quickly dove to the floor and threw back the hatch. The minute I did, a lot of ice and snow went out the hatch. Pete commented about it, and it wasn't 30 seconds until my water boiler started operating properly again. I think that's something you're going to have to be careful about when you're moving around inside there. I hadn't thought about it before the flight."

When Al opened the hatch, the water vapor in the cabin rushed out into the lunar vacuum, expanding, cooling, and freezing as it did so.

> **Interviewer** – "Pete, did you see the ice when it came out?"

> **Conrad** – *(laughing)* "Oh shit, yes! But, see, there's something I don't understand. Where the hell were the LEC straps all this time?"

Bean – "They're hooked on the side of the LM, I think."

Conrad – "Yes, but they went clear up into the cabin. So my problem is, how the hell did the door get all the way closed if the strap was hanging out, other than it may have warped a little bit. That's interesting."

Bean – "You sure it was going into the cabin? I kind of remember it going into the cabin."

Interviewer – "Well, we certainly see it hanging down from the porch in the TV at this point. Pete brought it out the door."

Bean – "I guess it has to come out the door, otherwise, how would I ever reach it?"

Interviewer – "The pressure that built up in there from the sublimator wasn't very much."

Conrad – "Yeah, it didn't take much."

Bean – "I'll bet the door just warped around the straps a little bit. Just the reverse of that peeling we did earlier."

Bean: That's what had happened to the PLSS.

Conrad: What's that?

Bean: Oh, the door had swung shut, like it did before; and probably bothered the sublimator, 'cause it wasn't in a good vacuum anymore.

Conrad: Uh-huh.

Bean: So I opened the door and it's probably going to start working in a minute.

Conrad: I should hope so. When you opened the door, that thing shot iceballs *(laughing)* straight out the hatch.

Bean: Yeah. There's probably all from my ... Never thought you'd have to do that. Hey, you bent the outside of that front hatch on the way out. You tore some of the skin.

Conrad: How did I do that?

Bean: I don't know. Must have hit your PLSS there. It's got a nice scrape mark right along the way out.

Conrad: Sorry about that. Trying to be gentle. *(pause)*

Bean: Houston ...

Gibson: Al, It's looking better.

Conrad: *(to Al)* Did you ...

Bean: *(responding to Gibson)* Yeah. I didn't realize that the hatch could close quite so tightly like that, because when I was working on the other side [Pete's side] of the cabin, the hatch went closed and I didn't notice it. And apparently it quit holding a good vacuum in here. My H_2O 'A' flag is off, now, so everything is copasetic.

Gibson: Roger, Al. It [meaning Al's PLSS telemetry] looks good down here.

Bean: Okay.

Conrad: Okay. I've got both [PLSS LiOH replacement] canisters, Al; both batteries. As soon as I get them in here [in the ETB], I got to pack the contingency sample.

Bean: Okay. Sitting tight, waiting for you, babe.

They are about 30 minutes into the EVA and about 10 minutes from the time Al is supposed to have started out the hatch.

Conrad: Okay. I tell you, you really can't move as fast as I thought you could. You got to take it real easy. Get the feeling that I'm most spiffy on the balance up here. *(garbled)* Gonna have to ... *(long pause)*

Conrad: Oh, man, did I get dirt all over myself. *(pause)* This is what is known as dirt dirt. *(pause)*

Bean: Let me know when you start heading back out there [west of the ladder] to do the ETB, Pete. I'll get a good shot of you, babe [with the Hasselblad].

Conrad: Getting ready to do it in a second, Al. Just as soon as I get the bag, [the ETB, hooked to the LEC]. I got the contingency sample in the bag.

Bean: Okay.

Conrad: I've got everything else: PLSS batteries, LiOH canisters, and I just got to hook up [the ETB to] the LEC. Bo bo bo bo; dee dee; dee dee. *(pause)* I tell you one thing, we're going to be a couple of dirty boogers. *(pause)* Come on hook. That a boy. Why don't you take up a little slack, Al?

Bean: All right.

Conrad: Just a little.

Bean: In work.

Conrad: I tell you, this is dirt dirt. *(pause)*

Bean: *(laughing)* That's the greatest.

Conrad: What's that?

Bean: My end of the ETB [means LEC] just came out of its metal slot. Somebody that made this pin didn't make it like all the training units; they made it littler, so it came out. I caught it as it was going by. You know this metal pin that keeps it from sliding all the way through?

Conrad: Yeah.

Bean: Unfortunately, it's smaller than the metal holder. Kind of interesting. *(pause)*

> **Conrad** (from the 1969 Technical Debrief) – "Our two rock boxes weighed out, Earth weight, at 44 pounds and 52 pounds, if I remember correctly. Neither of those boxes, which obviously were the heaviest things we sent up, presented any problem. The only problem was one that was already mentioned. We knew it was going to happen anyway, and I really don't see a heck of a lot you can do about it. This problem is that the lower end of the strap got completely covered with dust and I got dust all over my hands and over my suit arms from handling that strap. I don't see anything you can do about it."

The eventual solution was to carry gear up and down the ladder by hand and not bother with the LEC.

> **Conrad:** Wait; wait; wait; wait; wait; wait.
>
> **Bean:** Okay. *(pause)*
>
> **Conrad:** And, nasty booger, come on. That a boy. Wait just a minute. *(pause)* Okay; now … Hold the phone a second. *(pause)*
>
> **Conrad:** You can … Wait; wait. Here I go. Dee dee; Dee dee. *(hearty laughter)* Wait until I get in this shadow. Because I can't see what I'm doing looking right into the Sun.
>
> **Bean:** Okay.

Pete had planned to stand directly out in front of the ladder. Had he landed with the ladder pointed due west, he would have been in the LM shadow throughout the LEC operations. However, because the spacecraft is rotated right, he is standing north of the shadow and in full sunlight.

> **Conrad** – "I couldn't see anything. I was looking right smack into the Sun. So I had to move over and get in the shadow. And I couldn't really look down while I was working the LEC real easily, either. I was holding onto this thing, so I'm moving sideways and then I step in a crater hole. I stepped in a couple of them."
>
> **Bean** – "That's right. When you look up [at the Sun], then you don't see for a little while."

Conrad – "And the other thing was we had a porch rail on either side. I had to get the ETB lined up with the center of the hatch, too."

Bean – "That's right. It was hard to do, because you couldn't get in your normal position and still see."

Conrad – "Once we got this figured out, the rest of the time we ran stuff up and down, no sweat. We were on the learning curve, here. It didn't work the same in 1/6 g as it worked in 1 g."

Bean – "I remember it flying all over [that is, bouncing around]."

Conrad – "Yeah, in 1 g there was enough weight on it that you could pull it down and it was almost damped. But you start pulling on it in 1/6 g, it started doing this and then this."

Interviewer – "Bouncing up and down and rotating."

Bean – "We worried it was going to break the camera. We had a camera in there."

Conrad – "If you think about it, the batteries and the LiOH canisters, that was probably 10 [or] 12 pounds. And so, up there it was like 2. And those were nylon straps or whatever they were and they got a little elasticity in them and they got going!"

Bean – "And another thing was that, when you pulled on the end of the LEC and it took the bag up in the air, it just threw dirt all over the place. It was swinging around. And I guess we were thinking, 'Gee, this thing could bang into the spacecraft'."

Conrad – "Yeah, we worried about that, too. I remember that."

Bean – "I don't think it had any damping because of that light gravity and it was acting weird."

Conrad – "Yeah, especially once it started up."

Bean – "It started swinging around, scraping on the inside, and getting dirt all over you."

Conrad: I'm about to fall down this little crater hole. Oops. Boy, it really does get ... I gotta get over here in the shadow. I'm down in a – oops – another crater hole. *(laughs)*

Bean: It's a regular obstacle course over there.

Conrad: Man, am I going to get dirty! Hold it; now. [I'll] back up a little ways. *(pause)* Tell me if it's [meaning the LEC] clear of the porch rail, huh?

Bean: It is now.

Conrad: It is? I can't see a thing, looking into the Sun. Pull.

Bean: Okay; I'll bring her in. *(pause)*

In order to get the LEC and bag properly aligned between the porch rails and the hatch opening, Pete has to get out into the full Sun.

Conrad (from the 1969 Technical Debriefing) – "I guess the biggest note I'd like to make – and I think Al and I agree on this – is that the side visor, the side blinders, were excellent. But you also need a top one. We had a low enough Sun angle that, anytime you put your hand up, looked directly up-Sun, and just blocked the Sun out, you could see perfectly up-Sun. It was only when the Sun was shining in the top of the visor that we had difficulty. So, I think we need to modify the visor so that you have a center-top shield that you can pull down and blink the Sun out. If you have that, you can turn 360 degrees and see perfectly in any direction. It will also allow you to look in shadows. The only other time [than when you are looking directly up-Sun that] you have difficulty seeing in a shadow is when some other object is reflecting sunlight into your visor when you're trying to look in the shadow. Once you're in the shadow, you can see well. This is nothing new; Neil already pointed that out."

Bean: Comes in easy.

Conrad: Okay.

Bean: Good rig.

Conrad: I can't ... Wait a minute. Wait; wait; wait; wait. *(moving over into the LM shadow)* That a boy; now I can see. *(pause)* Pull. Keep going. *(pause; laughing)* Am I getting dirty. Whee! Got it.

Bean: Okay. I'll have this stuff right back out to you in a flash.

Conrad: Okay; let me see. While you're doing that, what was I supposed to do? Oh, I know: "Possible TV deploy." I'll go work on the, yeah, *(pause)* tripod.

Bean: Okay.

Conrad: Dum dum, da dee da dee dum. Trying to learn to move faster. *(pause)* Pretty good. Hey, I feel great.

Interviewer – "What gait did you use? Did you try the kangaroo hop at all? And there's a skip stride that some folks used. And then there's the loping, foot to foot stride."

Conrad – "That's what we used."

Bean – "We tried the bunny [hop], but it was too much work. The loping was the easiest. For short distances, sometimes you just walked, you just kind of bounced over. Where you did the loping was for long distances."

Conrad: How long we been out, Houston?

Gibson: Pete, you're 34 minutes into the EVA; and you're right on the nominal timeline. *(long pause)*

Bean: That contingency sample is black.

Conrad: You'd better believe it. *(pause)* I may have filled the bag too full. *(pause)* Oh, I know what it is [that's keeping the TV camera from coming loose]. Dee dee dee.

Bean: *(putting the Hasselblads in the ETB)* One camera; two cameras. *(pause)*

Conrad: Come on, little fella. *(straining a little as he tries to remove the locking pins that hold the TV camera in the MESA)* [Here] comes one TV camera.

Bean: Okay; ready for you, Pete.

Conrad: All right. Ah, shoot. Ah, all right. I've got to stop what I'm doing. Let me come over here and get it. Here I come.

Bean: Okay.

Conrad: Dee dum dee dum. I feel like Bugs Bunny. *(pause)* *(giggles)* *(pause)*

Bean: Take it away whenever you need it.

Conrad: Okay. I'm going right, now. If I fall over I can ... *(pause)*

Bean: Sure goes out easy.

Conrad: Yeah. Yeah. *(pause)* Let her keep coming. Let me get over to this side. Let's get it over the handrail. Hold it, hold it right there.

Bean: Okay.

Conrad: Now, I can't see it on account of the Sun, so tell me when it's over the handrail.

Bean: It's over the handrail now.

Conrad: All right. Just lower it real slow. That a boy. Easy. Hold it right there.

Bean: Okay.

Conrad: Just stay put. *(pause)* Okay, let her go.

Bean: That's it.

Conrad: All right. Now, just hold it right there. Hold it; hold it; hold it. *(removing the cameras from the hanging bag)* One ... That is – dad-da-dee-dee – two. *(pause)* Okay; let it go. All the way.

Bean: Okay.

Conrad: All right: *(reading his checklist)* "LEC; possible TV deploy; LMP egress; contingency sample area [photos]; deploy color chart and place 70 millimeter [camera] on the MESA." So I get me a camera [to photograph Al coming out of the LM].

Bean: Okay; be out in a minute. Got to set the [sequence] camera, and I'll be right out.

Conrad: All right. Let me know so I can photograph you.

Bean: Okay. *(pause)*

Conrad: Okay, contingency sample 8; f/8. *(pause)* Eight and five. [I've got to] step back. *(long pause)*

Conrad: *(to Gibson)* We sampled in quite a few places, Houston, so I'm taking a bunch of pictures.

Gibson: Roger, Pete. *(pause)* Pete, for your information for those photos, your shadow length right now is about 45 feet on a level plane.

Conrad: Okay, very good. "[Photograph] contingency sample area" I got. "Deploy the color chart (on an undisturbed surface)" Ho ho. Take your time, Al. *(pause)* Hey, I'm learning to do it. *(pause)*

Pete may be learning how to run or, at least, to take advantage of lunar gravity by using bouncing steps and hops as he moves. Houston's estimate of a 45-foot shadow suggests a Sun elevation of 7.6 degrees. Because the Sun moves about 0.52 degrees per hour in the lunar sky, this estimate is consistent with our earlier estimate of 5.7 degrees.

Bean: *(Pete belches)* Houston, how does the LM look? I'm getting ready to go out the front door.

Gibson: Roger, Al. Stand by on that. *(long pause)*

Conrad: Dum dee dum dum. *(pause)* Whoops. No way I'm gonna . . . I wonder if I can get in the bottom of this crater hole?

Pete may be getting in a small crater so that he can set the chart on the Sun-facing inner slope.

Gibson: Al, Houston. The LM is looking good. You're Go for egress. Pete, you're at 40 minutes into the timeline, and you're about 4 minutes ahead.

Bean: Okay.

Conrad: Ho, ho. Oh dear. *(pause)*

Conrad – "I dropped that color thing [a color chart that they will photograph to help the photo processors back on Earth] and I got dirt all over that. And then I thought I was going to shake it off and it didn't come off, which is what was really weird because, later on, in the spacecraft, all that stuff floated out when we got to zero-g."

Interviewer – "Could you have kneeled down to pick it up?"

Conrad – "No, you can't kneel."

Interviewer – "I ask because, in the J-mission [Apollo 15/16/17] suits, you could kneel, especially if you had something to hold on to."

Conrad – "Yeah, because you can bend over [in the J-mission suit by bending at the waist, a feat that was all but impossible in the Apollo 11/12/14 suit]. I can't bend over, so it doesn't do me any good to kneel. If I kneel, I'm not getting any closer [to the ground] … Well, I'm getting closer, but I can't bend over and I can't get my arms down past my knees, so it's still not doing me any good."

Interviewer – "You said that you dropped the color chart."

Conrad – "Yeah. What should have been something very simple turned out to be a real pain in the ass. It got all fucked up."

Bean – "Did it spear in or something?"

Conrad – "Well, I thought it would just drop flat. And, not thinking about it … The reason it would drop flat [on Earth] if you let go of it any way at all is it would go to its flat-plate area (that is, air resistance would tend to orient the falling chart parallel to the ground). So, when I dropped it, it just went straight and it knifed in. And, of course, dirt flew up and came back down all over it."

Conrad (from the 1969 Technical Debriefing) – "One other item in the first EVA – the color chart – I took out because I could not bend over, and there was no reasonable way to stick it in the ground. I tried to work it into the ground so that it was perpendicular to the Sun. It didn't work because of the soft dirt. It fell over and became covered with dust. I got it back up and tried to brush it off, but it was impossible. I just made a complete shambles out of it. The dust clung to it so badly that we didn't get a color shot of that."

Bean – "Well, you do have pictures of it [from EVA-2]."

Conrad – "Yeah, I know. But it's pretty dirty."

Bean – "Yeah. We were beginning to understand how the dirt got on everything. The reason, I think, is because all the teeny, teeny, fine dirt that isn't on Earth any more. On Earth, it's been washed away or made into mud, and just the bigger pieces of dirt are left. And, yet, up on the Moon, none of that ever blows away. The wind doesn't take it away."

Conrad – "We got to where we rubbed it in the spacecraft."

Bean – "It was kind of greasy."

Conrad – "Well, we thought it was like graphite. It was almost a lubricant."

Bean – "Yes, and it was tenacious as hell. 'Cause, remember when I put that little automatic camera thing in the bag and couldn't find it? It was chrome and I still couldn't find it in that bag. *(Pete chuckles knowingly)* That was one of the mistakes we made."

Conrad – "Yeah. You mean the timer."

Bean – "The auto timer. We would have had a great picture. We still would have, see. We should have done it right in front of the LM. It would have been even a better picture."

Conrad – "Yep."

One of the perils of doing lengthy, detailed interviews of this kind is that, sometimes at the worst possible moment, the interviewer's attention drifts away from what the interviewees are saying – confident that the pearls are being recorded on tape. In this case, I (author Jones) was too busy thinking about dust in the context of other missions and didn't realize that Pete and Al were trying to tell me the wonderful story of an illicit photographic timer which, had they been able to find it in the rock bag, would have allowed them to take a picture of the two of them, standing together next to the Surveyor. Fortunately, they later raised the subject again.

Bean: Okay, Pete, here I come.

Conrad: Wait; wait; wait; wait.

Bean: You ready now?

Conrad: No, no, no, no. Let me come. Dum dee dum dum. Got to run through this [small] crater. Here I come. Now, wait a minute. *(reading)* "LM[P] egress at [f/]5.6 at 15 [foot focus]." I made a shambles of that color chart. I tried to throw it in the ground; and, naturally, it went in sideways and got itself so covered with dirt, you wouldn't know what color it was. *(pause)* Okay; I'm ready for you.

Bean: Okay, you might want to give me some directions, too.

Conrad: All right, yeah.

Bean: Bumping anything? *(pause)*

Conrad: Okay. You're coming straight out and the further you can bend over the better. All right; move to your right.

Bean: Okay.

Conrad: That a boy. Down. *(pause)* That's it; get your knees down. That a boy. Good shape; good shape.

Bean: Okay; I'm pulling the hatch closed here.

Bean – "We closed the door for thermal reasons. They didn't want the temperatures in the LM affected by looking at either the dirt or the sky. You're getting all that radiant stuff going on in that vacuum. You have to be careful."

Conrad: Okay. Don't lock it. Okay, you're right at the edge of the porch.

Bean: Okay. *(long pause)*

Conrad: Man, if I'd landed 20 feet behind where I landed, we'd have landed right smack in that [Surveyor] crater. *(pause)* [How are you] doing?

Bean: Oh, it's kind of hard to move the door. I was just getting in and trying to get it. *(pause)*

Bean (from the 1969 Technical Debriefing) – "Both times I egressed the LM and tried to close the hatch, it took 45 seconds or so to find something on the hatch I could pull. I think it would be worth the effort to put some sort of hook or something as a permanent fixture on the outside of the hatch so that, when the last man gets out, he can pull the hatch closed without having to grab the protective doors over either the handle or the vent valve. It would save 45 seconds or so each time."

Conrad: There you go.

Bean: I'm going to try to keep the door open for us, there.

Conrad: Okay. *(pause)*

Bean: Pretty good; I'd better get my visor down though.

Conrad: Yes, sir. My, that Sun is bright. *(pause)*

Bean: Boy, the LM looks nice on the outside.

Conrad: Houston, let me ask you a question. How important is that color chart? I tried to spike it in the ground, you know, so it was perpendicular to the Sun; and it just didn't do that, and it's all covered with dirt.

Conrad: I can go back and re-salvage it, if you want to take the time.

Gibson: Press on with what you're doing there, Pete, and we'll get an answer back to you.

Conrad: *(to Gibson)* Okay.

Conrad: *(to Al)* Okay; ...

Gibson: Pete, press on. No problem.

Conrad: ... turn around and give me a big smile. *(responding to Gibson)* Okay. *(to Bean)* That a boy.

Bean: Okay.

Conrad: You look great. Welcome aboard. Okay. *(reading)* "Place [70-mm camera on MESA]" ... Wait a minute. The chart I didn't get. "Deploy color chart on undisturbed surface." Didn't make it. "Contingency sample area [photos]" I got, and "LMP egress" I got. I'm off for S-band antenni *(sic)*.

Bean: Okay. My, that Sun is bright.

Conrad: Yeah. Take it easy.

Bean: *(moving slowly off camera to the right)* It feels good.

Conrad: Yeah, you really do begin to adapt.

Bean: If you hop it a little bit ...

Conrad: If you turn around and walk over to your right a little bit and look over that crater, you're going to see our pal [Surveyor III] sitting there. And that's one steep slope it's on. *(pause)* Okay; now what have you got all over your boot? Stop. You picked up a piece of landing gear insulation.

Bean: Okay. Here we go.

Conrad: That a boy. Okay, I'm going after [the S-band antenna] ...

Bean: Hey, you've got to watch it in these shadows.

Bean (from the 1969 Technical Debriefing) – "It just takes about 5 minutes to get used to walking around, and this time should be allowed in the first EVA. Once you learn that, you can start easing over and doing your job. I noticed no effect of our movements. I leaned forward and backwards about the same as I expected. I got used to it very quickly."

Conrad: Yup. You can't see what you're doing. Come over here where I am. See that Surveyor sitting there?

Bean: There that thing is! Look at that!

Conrad: Will you look how close we almost landed to that crater! *(pause)*

Bean: Beautiful, Pete.

Conrad: Look at the [LM] descent engine. It didn't even dig a hole! *(pause)* Okay.

Gibson: Pete; Houston.

Conrad: Houston, I'm in the process ... *(stops to answer Gibson)* Go ahead, Houston. *(pause)* Hey, Al? We've got to save this *(garbled under Gibson's voice)* ...

Gibson: Pete, will you give us status on the LM and also some comments on your boot penetration?

Conrad: Okay. My comments are exactly the same as Neil and Buzz. In fact, every time I get down in one of these little craters, I sink in a lot further. I'd say our footsteps are sinking in ...

Bean: What do you want to do with it [a thermal blanket that was covering the S-Band antenna that Pete has just handed to him].

Conrad: Put it over by the Y-gear [the north landing gear]. I think I cover that rock box with it later. Remember, [it was a] last minute change. And ... Where was I?

Bean: *(Garbled)*

Conrad: Well, I tell you. I think it's pretty much the same as Neil and Buzz found, don't you, Al?

Bean: I do. One thing I've noticed: it seems to compact into a very shiny surface [where you step on the surface]. I guess the particles are very small and very cohesive; so every boot print, as you look at it, looks almost like it's a piece of rubber itself. It's so well defined, you can't see any grains in it or anything.

MONITORING THE MOON – SETTING OUT THE ALSEP

Conrad and Bean are both on the lunar surface, have finished some of their house-keeping chores, and are about to unload the equipment that is to be left on the Moon.

Bean: The experiment bay looks real good.

Conrad: Yup.

The ALSEP experiments were contained in two compact packages. Pete and Al removed them from the SEQ Bay on the southeast face of the LM, and attached them to the ends of a carry bar so Al could take them out to the deployment site some 300 feet west of the LM. Details of the off-loading procedure were printed on a decal, which was fixed to the back wall of the SEQ Bay. As on Apollo 11, the pieces of equipment were mounted on booms and, once they were extracted from the Bay, they were lowered to the ground with pulley-mounted lanyards. Pete extracted Package 1 and Al extracted Package 2, which contained the RTG, a small, plutonium-powered electrical generator. Al will fuel the RTG after it is on the ground.

Bean: The LM exterior looks beautiful the whole way around. Real good shape. Not a lot that doesn't look the way it did the day we launched it.

Conrad: *(possibly pulling a lanyard to open the SEQ bay doors)* Light one. *(pause)*

Bean: Okay. Here we go, Pete. Ohhhhh, up they go, babes. One ALSEP. *(pause)*

They have raised the doors that cover the cavity where the ALSEP packages are stowed.

Conrad: There it is.

Bean: There it is, is right. [Now we] just lay it on the lunar surface. *(pause)* Better go to intermediate cooling, get good and chilled down. *(pause)* Okay.

Conrad: Wait a minute. *(garbled)* You've got to go easy.

Bean: Sure do. *(pause)* Here it *(probably the first package)* comes.

Conrad: Coming right out.

Bean: And just about right. Riding right out on the boom, Houston. Sure looks pretty.

Gibson: *(making a misidentification)* Roger, Pete. We copy. *(long pause)*

Bean: *(wanting to take a picture)* Look at me, Pete. *(pause)* It's a good shot, babe. The LM and everything's reflecting in your visor. *(pause)*

Bean – "I don't know it they ever tried to see if people could just pull the packages out and stand them on the ground. I think some of the things we did like this, where we had these nice, wonderful things that would deploy them and let them down, well, we didn't really need them. Now, I think one of the reasons they claimed we needed the rails and pulleys was, if you stroked the front gear, then it would be too high to reach in there. But I think the first thing a person ought to do when they're designing these things is see if – like taking stuff up the stairs [that is, the ladder] – if you can just grab it and do it. And then, if you can't come up with the most simplest thing in the world to do it. Nothing that's quite this fancy. Even if you had that high attitude, if you just had tabs pulled down, you could have pulled on the tabs and pulled it out and reach your hands up and got it. My comment would be to try to do it always the easy way first. And, if there's any way to do it the easy way, then don't build all that other stuff."

Bean: Lay her on the ground. Okay, I'll get mine out. *(long pause)* Right out on the boom, just like advertised.

Conrad: Wait until I get this [meaning the first ALSEP package] out of your way.

Bean: Okay. *(long pause)*

* * *

Conrad: It's going to look like ...

Bean: Say again.

Conrad: Going to look like the back of the flight crew training building in a minute.

Bean: Uh-huh.

Conrad – "It was where we practiced off-loading the ALSEP and deploying it."

Bean – "There was a lot of hardware hanging around. Well, look at it in the pictures. You can see it kind of starting to look stacked around there."

Interviewer – "And that was in Houston?"

Conrad – "No, at the Cape. It was also where the Surveyor [mock-up] was."

Conrad: Okay. SEQ Bay door is coming closed. *(pause)* Uh-oh. It's all right. Push that door ... *(pause)* What am I hung up on?

Bean: I don't know. *(long pause)*

Conrad: Hey, Al, *(garbled)*. I have something *(garbled)*

Bean: Let me look, Pete. *(pause)* Here's one here. Let me get it off for you.

Conrad: Thank you. *(pause)*

Bean: Okay, lift your left foot up and you're okay. *(pause)* Okay.

Evidently, Pete has gotten his foot caught in a cable, a common problem given the near impossibility of seeing one's own feet.

Conrad: Thank you. *(pause)*

Gibson: Yankee Clipper, Houston. High Gain Antenna: Pitch, minus 13; Yaw, 225. *(long pause; the communication improves noticeably)*

Bean: Hey, Houston, do you hear this constant beeping in the background?

Gibson: That's affirmative. We've heard it now for about the past 45 minutes.

Bean: What is it? *(responding to the second half of Gibson's communication)* That's right, so have we. What is it?

Gibson: *(static returns briefly)* Intrepid, we've tried to isolate it. It appears it's something on the downlink coming from the LM. *(long pause)*

Conrad: Hey, Al.

Bean: Yes, sir. And there's the lunar tools all set up for you, fella.

Conrad: Okay.

Al is about to fuel the RTG. To do so, he rotates the cask that contains the plutonium fuel element down 90 degrees to a horizontal position. He then removes a protective dome and, finally, removes the element and inserts it into the RTG. Pete will hand him the appropriate tools. The fuel cask was designed to survive re-entry into the Earth's atmosphere just in case there was a launch accident. On Apollo 13, the LM was used to get the crew and their crippled CM back to Earth and was abandoned just prior to re-entry. The fuel cask survived the journey and currently resides at the bottom of the Tonga Trench in the western Pacific.

Bean: This ALSEP's doing okay. *(pause; static has cleared)* Old Chuck Weatherred will be happy to know we're throwing it up for him, here. That's all right.

ALSJ contributor Harald Kucharek notes that Chuck Weatherred is an honoree in the National Air and Space Museum Laureates Hall of Fame 1969. He worked in the Bendix Space Systems Division and was program manager for the EASEP and ALSEP scientific instrumentation stations deployed on the Moon by the Apollo 11 and 12 crews.

Conrad: Here you go. Now, what do you need of these?

Bean: I need anything you've got. *(garbled)* ...

Conrad: There you go.

Bean: ... that tool.

Conrad: How about this? You need that one?

Bean: How about the ... Here, let me put that on. No, no. You put that in the package 2, and I'll pick that up later.

Conrad: It's already in package 2.

Bean: Then one must be yours.

Conrad: Okay. *(pause)*

Figure 2.4. Al Bean is fitting the fuel-element extraction tool onto the plutonium fuel element in its cask. The task wasn't easy. The fuel element was stuck and only after Conrad beat on the cask with his hammer did it come loose. The fuel had to be inserted into the grey, finned radioisotope thermal generator shown on the ground to the right of Bean. The generator and the geophysical instruments were set up at a distance from the LM so they would not be affected by the exhaust and lunar dust kicked up when the LM lifted off. The hand tool carrier is behind Bean on the left.

The items in question may be the Universal Handling Tools (UHT), the long-handled tools they will use to release the bolts holding the experiments onto their pallets. Or, they may be discussing tools associated with the fueling operation. Package 2 is the RTG package.

Bean: Okay. I put that there. *(pause)* Excuse me, Pete, I'll move it [possibly the Hand Tool Carrier, HTC] over and plant it.

Conrad: Wait a minute.

Bean: All right.

Conrad: Got to put this together right. Where's the arrow? *(pause)*

Bean: Shall I let down the cask while we wait?

Conrad: Wait a minute. Here you go.

Bean: Okay.

Conrad: [I want to] put this together. *(pause)*

Bean: Houston, we're going to go ahead and pull down the fuel cask right now, and then I'll take the element out of it.

Gibson: Roger, Al. Copy. You're working with the fuel cask.

Conrad: *(to Al)* Wait, wait, wait, wait, wait! *(laughs)*

Bean: I knew it. That's a bad place to put it [the HTC?], Pete.

Conrad: Huh?

Bean: That's a bad place to put it.

Conrad: Yeah. *(pause)*

Bean: Fuel cask comes down beautifully. In position. Came down just right.

Conrad: Oh, I know what I have to do, Al. And I'm standing here not doing it.

Pete is supposed to be removing the SIDE from package 2 – as per the ALSEP deployment decal, but not his cuff checklist – so that Al can fuel the RTG.

Bean: Okay, maybe I need to move the SIDE on.

Conrad: Yeah. *(pause)*

Bean: We're moving right along, Houston. We're catching up. *(long pause)*

Gibson: Yankee Clipper; Houston. Go to Wide Dead band.

Conrad: Houston! You can log me for my first Boyd bolts on the Moon. *(pause)*

Bean – "A questionable honor."

Conrad – "Boyd bolts were a real pain in the ass."

The various experiment modules were attached to the two ALSEP packages by Boyd bolts. In order to remove the Boyd bolts holding down the SIDE, Pete inserted the head of the UHT into a guide sleeve that covered a bolt, engaged the bolt, and turned the UHT about a half turn to release the bolt. Despite having the little alignment tubes to help guide them, the astronauts found that getting the UHT properly positioned could be a major problem. The bolts were often in deep shadow and difficult to see, and dust sometimes got into the sleeves, increasing the difficulty.

Bean: Hey, Houston.

Gibson: Pete, go ahead.

Bean: It's real interesting, as we put out this ALSEP right ... *(hearing Gibson)* There's one thing that's pretty obvious, as we're setting out the components of the ALSEP here. [And that] is: I just hope that these thermal coatings don't have to stay as white as they are right now, because there's just no way [of keeping them absolutely clean]. With all this dust, there's just no possibility ...

Conrad: *(garbled)*

Bean: ... of not getting them a little bit dirty. And ...

Conrad: Little bit dirty isn't the word for it.

Bean: I know it. This is going to be a real problem, I guess, if thermally they've got to maintain that [clean] coating. Because there's just no way you can do it. Everything that touches the ground picks it up. Your suit is about half dirty because the [LEC] strap landed on it. Uh-oh. It's a little dirty even on top of the fuel element, there. Okay.

A perfectly clean, white covering will reflect most of the sunlight falling on it and keep the equipment inside relatively cool. A coating of dust will increase the amount of heat absorbed and, if the dust covering is heavy enough, will let the internal temperatures rise enough to cause damage.

Conrad: Wait just a minute now before you get turned away.

Gibson: Al, we copy your comments.

Conrad: There. Okay.

Bean: I guess we'll just have to make allowances for things like that when we build them. They sure are going to get dirty. *(pause)* Okay, I'm unlocking the cask dome, right now. It unlocked perfectly. Shaking it down, trying to get it off.

Conrad: There you go.

Bean: It came off beautifully. [I'll] put the tool and the dome aside.

Conrad: Very nice.

Bean: Okay. I'll get out the cask removal tool. *(pause)* Why don't you stand over there, Pete?

Conrad: "I think you wanted me to move around so I could take a picture of you pulling the element straight out."

Conrad: I've got to go back to Min cooling. I'm about to freeze to death. *(garbled)*

Bean: Yeah. Okay. *(pause)*

Conrad: Go ahead.

Bean: I am.

Conrad: Oh.

Bean: Yeah; it's not ... *(pause)* There you go. *(pause)* Sliding right in there. Okay, tighten up the lock *(garbled)*. Hold it. *(long pause)* You got to be kidding.

Conrad: Make sure it [the removal tool]'s screwed all the way down. *(pause)*

Bean: That could make a guy mad, you know it?

Conrad: Yup. *(pause)*

Bean: Let me undo it a minute, and try it a different way.

Conrad: Yup. *(pause)*

Bean: It can really get you mad.

Conrad – "If I remember right, basically that cask had two steel rings inside it that were imbedded in the carbon. That's a carbon cask. That's a re-entry cask. And there are two rings inside. And the plutonium rod fit in and seated in those two rings. Then it had a thing on the top that you screwed the tool in. And there was something on the top ..."

Bean – "Like three jaws. You squeeze them in, supposedly. As you tightened it, something happened to them, and then you pulled it out. It sounds to me like right here I started to pull and we could see it isn't pulling and we started talking about 'it makes you mad.' Then we decided to unlock the tool and relock it. Because the tool had little bitty pins and we were afraid if we pulled, we'd break it and then we really would be sunk. So we took it off, put it in a different way and locked it down again to see if, somehow, it would work better. And it didn't work any better."

Conrad: Houston, Al put the tool on, screwed it all the way down, and the fuel element would not come out of the cask. He's taking the tool off, and he's working her again.

Gibson: Roger. We copy. *(long pause)*

Conrad: *(to Houston)* [You] guys got any suggestions?

Bean: It really kind of surprises me. *(pause)*

Conrad: *(garbled)* come over and look.

Bean: I tell you what worries me, Pete. If I pull on it too hard, it's a very delicate lock mechanism. Maybe [I should] not push the pins in quite so far, and wiggle it a little. I just get the feeling that it's hot and swelled in there or something. Doesn't want to come out. I can sure feel the heat, though, on my hands. *(pause)* Come out of there! Rascal. *(pause)*

Al has correctly diagnosed the problem just from feeling the heat.

Conrad – "Now, the problem turned out that they never calculated the amount of time the plutonium was in there. It heated those two rings to where they expanded and were just snug on there. I guess by tapping the hammer on it, I gave it just enough bounce that the load Al was putting on it would overcome the friction holding it. And he'd get it a little bit more and little bit more."

Conrad: Suppose it outgassed or something? *(pause)*

Bean: Say, Houston?

Gibson: Go ahead.

Bean: Okay. We've really got a problem, I guess. I've tried using different pins … You know, it's got a three-pin removal tool, so I tried using different pins in different holes. That doesn't appear to have any effect. You know, everything operates just exactly like it does in the training mock-ups and up at GE [General Electric Corporation]. The only problem is, it just won't come out of the cask. I am suspicious that it's just swollen in there or something and friction's holding it in. But it's such a delicate tool, I really hate to pull on it too hard. I think what we can do …

Conrad: Hey, I'll be with you.

Bean: … Go get that hammer and bang on the side of it.

Conrad: No. I got a better idea. Where's the hammer?

Bean: That's what I said.

Conrad: No, no. But I want to try and put the back end in under that lip there and pry her out. Let me go get the hammer. Be right back. Where did you put it?

Bean: Huh? What? Hammer's on the MESA.

Conrad: Okay. *(pause)*

Bean: Let me get the tool off; it's starting to warm up.

Conrad: Okay. *(pause)*

Bean: Can't figure that out. *(long pause)*

Gibson: Al, when you're working on that, try to make sure you've got the pins all the way in. Tighten up on it; then you can try pushing down on it a little, before you pull it out.

Bean: Okay. *(to Pete)* Don't touch these needles [the pins on the removal tool]; if these break off, that's all she wrote.

Conrad: Yeah. I understand.

If the pins break off the tool, they will have no chance of getting the fuel element out.

> **Conrad** – "There was no problem with radiation. We weren't going to have it out that long."
>
> **Bean** – "We had to use the tool because we didn't want it to touch anybody's suit."
>
> **Conrad** – "Right. It was too *[thermally]* hot. It would burn your suit."

Bean: And don't pound on anything *(garbled)*.

Conrad: No, no. I'm not going to.

Bean: Okay. *(pause)* We'll try again. Rotate and try it this way. *(pause)* [The removal tool] drives in there just like it's going to do the job. Only it doesn't do the job.

> **Conrad** – "It was really stuck in there."
>
> **Bean** – "Yes, but we didn't know it."
>
> **Conrad** – "It didn't have to do with the latch."
>
> **Bean** – "Which we said earlier. Everything seemed to operate, but it wouldn't come out."

Conrad: Oh man, look at this dust fly. *(pause)*

Bean: Just a minute. Just a minute. [Let me] get those pins in there again. *(pause)*

Conrad: You're not getting those pins all the way in.

Bean: They're not in now because I'm lining them up. Just a damn minute. Now they are all the way in. They're all the way ... Not quite. That bottom one down there's ... *(pause)* Now, my recommendation would be pound on the casket, then ... you know.

Pete begins hitting on the side of the cask with the flat of the hammer.

Bean: Hey, that's doing it! Give it a few more pounds. *(pause)* Got to beat harder than that. *(pause)* Keep going. It's coming out. It's coming out! *(pause)* Pound harder.

Conrad: Keep going.

Bean: *(laughs, cheering him on)* Come on, Conrad!

Conrad: Keep going, baby.

Bean: That hammer's a universal tool.

Conrad: You better believe it ...

Bean: There, you got it!

Conrad: Got it.

Bean: Got it, Houston. *(Pete giggles)* That's beautiful. That's too much.

Gibson: Well done, troops.

Conrad: I gotta go put the ...

Bean: We got it, babe! It fits in the RTG real well! It's just the cask was holding in on the side.

Gibson: Yankee Clipper, Houston. One minute to LOS [loss of signal].

Here, Gibson is talking to Dick Gordon, CMP of the Yankee Clipper, which is about to go behind the Moon.

Bean: Don't come to the Moon without a hammer. *(pause)* That's it, Pete.

Conrad: *(laughing)*

Bean: Outstanding!

Conrad: *(laughing; pause)*

Conrad (from the 1969 Technical Debriefing) – "[When we started to off-load the ALSEP], the first thing we noted was that, as soon as we put the packages down on the surface, they began to accumulate dust. Everything went as advertised until Al screwed the cask removal tool on the cask [he means the fuel element in the cask] and [it] would not budge. We got the normal fix-it: the hammer. While I beat the blazes out of the side of the container, Al managed to start the [element] out. He'd get a little notch of it every time I'd hit the container. And I really, I guess, started cracking the container. We finally got the element out and the generator fueled."

Bean (from the 1969 Technical Debriefing) – "It looked to me like the part that was sticking was about the first inch or so, because Pete would beat on it, and it would move out 1/8 inch or so until about an inch of it extended from the cask. Once the element was about an inch out, it suddenly came free and came all the way out, if that will be any aid to whoever is designing the equipment. Something was holding it that first inch."

Bean: Let's make our move [and carry the ALSEP out to the deployment site].

Conrad: I'm ready. *(pause)*

Bean: That cask's going [in?]. Okay, Houston. The fuel element is in the RTG. I can feel it radiate heat already! *(to Pete)* Put your hand over here.

Gibson: Copied that, Al.

Conrad: Wait a minute; don't move . . . No, wait a minute, Al. Have you got a strap around your boot? Let me look and see.

Bean: Okay.

Conrad: No. You're all right.

Bean: All right?

Conrad: Yeah.

Bean: Boy, this thermal coating doesn't mean a thing here [because it gets dirty so quickly].

Conrad: Uh-huh. That thing is really getting covered with dirt.

Bean: Gosh! I hope they made allowances for it.

Conrad: Me, too. Okay.

The thermal coating that Al refers to is the reflective, white-cloth covering on the packages. As discussed earlier, any dirt that gets on the covers will increase the amount of sunlight absorbed, raising the internal temperatures of the various pieces of equipment. He is hoping that the designers took into account the possibility that the equipment was going to get dirty and would get hotter than if it stayed clean.

Next, Al attaches the ALSEP packages to the carry bar. Getting the packages properly locked onto the carrybar was always a problem and, on Apollo 16, one of Charlie Duke's packages came off the bar while he was carrying it. It fell 4 or 5 feet to the surface and bounced and rolled after it hit. There was, however, no apparent damage to the equipment.

Conrad: You're in, pal.

Bean: It doesn't look like it [is properly attached]. Let me look at that just a second.

Conrad: Okay.

Bean: It's the right way. *(pause)*

Conrad: That's it.

Bean – "The trouble was in lining it up. You had to line up the package perpendicular to the shaft to the nearest degree and you couldn't ever do it. That was a hard job to do. We needed a better alignment way to click those in. In fact, when I carried it out, it fell off once, didn't it? They had little over-center holders and you'd hit a bump and it would come right out. 'Cause the thing was [flexing and] bending. When I carried that baby out, it was like carrying springs. In training, we just carried makeshift ones and the pole was stiffer."

Bean: Hey, feel the heat off that machine. That's amazing.

Conrad: 1400 degrees [Fahrenheit – or about 760 Celsius]. *(in a conspiratorial tone)* Almost as hot as the Sun! *(chuckles)*

Conrad – "I must have taken this [picture] talking about the heat, because your hand is right next to the radiator."

Bean: Hey, do me a favor.

Al is ready to start carrying the ALSEP and wants Pete's help in getting his hands properly positioned. The problem is one of not really being able to see something so low and close to the body while, at the same time, trying to get in a position to grab the bar. On Apollo 16 and 17, Duke and Schmitt, respectively, generally bent forward to grab the bar and then, as they lifted the packages, ran forward to get it under control. Al may be trying to do a more controlled, static lift.

Conrad: What do you need? No, go lower. There you go. You got it.

Bean: Okay.

Conrad: Okay; let me go scout ... smoke over the area.

Pete is saying that he'll go out west of the spacecraft and look over the potential deployment area. They need to be at least 300 feet from the LM to minimize the amount of dust blown onto the experiments as they lift off from the lunar surface. They will do the deployment west of the LM because shortly after lift-off, the spacecraft will tilt toward the west to catch up with the orbiting CM and, as a result, will blow most of the dust toward the east.

Bean: Okay. Got everything you need?

Conrad: *(examining his cuff checklist entry)* All right, there's no TV, so I got the SIDE [subpallet] and the picker-uppers for the rocks [meaning the tongs]!

Bean: Okay.

Conrad: Okay. Let's go right off to our little mound over there; how does that grab you?

There are two conical mounds of dirt near the ALSEP deployment site, which, evidently, the crew has already noticed. They did not, however, mention the mounds during the site description they did before the EVA preparations.

Bean: Okay. Now, something's wrong.

Conrad: What's the matter? *(long pause; probably while they make sure the packages are secure on the carrybar)*

Bean: Well, it's kinda ... The thing doesn't ... Well, let's try it.

Gibson: Pete, we copy. You've got the UHT, tongs, and [SIDE] subpallet.

Bean: We're making our move, Houston. I can tell this is going to be a workload. I'll take it easy.

Conrad: *(to Houston)* How long did you say our shadow was ... [that is], the LM's shadow, 150 feet?

Gibson: Stand by, Pete.

Bean: No, that isn't any 150 feet. *(pause)*

If the shadow is 150 feet, they'll only have to go twice that far to get to the 300-foot minimum distance.

Conrad: Take your time, Al; I'll just go on out ...

Bean: That's what I'm doing. What's the hurry? We got it made.

Conrad: No, I'm just going out to scout the area, that's all.

Bean: Okay, I'm going to set it down and rest ...

Conrad: Okay.

Bean: ... and go to intermediate cooling. *(garbled)* *(pause)*

Al is carrying the ALSEP low, with his arms hanging down.

Bean – "It was hard to carry that way; it kept bumping your knees. And then your knees would knock and it would rattle and jump."

Conrad – "Well, it weighed about 300 Earth pounds, so it had a lot of momentum."

Bean – "On Earth, when they gave us one that weighed like it did on the Moon [that is, a 50-pound mock-up], it had the momentum of the light masses. But you get up there with those big masses, even though it weighed the same, it wobbled different."

Later, Ed Mitchell (Apollo 14) discovered that it was easier to carry the ALSEP with the carry bar cradled in the crook of his elbows. Jim Irwin (Apollo 15), Charlie Duke (Apollo 16), and Jack Schmitt (Apollo 17) also carried their ALSEP high, either with their hands at shoulder height or in the crook of the elbows, and consequently had less trouble than Al did with this task.

Bean (from the 1969 Technical Debriefing) – "After we mated the ALSEP and got ready to carry it out, the workload carrying it out was about the same as I had guessed from working on Earth. The hard part is holding that weight in your hands. Even though it is not much [about 300 pounds on the Earth, but only 50 on the Moon], the combination of the weight, the fact that you're moving along, and the fact that your gloves don't want to stay closed [against the internal pressure of the suit] tends to make it a fairly difficult task. I would say that it would be acceptable to carry it this way for distances of up to 500 feet; but at distances greater than that, I don't think you want a hand carrier arrangement. You will want to have a strap that fits over your shoulder, or something like that. It's not your legs that get tired; it's a combination of your hands and arms, and it just makes you tired. Another thing that occurred that we hadn't seen on Earth is that, as you bounce along at 1/6 g, the RTG package tends to rotate. The c.g. is not exactly lined up beneath the crossbar. It tended to rotate and unlock the two-piece crossbar. This was disturbing because I would have to stop once in a while and relock the crossbar. I would recommend that we definitely put some sort of snap lock on that cross piece so that, when you put it in position and it rotates to the carry position, it locks there. If it had opened up as we carried it, we might have dropped the gear and broken some of it on the lunar surface. It's funny that never showed up in any of our 1/6-g work in the airplane or anywhere else, but it was a continual problem on the lunar surface."

Also during the 1969 Technical Debriefinging, Pete and Al discussed the rest they could get by just standing still, arms hanging down and leaning forward to put their center of mass over their feet.

Conrad – "The position described is a very comfortable position. I never got tired. It's just a normal position to rest in. You can stand perfectly still in that position and rest. Did you *(to Al)* feel the same way about resting? Did you just stand?"

Bean – "I never remember doing anything but standing there; and never seemed to get tired. As you said earlier, you could work 8 hours out there; if you got tired, you could probably stand against something or just stand there, cool off, and press on. At the end of the EVA, I was feeling as good, particularly in my legs, as I was at the start."

> **Bean:** You know, they ought to build this equipment for lunar operations some other color besides white.
>
> **Conrad:** *(laughs; pause)* I'm going to go right up to the Head Crater, I guess.
>
> **Bean:** Well, if you're going to do anything ... *(pause)* Okay; let me move another couple of ...
>
> **Conrad:** Got any direction you want me to go?
>
> **Bean:** Well, it looks to me like either the direction you're headed is good, or the one a little bit more to the right. You're going to have to go far enough so we don't end up in one of the craters when we furthest south deploy.
>
> **Conrad:** Okay. *(pause)*

They will deploy all of the experiments south of the Central Station, so they need to leave themselves enough room between the Central Station and Head Crater to get everything on level ground and in the proper relative position.

> **Conrad:** But I want to go 10 degrees off our takeoff angle; and I think I'm headed out about that way now.
>
> **Bean:** Okay. *(pause)*

Presumably, they want to get far enough off the direct LM track to prevent any significant interactions with the engine exhaust.

> **Conrad:** You just stay back there and take your time. I'll go out here and scout the area.
>
> **Bean:** You're getting pretty far out.
>
> **Conrad:** Huh?
>
> **Bean:** Getting pretty far out.
>
> **Gibson:** Pete and Al, your LM shadow should be about 110 feet.
>
> **Conrad:** Okay. I'm looking for ... I'm dying to find out what this mound is over here anyhow, Al. *(pause)* We got a very peculiar mound sticking up out of the ground, Houston. I want to go look at it. As a matter of fact, I think I'll go take a picture of it. No, my ...

Gibson: Roger, Pete. Could you give me your position and distance with respect to the LM?

Conrad: Wait one [minute].

Bean: Go ahead, Pete. Do what you're doing. *(to Gibson)* Pete's about ... I'd guess, about 300 feet at 12 o'clock [from the LM] in the bottom of a shallow crater that you're bound to see on your map. It's sort of a doublet. *(pause)*

Conrad: I'm headed to the right-hand edge of the Head Crater.

Gibson: Roger. *(pause)*

Conrad: Hey, Al. Here's a neat spot to put it out up here.

Bean: Is it flat for a good piece?

Conrad: Oh, you'd better believe it.

Bean: Okay. It's a good long ways away too; it must be at least, what, 500 feet from the LM?

Conrad: I don't know.

Bean: 600?

Conrad: It's the world's most peculiar ... I got to photograph this thing. I can't imagine what it is. The mound's sticking up; and I can't imagine how it got there or what would make it. *(pause)*

Bean: Boy, you can cover the ground on this lightweight ... at 1/6 g. Really move.

Conrad: I got to get them a stereo of this thing. It's really fantastic! Oops. *(pause)*

Conrad: How's our timeline going, Houston?

Gibson: Pete, at 1 plus 48 into the EVA, you're looking good. Looks as though you're right on there, if you've just about completed your traverse.

Conrad: We have, Houston. Yeah. Now, look, Al, look over here.

Bean: We're way out from the LM.

Conrad: Over here, Al.

Bean: Okay.

Conrad: See where I'm headed. This great big flat area.

Bean: That's a good ... Hey, there's another one of those mounds over there.

Conrad: Where? Hey, you're right! What do you suppose they are?

Bean: I don't know, Houston, what they are; they're just sort of mounds. Looks like – don't take this the wrong way – it looks like a small volcano, only it's just about 4 feet high and, at the top, it's about 5 feet across; and it then slopes from the top on down to the level with the terrain, and that diameter – that circle where it finally becomes level with the terrain – is about 15 or 20 feet. So it looks sort of like a small volcano. There's a couple of them out here. They look like they were formerly made out of mud or something.

Gibson: Al, Roger. We copy.

Al's comment "Don't take this the wrong way" is probably related to a story that Jack Schmitt told me (author Jones). During Apollo 11, Buzz Aldrin said he saw bright flashes of light at places in the soil – like biotite (a form of mica), he said. He wasn't saying that the material was biotite, only that it flashed as biotite would. According to Schmitt, some geologists were highly critical of Buzz's use of the word – a criticism that Schmitt rejects as being completely unwarranted. Furthermore, Schmitt said that Pete and Al had decided to avoid such controversy by not using technical terms. Pete and Al deny that they made any such decision and, whatever the truth may be, Al's comment here indicates that he was sensitive to the potential for such criticism. By clearly indicating that he doesn't believe the mounds to be volcanic vents, he then uses the term "volcano" to convey a very accurate and useful impression of the shape of the mounds.

Gibson: Is there any hole or central vent?

Conrad: I don't know. I got to go over … We'll go over after we get the ALSEP out. There's a couple of them here. *(pause)* This is a … We couldn't ask for a better spot to put this ALSEP down.

Bean: No. This is nice. Hey, [there are a] lot more rocks up here [than there are around the LM].

Conrad: *(to Houston)* Listen, there … We could play geologist for two days and never get any further than we are right now. Seeing all different kinds of things.

Bean: Hey, here's a different one.

Conrad: Yeah. It's really neat. Better than any of the geology field trips. Look at that thing. *(laughs)* I'm getting a quick pan of the area here to tie down the ALSEP deployment. There you go.

Bean: Okay. Man alive!

Conrad: Tired?

Bean: No, I'm not so tired. That handle, you know, when you carry this thing around in 1 g, the ALSEP tends to hang down. But you carry it around up here at 1/6 [g] and, particularly, the RTG tends to rotate the whole pallet. So, in a few minutes, you've got one of them kind of half way up in the air and the other one's down by your left leg, being afraid that the handle's going to come undone. You know, the handle doesn't lock. Okay. *(pause)*

Bean: What happened to all those … Hey, that's slick. You know these neat little decals we got on here?

Conrad: Uh-huh.

Bean: You can't hardly read them in the sunlight because they don't have enough contrast to them. They're so bright. *(pause)*

Bean (from the 1969 Technical Debriefing) – "One thing that turned out a little different than I imagined – I had difficulty reading the decals that we had put on the ALSEP. As I recall, looking at it on the surface [he means 'during training'], it looked as if there were black writing on a silver background. When I tried to read this on the lunar surface, it was very difficult. The brilliant light reflected off the silver and you couldn't see the black. We had the sequence of laying that ALSEP down pretty good, and I would recommend on the next one, they use a black on orange – or something like that – to decrease the amount of reflected light off the decals. It's going to be needed."

Bean: Okay, let's move them. *(pause)* Okay. Let me make sure now that we're not going to run out into some holes. *(pause)* That's pretty good. Pete, I'm going to move just a little bit further to the east – or, correction – to the north, so that I won't end up over in that hole with the SIDE. Okay?

Conrad: All right.

Using his training experience, Al is visualizing where each piece of equipment will go.

Bean: And I think it would be a real good spot. *(long pause)* Okay, I'm …

Conrad: Awfully frustrating.

Bean: And I think this is the spot, Pete, right here. *(pause)* Let me look and make sure, now, that we're going to have a good place for everything. Yeah, we will. Magnetometer can sit over there and the seismometer will sit in a good flat place. Although, the trouble with the seismometer [is that] we don't have any good solid bedrock or anything to set it on. All we've got is this dirt. And I don't see any area around that has any rock.

Gibson: Roger, Al.

Interviewer – "Were you being literal about rock to set the seismometer on?"

Conrad – "I think they wanted as hard a surface as possible."

Bean – "And then we didn't find any. 'Cause we thought we'd maybe see some dirt and some rocks and things, and we were going to set it on a flat piece of rock. But now we're telling them, 'forget it, all we got is dust up here.'"

Bean: I'm afraid we're just going to have to take what we can get on this seismometer. *(pause)*

Conrad: You know, there must be some thermal expansion or something. I'm having a heck of a time getting this UHT in this SIDE. *(pause)* It just flat won't go in there.

Bean: Well, just pick it up with your hands. I always handled the SIDE with my hands.

Conrad: I can't bend down that far.

Bean: Okey-doke. *(pause)*

Conrad: *(garbled)* free. *(long pause)*

Interviewer – "In 1 g you could bend down far enough to pick it up?"

Conrad – "Sure, you could get a little help [from Earth's gravity]. Especially with a 70-pound PLSS on your back and everything, you could get down. But up there, you strain ..."

Bean – "And you're still way up off the ground."

Interviewer – "Could you have done the ALSEP deploy solo?"

Conrad – "Yep."

Bean – "Sometimes, when you were training, you'd try it. It wasn't that hard. All you had to do was to be able to read the labels – unscrew the Boyd bolts and read the labels. That was about it."

After he gets the SIDE free, Pete will deploy the legs that will keep the experiment off the ground, extract the cable reel, and connect the experiment cable to the Central Station. In the meantime, Al will position the RTG.

Conrad: Where'd you go, Al?

Bean: I'm right over here, babe.

Conrad: *(turning)* Oh, you're miles away.

Bean: Yeah, I am. I moved over here.

Conrad: Oh, son of a gun!

Bean: I had to do it, Pete, so that I'd have a good place for the *(garbled)*. *(pause)* Okay, there's a good spot. Looks good. *(long pause)*

Bean – "It seems to me I set the ALSEP [Central Station] down and had to go move, for some reason. I don't remember for sure. But that may have been what happened here. It seems to me that, somehow, things didn't look right and I had to move it about 3 feet."

Gibson: Al, when you deploy that PSE stool, it would help if you tamp that ground down as well as you could before putting the stool down.

Bean: I'm going to try to do that. This ground, you know, it doesn't get hard as you move down a couple of inches. If you tamp it, and that's what I'm planning to do … But I don't know. We'll just have to see what happens. I'm worried about it getting a thermal short myself.

Gibson: Roger, Al. *(long pause)*

Bean (from the 1969 Technical Debriefing) – "The first experiment I put out was the passive seismic. I had two anomalies that I know. One was the skirt [which is discussed below] … the second anomaly [the first in actual sequence] was that the little dish the passive seismic sits in needs to have a solid bottom so when it is placed on the ground, there is no danger of dirt easing up through the center of it, as there was in the case of our dish, and touching the bottom of the passive seismic itself and causing a thermal short that would ruin it. We spent quite a bit of time tapping out a nice clean hole so this wouldn't occur. Really, I think the fix should be to put a solid bottom on that dish."

Conrad – "What you wanted to avoid was making [thermal] contact with the ground. Don't forget, the only way you're going to pass heat one way or another is radiate or conduct. To conduct anything that way, you've got to make a good contact. Remember those things had those skirts around them."

Interviewer – "To keep the Sun off the ground right around the instrument."

Conrad – "Sure, because then you're going to get nice and cool underneath."

Bean – "I remembered that skirt being there to keep the ground from making thermal noises that was picked up in the seismometer. Now, that may be wrong."

Conrad – "That could be. But it's really kind of the same thing."

Bean – "Stabilizes the [ground] temperature so it isn't doing things that the seismometer picks up as background noise."

John Saxon, who served as Operations Supervisor at NASA Honeysuckle Creek Tracking Station, Australia, during Apollo writes, "I should share one of my best ALSEP tracking stories. One of the nice things about the Manned Spaceflight Network ground equipment was that we could decommutate the Telemetry and display spacecraft – and astronaut – parameters on strip chart recorders and other devices. After the excitement of the manned sections of the Apollo missions, supporting the

ALSEP packages could seem rather routine – so we often checked temperatures, seismic activity, etc. One quiet evening we noticed that one of the usually fairly dull seismometers was outputting apparently random, short, sine-wave activity super-imposed on the background activity. These were obviously artificial artifacts, but no one on site could work out what might be causing them. So I checked with the on-duty ALSEP controller at Houston, who came up with the answer immediately. 'Oh – what you are seeing is the Mylar heat insulation which is spread around the seismometer. The sun is just rising in the lunar dawn at the site and the insulation is warming up and pinging as it changes shape.' A bit like a biscuit tin lid when you flex it. So that was an unintended temperature-change sensor."

Bean: Boy, I'll tell you.

Conrad: Man, are you dirty.

Bean: Boy. Sure. [It's from] carrying that ALSEP. Look at the ALSEP.

Conrad: I know.

Bean: Ridiculous *(laughs)* I remember how they took care of this white paint. You had to have gloves to touch it. *(Pete laughs at the thought; pause)*

Bean (from the 1969 Technical Debriefing) – "I think we're kidding ourselves if we think there is any way to deploy this experiment [that is, the ALSEP] without getting a lot of dirt and dust on it. The pictures are going to show this. They just have to be designed to accept dirt and dust. If they can't accept the dirt and dust, then they are going to have to be packaged in some way so they can be deployed completely, and then the last act would be to pull some sort of pin and flip off the covering that would have all the dirt and dust on it, exposing the nice clean experiment."

The only Apollo experiment that rapidly failed from overheating was the Surface Electrical Properties Experiment that was flown on Apollo 17. The SEP rode on the back of the 17 Rover and was housed in a cloth bag that was supposed to keep a radiator on the top of the instrument clean while the crew was driving from one geology stop to the next. At each stop, they would expose the radiator and let it cool the instrument. Unfortunately, the glue failed on the Velcro patch with which the top flap was secured. The radiator accumulated a fine layer of dust and, despite repeated efforts to clean the radiator and secure the flap with tape, the instrument overheated.

Conrad: I got a kind of a problem here. *(pause)* I'm going to have to do this myself. I'll have to do this backwards; but it's going to work right. *(pause)* Oops. *(pause)*

Bean: Do me a favor, Pete?

Conrad: *(garbled)* I'll be with you in a minute.

Bean: Okay. *(pause)*

Conrad: You can relieve me in this SIDE right here while I *(garbled)*, because it looks like it might ...

Bean: Yeah. Give me that thing. Let me hold it.

Conrad: Easy, easy. That a boy.

Bean: I got it.

Conrad: That a boy. Now, wait.

Bean: Watch for that *(garbled)* cable.

Conrad: Yeah, but let me get it plugged in.

Bean: They [the cables] sure splay around under this 1/6 g, don't they? It's moving around a bunch. *(garbled)* Watch it, watch it.

Conrad: Where's the orange stripe?

Bean: Should be [the] same as on that other one.

Conrad: Obviously, it's on the side I can't see.

Bean: Yeah. Blue to blue. *(long pause)* You know, that noise – that whistling – is coming from you, Pete, because every time you move around, it whistles. When you stop, it quits.

Conrad: Is that right?

Bean: Yeah.

Conrad: There it is.

Bean: Is it locked good?

Conrad: Yes, sir.

Bean: Okay, here's ...

Conrad: *(garbled)*

Bean: Here's yours.

Conrad: [We don't want to] drop that baby [the SIDE].

Bean: Okay. You got it?

Conrad: Let me have the cable right there.

Bean: All right. Oops! That came out.

Conrad: That's what I wanted.

Bean: Okay.

Conrad: That's what I wanted.

They may be referring to a lanyard on the SIDE, which Pete will pull to release the legs.

Bean: All right. Those legs came out nice.

Conrad: Yup. *(pause)* The next thing I got to do is head this way with it, right?

Bean: No. Before you do that, come here a second, would you, please?

Conrad: Yeah, but let me set it down one moment. Uh-oh; the top popped open.

Bean (from the 1969 Technical Debriefing) – "When we were deploying the suprathermal ion detector, the lid came open a couple of times. This lid was supposed to be deployed [by] ground command after we had left the area [after launch] so that the exposed, mirrored surface would be nice and clean and the two detectors would not get dust in them. I'm pretty sure that we did get some dust on the top of it. I hope it's not enough to bother the operation."

Conrad: I[s it] supposed to do that?

Bean: No. Not supposed to do that. You can close it, though, before it gets any dirt in it. You can just reach up there and close it.

Conrad: Okay. You better ... You better ... *(garbled under Al's voice)*

Bean: *(garbled under Pete's voice)*

Conrad: ... close it while I'm holding it.

Bean: Okay.

Conrad: Before we get any dirt on it. Son of a gun. Why did that happen?

Bean: Never saw it do it before. *(garbled)* We could still salvage that, though.

Conrad: Wait a minute. The little ... The little ...

Bean: I see it. I see it.

Conrad: That a boy.

Bean: Now kind of turn it over with your other hand, Pete. So ...

Conrad: Wait a minute, *(garbled)*.

Bean: Wrong way.

Conrad: Where do you want to go with it?

Bean: Well, I want to see the ... Let me get this. Okay. Now, let's see the other side.

Conrad: This way?

> **Bean:** That's a boy. Yeah; this won't hurt a thing. Now ... Wait a second. *(pause)*
>
> **Conrad:** That's it. You got it.
>
> **Bean:** But just don't touch that. It'll be okay.
>
> **Conrad:** Okay.
>
> **Bean:** Didn't hurt a thing.
>
> **Conrad:** Let me just put this down here.
>
> **Bean:** Okay.

Clearly, Pete and Al are having more than their share of difficulties with the ALSEP deployment – far more than during any other mission. It is important to keep in mind that Neil and Buzz only deployed two scientific instruments: a passive seismometer and a laser reflectometer. Neil and Buzz also had some problems with their deployment – for example, a BB leveling device that refused to operate properly for a while – but Pete and Al are the first crew to do a full ALSEP deployment and, not surprisingly, are discovering a number of design and procedural flaws that didn't manifest themselves in 1-g training. As a result of the astronauts' real-time evaluations and later interview comments, design and procedural changes will be made that will give later crews a much easier time of it. Engineers are able to implement extraordinary improvements in the scientific packages over the scant 6 years between the first and last ALSEP deployments.

> **Gibson:** Pete, before you set the Central Station down for the final time, if you would also tamp that ground down, it would help in keeping the dirt off of the thermally sensitive areas.
>
> **Conrad:** *(to Al)* Oh, I did it, I did it. *(laughs)*
>
> **Bean:** You set it down.
>
> **Conrad:** *(laughing)* I set it [meaning the SIDE] down and it didn't fall over. I can't believe it.

Pete has put the SIDE out of the way so that, later on, Al can carry it out 60 feet SSW of the Central Station for final deployment. According to the *Apollo 12 Mission Report*, the three legs on the SIDE were too close together for good stability and, consequently, the instrument was easy to tip over.

> **Conrad:** Here I come. What do you need, Al?
>
> **Bean:** Use your tongs to hold this up a minute. It [the RTG]'s a little hot, and I don't want to touch it. Watch [out for] this.

Bean (from the 1969 Technical Debriefing) – "We had no trouble putting down the RTG. I did notice, however, that you could feel the heat radiating from the RTG. When I removed the bracket that carried the power cable that ran from the RTG to the Central Station, it felt warm to the touch. I didn't want to keep my fingers there too long, so I handled it with the ALSEP tool [UHT] as opposed to just my gloved hands, as I had been doing in practice. Apparently that bracket can get pretty hot, although we only had the element in it a short time."

Conrad (from the 1969 Technical Debriefing) – "I guess the point is, when you fuel that generator, you had better get on the road and get going to wherever you are going to take it. You should get those parts off the fuel element as soon as possible, because they heat at quite a high rate."

Anything that can be felt as "warm" through the EVA suit's gloves must be quite hot and thus a potential threat to the gloves. At this point, it has been about 17 minutes since they fueled the RTG.

Conrad: Where are my tongs?

Bean: Right there in your ... There you got them. Now if you'll just hold it steady, then I can do the rest. *(pause)* There, now. That's good. *(pause)* Thanks, Pete. Got it.

Pete may be using the tongs to hold one part of the equipment while Al is using his UHT. He is wearing his tongs at his left hip, attached to a spring-operated, self-rewinding device appropriately called the *yo-yo*. The yo-yo operates very much like a rewinding tape measure and lets him carry the tongs without having to use his hands and, then, when he wants to use them, he just reaches across with his right hand and pulls out as much cable as he needs. When he is done with the tongs, he simply releases them and they spring back onto his hip.

Conrad: Okay. I'll go get the rest of the stuff over here. Where we at?

Bean: Okay. Now we're going to get the easy part.

Conrad: Okay. Take your time.

Bean: Thing's coming out real well.

Conrad: The name of the game is to get the ALSEP here. *(running or, at least moving quickly)* Whoooom! Up through one crater and over another. *(hearty laugh)*

Bean: Oh, man.

Conrad: Does that look as good as it feels?

Bean: It does. Hey, I'll tell you the way to do it. Pete! Tend to rock from side to side as you run. Like that. There you go.

Conrad: *(joyous laugh)*

Bean: You really move better that way.

Conrad: Here I come, ready or not. *(pause)*

Bean: Houston, I'm not kidding. We are really getting dirty out here. There's no way to handle all this equipment with all the dust on it. Every time you move something, the dust flies; and, in this low gravity, it really takes off. Goes way up in the air and then comes down and lands on you.

Conrad: How far do you estimate we're from the LM? 600 feet? 700 feet?

Bean: At least.

Conrad: I think you're right. 6[00 or] 700 feet.

Pete and Al actually deployed the Central Station about 450 feet from the LM.

Bean: Way to do it.

Conrad: Here I come. Dee dee, dee dum. *(long pause)* Dum dum, dum dum. Coming after the [Central Station] antenna mast [which had been serving as the carry bar].

Bean: Okay. *(long pause)* Okay. Let me plug this thing [probably the RTG] in, Pete.

Conrad: I can't get down low enough to get it [the antenna mast] off.

Bean: I'll get it for you, if you can't.

Conrad: Okay. Wait a minute. Let me move that dust cover for you.

The dust cover in question is probably one that protected the SIDE cable, rather than the one on the instrument that is protecting its innards. After Al picks it up – undoubtedly with the tongs – Pete will throw it, just for the fun of seeing how far it will go.

Bean: How's that strike you? Okay. *(garbled)* the stripes ... Hey! Look at that *(dust cover goes)*.

Conrad: Strikes me pretty good. I'll be tamping the dirt around this Central Station.

Bean: Okay. *(pause)* You're going to have to design things with little legs on them or something, so you can put them down and they stand off the soil. *(laughs)*

Conrad: If we had that suit we could bend over in, we'd have the job done by now.

Bean: Hey, can you push down on this side of that?

Conrad: Yes. My biggest sweat is trying to bend over. Wait a minute. Ready, get set, push. That on, or not?

Bean: I don't think so.

Conrad: I don't think so.

Bean: No, let me look at it a second. Here we are.

Conrad: Which way do you want this?

Bean: Under there.

Conrad: Under. *(pause)* I had to really push that other one [that is, the SIDE cable connector]. Look at all that dust! *(pause)*

Bean: Okay now … *(pause)* Wait a second. Wait a second. Let me get it with you. *(pause)*

Conrad: Ready? *(pause)* That got her.

Bean: That did it, now.

Conrad: That did it.

Bean: Now we won't touch the shorting button.

The shorting circuit allows current from the RTG to flow through some resistance without going through the experiments. Only when the experiments are all connected will they push the shorting switch and divert the current flow out of the shorting circuit.

Bean: *(to Houston)* Okay. We've connected the RTG to the Central Station, Houston. And we're ready to go to work deploying the experiments.

Conrad: I['ve] got the antenna mast.

Bean: All right.

Conrad: Dee dum, dee dum. *(long pause)*

Gibson: Pete, did you experience any trouble with the dust cover on the SIDE?

Conrad: It opened, Houston, and we closed it again, and we got no dirt on it, fortunately.

Gibson: Roger, Pete. Well done. *(long pause)*

Conrad: Okay. Let me tamp the dirt down underneath the Central Station. Hey, easy does … Whoops! *(laughing)* That's me.

Bean: What happened?

Conrad: Oh, I started to walk away with the Central Station attached to the UHT via the little gizzy.

Bean: Okay. The RTG's down and cooking.

Conrad: Okay. I'm tamping the dirt, here.

Bean: Okay.

Pete is probably still tamping down soil to form a base for the Central Station.

Conrad: Have I got it tamped pretty good? Look over where I'm tamping. How am I doing?

Bean: Looks good. Need a little bit more this way, I think. The problem with tamping, Pete, it looks like every time you tamp that, as your foot comes up, you know it ...

Conrad: Yeah, *(garbled)*

Bean: ... it redusts the area.

Conrad: Yeah. Okay. *(pause)* Here we go. Al. *(pause)* Now, take your UHT, there ...

Bean: And move that.

Conrad: ... and keep the SIDE cable clear.

Bean: Sounds good.

The ALSEP package 1 has been standing on edge. The base of package 1 is the Central Station. After they remove the experiment packages, Pete will configure the Central Station by raising the top and, with it, a pre-attached thermal curtain. He will then assemble and aim the antenna. For the moment, Pete and Al are tipping package 1 over onto its base so that they can remove the experiment packages.

Conrad: All right.

Bean: Go, babe.

Conrad: Here she comes. Over she goes. *(pause)* Get it right down the Sun mark.

Knowing where the Sun is supposed to be at this point in the mission, they can use a shadow device to align the package in a predetermined direction so that, later, Pete's antenna alignment job will be relatively straightforward.

Figure 2.5. Powered by nuclear energy, the ALSEP is deployed and ready for work – measuring everything from moonquakes to the solar wind. Al took this photo to show the location of the magnetometer, the instrument in the foreground with three booms. Pete is in the background at the Central Station, where he is installing the antenna. The SIDE and RTG packages are to the left of the Central Station.

Bean: No. There you go.

Conrad: How's that?

Bean: That's good.

Conrad: Okay.

Bean: Ready to go to work.

Conrad: Go. Have at it. *(pause)* Okay; what have I got? Let's see. I've got to get the solar wind, huh?

Pete's next task is on the checklist page that starts with "1 + 58, SWE Deploy." Apollo 12 carried two solar wind experiments. The first was the Solar Wind Collector (SWC), the foil "window shade" that they deployed near the LM at the start of the EVA. That experiment captures solar wind atoms for return to Earth, where they will be used to help determine the relative abundances of the various constituents. The experiment Pete is about to deploy is the Solar Wind Spectrometer (SWS), which uses a device called a Faraday Cup to determine the flux (flow rate) of the solar wind as a whole.

Bean: Solar wind. I'm going to do the ... *(pause)* I better ... *(pause)*

Conrad: Houston, how we doing on the timeline? *(no answer; long pause)* Do you read me. Al?

Bean: Sure do.

Conrad: Hello, Houston. How we doing on the timeline? *(no answer; pause)*

Bean: They may be talking to Yankee Clipper. [not true] *(pause)*

Gibson: Al, did you have a reading on the shorting amps?

Conrad: Wait a minute. I'll get it.

Houston wants to know how much current is flowing through the shorting circuit. Al's checklist calls for such a report, but he neglected to do it.

Conrad: Hold the phone. Do you read me, Houston?

Gibson: Sure do; go ahead.

Conrad: Yeah; I was asking you how we were doing on the timeline? *(no answer; long pause)* Can't see that [shorting amp] needle anywhere. Can you, Al?

Bean: Let me see it. Let me help you. Doesn't it show any amps?

Conrad: Place is covered with dust.

Gibson: Pete and Al, at 2 hours and 7 minutes into the EVA, you're about 5 minutes behind.

Conrad: *(responding to Gibson)* Okay.

Bean: Hold me a little bit.

Conrad: Wait a minute. Going back to Min cooling. What do you need?

Bean: Hold me and I'll look.

Conrad: Oh, okay. Easy, don't ... *(pause)*

Pete may be holding Al's hand while he tries to get close enough to the meter to see the needle. The meter is on the Central Station base and is only a couple of inches off the ground.

Bean: I don't even see a needle in there.

Conrad: I don't either. That's what's bothering me. It's not reading zero, and I don't see a needle.

Bean: I don't even see a needle in there. Houston.

Conrad: Yeah. Okay. Let's quit screwing with it. *(pause)* [Let's] get on the timeline.

Bean: Wonder where it went. *(long pause)*

Conrad: Houston, did you get our last comment?

Gibson: Negative, Pete. Go ahead.

Conrad: We can't see a needle in the shorting amps [meter] anywhere. It's not at zero. It's not in sight.

Gibson: Roger. We copy. Go ahead. *(long pause)*

Bean: What I've done for the seismometer, Houston, is I've sort of dug a little crater so that, essentially, the hole in the center of the stool has full clearance between it and the ground. Hopefully, this will keep the SIDE [means the seismometer] from getting in the ground at all.

Gibson: Roger, Al. That's good. Go ahead.

Bean: I mean the seismometer. *(to Gibson)* We'll see how it works. It looks like it might work. I tamped it also in the small crater there. *(pause)*

Conrad: Okay. I've got the solar wind deployed here.

Bean: Okay. *(pause)* Just a minute. *(pause)* Boy, you really have to be careful of these cables, don't you? *(pause)* Okay. Looks good. Looks good.

Conrad: Okay. *(pause)*

Conrad (from the 1969 Technical Debriefing) – "The only [experiment] I deployed was the solar wind and it went exactly as advertised. I checked the four legs down, took it out the proper distance, aligned it, and turned her loose."

Conrad: *(consulting his checklist)* Okay. After the solar wind, I get an EMU check and I'm down to Min cooling anyhow, and it says "LSM [Lunar Surface Magnetometer] off-load, two Boyd bolts." Wait a minute. Let me by.

The EMU (Extravehicular Mobility Unit) is the combination of suit, backpack, and everything else they are wearing. An EMU chcck is, formally, a check of the suit pressure and the remaining oxygen supply. EMU checks are called out at several places in the checklist. On this and later missions, Houston sometimes asked for an unscheduled EMU check and, when they did so, were actually saying that the flight surgeon – who was monitoring their heart rates – thought they were working too hard and needed to slow down for a moment.

The reason that Pete asks Al to let him by is that they are probably both working around the Central Station and need to be aware of the possibility of bumping into each other. Pete is removing the magnetometer from the Central Station pallet and Al is removing the seismometer.

Bean: Wait a minute. Okay. *(pause)* Say, we have to be careful. Don't move sideways or backwards. You just don't know what's there. You've always got to move forwards.

Conrad: Yeah. *(pause)*

Bean – "You couldn't look down at your feet. And, if you started moving backwards, you're going to run into those cables and stuff."

Interviewer – "Your concern here was specifically about cables?"

Bean – "And the experiments. Actually, those things were [in] pretty close quarters when you got your suit and everything on. Because you're thrashing around up there and you're fairly clumsy. And if you started backing up, you'd go over something as John Young found out [when he caught a foot in an experiment cable and pulled it loose]. It's hard to miss all that and, if it gets dirt on it, you don't even see it. We were just lucky we didn't."

Interviewer – "How about out on open terrain, just running around?"

Bean – "I don't think it makes any difference. You do trip more going backwards. I tripped a couple of times going backwards."

Conrad – "Other than falling in a hole you don't see."

Bean – "Your balance isn't as good bouncing backwards. Actually, you're in pretty close quarters for two people and all those experiments, although it looked like they're far away [from each other]. All those cables are twisting and turning, and some of them are standing up. You know, they're not laying down nice and even."

Because of the weak lunar gravity and the length of time the cables have been folded or coiled up inside the experiments, they retain memory for quite a long time and don't lie flat on the ground.

Gibson: Pete, we copy an EMU check.

Bean: *(garbled)* this way a little bit.

Conrad: Shove it?

Bean: Just put your thing [UHT] in the holder, and move it over this way a little bit.

Conrad: I'm not sure I can do that; but I'll give it a try.

Bean: Watch it. There you go.

Conrad: How's that for one-footed la-dee-da?

Bean: Pretty good.

On the later missions for which there is good TV coverage, one occasionally sees an astronaut making one-footed, hopping turns, such as Pete may be doing here.

Conrad: [If] you make me knock my solar wind over, I'm going to be mad at you.

Bean: Okay. Here; just a second. Let me give you a hand. *(pause)* Let me show you ... You hold it there.

Conrad: Hey, don't touch that [magnetometer]. I've got one Boyd bolt off it.

Bean: *(garbled)*

Conrad: Huh?

Bean: Want me to help you move it that way?

Conrad: No. It's got one Boyd bolt out of that magnetometer.

Bean: Okay. Okay. Go ahead.

Interviewer – "Can you give me an estimate of how many times you did this deployment in training?"

Bean – "I'd say a dozen is a good estimate. We knew how to do this, but those Boyd bolts were always hard. One thing you found out about spaceflight, there's so many things you're doing that things don't always go right. Boyd bolts don't loosen. Sometimes the UHT doesn't fit in the hole."

Conrad – "You don't realize how much 1 g helps you, sometimes. You don't realize that 1 g's helping you. Putting that tool [UHT] in a Boyd bolt ..."

Bean – "1 g helps you lean over; it helps you push in."

Conrad – "Yeah. Helps you lean over; you get a better look at it. And that was another reason I don't think we felt we needed those [guide] collars. We'd have felt better, in 1/6 g, if we'd have been able to just be able to look at the thing, even if

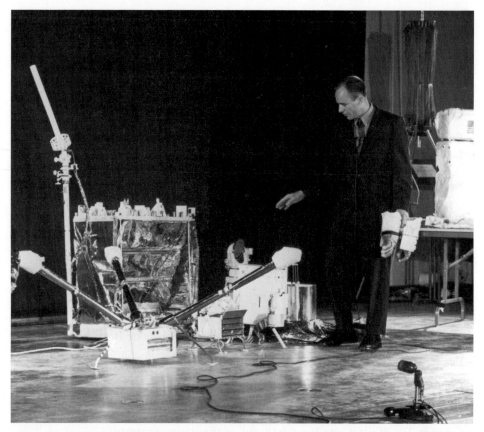

Figure 2.6. Alan Bean described the ALSEP for the press in October 1969. The LSM is in the left foreground, with the Central Station behind. Next right is the SWS, then the SIDE, with a mock-up of the Surveyor TV camera behind the SIDE. The shiny cylinder surrounded by a thermal skirt is the PSE. Al has a pair of EVA gloves in his left hand. A PLSS sits on the table at the right edge of the photo with the deployable S-band antenna behind.

you're coming at it from an angle, and then stick it in. You wouldn't have to worry about having it absolutely straight up and down."

Bean – "And then, [with the sleeve] you can't see in there."

Conrad – "Having it straight up and down was real easy in 1 g; it wasn't up there."

Bean – "And it was a lot darker inside those little collars. So, when you stuck it in the hole, you weren't sure if it was lined up to connect."

Conrad – "They were a real pain."

Bean – "And that's what's going on here. And I think it's just normal, because you're doing so much stuff. There's no way to make it all just perfect. It's like car repair."

Conrad – "The timeline they put together was our really best estimate of how long it was going to take to do something. And, as you can see, we were slowly falling behind. And I think you can pick out all the reasons."

Interviewer – "When you designed the timeline, was it based on how long it took you in 1 g? Or was there some padding (that is, some allowance for the unexpected) in there?"

Conrad – "Well, I think we put a little educated guess in there. But we used to run a 4-hour EVA, in 1 g, weighing 300 pounds. And there were places where weighing 300 pounds bothered us. And there were places where it was helping."

Bean – "I don't think it was quite 300, because the backpack was sort of a lightweight [mock-up]. [True] Everything else was normal weight, but they had a different kind of backpack."

Conrad – "For some reason, I thought they were weighted . . . Well, maybe not. Maybe that's another reason we were a little bit different standing up in 1 g than we were up there."

Bean – "Because I can remember they [members of the support team] had to follow us with the cooling hose."

Interviewer – "What kind of cooling did you have, air or water?"

Conrad – "I believe we had air."

Bean – "That's what I think, too. Later on, they changed it [to ice water that was circulated through the Liquid-Cooled Garment, starting with Apollo 16]. You just didn't have all that stuff [that is, the oxygen and cooling water and associated fans and pumps]. The guys had to carry it around. You couldn't carry all that weight; and, of course, the backpack [sublimator] wouldn't have worked on Earth, anyhow. So they had to carry it [the air-cooling gear] around. And I don't think they had water cooling for us. We didn't put that part of our suit on [meaning the LCG], and they just blew more air through those hoses, sort of like suit ventilators except they just kept doing it. They could pressurize you and [the training] was a lot of hard work. I think that's why the work [on the Moon] didn't bother us, except – like Pete pointed out – that the gravity makes a lot of difference, 'cause you can't bend over as easy, when you need to. The 1/6 g has got some advantage; but, if you have to bend over, then it's a disadvantage because you don't bend over so far or so easy."

Conrad – "The practice deployments were good physical training."

Bean – "Yeah. I can remember the first time we did all this; man, we were just really dragging. And then, as each time we did it – 'cause we'd do it about once a week . . ."

Conrad – "We got physically more adapted to it."

Bean – "We could do it better and we physically got better at it."

Interviewer – "Is this in the interval between 11 and your flight, or prior to 11?"

Conrad – "Well, we had started some of it before 11. Obviously we were going to have an ALSEP all along. I remember doing it unsuited a lot, in the beginning ..."

Bean – "To learn how to do it. Probably when we finished [as the backup crew on Apollo] 9, then we started doing these things, as best we could."

Conrad – "The thing that came late was the Surveyor."

Interviewer – "Apollo 9 was in March of '69. So you would have had 10 months to train as the 12 prime crew."

Conrad – "But we were on a schedule to go sooner than we did. Until July, we were on for September [instead of November]."

Had the first lunar landing not been accomplished on Apollo 11, a September launch of Apollo 12 and a launch of Apollo 13 in December would have given NASA two more chance to achieve the Kennedy goal of landing before the "decade was out." Thanks to the success of Apollo 11, NASA could lengthen the interval between missions to allow more preparation and the addition of new capabilities.

Interviewer – "It was a short training cycle."

Conrad – "Yeah, but we had a long cycle on 9. So all of that [spacecraft and flight] stuff we had down cold."

Bean – "All the LM ..."

Conrad – "It wasn't like we were starting cold."

Bean – "... doing the rendezvousing, so we were able to take those basic skills we'd developed and put it over on this. Nobody could have done this from a standing start. Just no way to learn it all. There's too much to learn. None of it's hard. There's just a lot."

Interviewer – "One of the advantages the J guys [the crews of Apollo's 15, 16, and 17] had is that they trained as backups to landing missions."

Conrad: All we're doing is making a mess. You'd be better off ... Let me ... Listen, can you ... Leave it; leave it sit. Let me hold that. You put that stool back here closer.

Bean: Okey-dokey.

Conrad: Am I over all the cables?

Bean: Okay.

Conrad: Huh?

Conrad: Okay. Take your EMU break.

Bean: I am.

Pete is telling Al to rest for a moment.

Bean: I'm trying to stay away from ... That's a nice job on that solar wind.

Conrad: Thank you.

Bean: You've got to be careful you don't kick dirt on them once we get them set down. I guess the way you could do it is have it some sort of package and the package'd get all the dirt.

Conrad: What are you doing anyhow?

Bean: Well, I had to get this ... *(pause)* Oh, I'm sorry.

Conrad: Here you go; move it up right here where my footprints are.

Bean: Okay. Wait a minute.

Conrad: That a boy. That a boy. Now, don't ... Doing good. There; now, kind of tamp it right around in there. That a boy.

Bean: They took off the central BB. *(pause)*

The PSE was one of two pieces of scientific equipment that the Apollo 11 crew had also deployed. On both flights, the experiment had a device on top to help the astronauts get it properly leveled. On 11, the device consisted of a BB – a small metal ball – in a small cup marked with concentric rings. Because of the low gravity field, the BB on Apollo 11's PSE wouldn't come to rest in the bottom of the cup while Aldrin was trying to get the instrument leveled, and it gave him considerable trouble. For Apollo 12, a bubble level was added so that Bean could see which method worked best.

> **Bean** (from the 1969 Technical Debriefing) – "The addition of the bubble level to the top of the passive seismic was a good one. I noticed it was really easy to level the experiment with that. While I was doing it, I kept an eye on the little BB in the bowl-leveling scheme. It was just rolling all over the place, as it had for Buzz when he tried it on Apollo 11. I don't think that's the way to go for any other leveling. I think this bubble works real well, and it works pretty fast."

Bean: *(positioning the PSE stool)* Now, let's see what happens if we set it right ... *(pause)*

Conrad: That's good.

Bean: That looks like it'd be okay, doesn't it?

Conrad: Yeah.

Bean: Just a second. Let me do a couple of things. *(pause)* Do this; back this out. I take it and knock a little bit out of there.

Conrad: Pull it over a little bit more and level it up.

Bean: Yeah, this doesn't need to be level. Okay. That's got kind of a deeper hole there in the center.

Conrad: Yeah. You're in good shape. That doesn't have to level, huh?

Bean: No, because the other one levels on this stool.

Conrad: Okay. Push it down a little bit.

Bean: Okay. That's a good idea.

Conrad: Seat it. *(pause)*

Bean: Looking good. Looking good.

Conrad: Just give me the tongs, and I'll put them away.

Bean: Okay.

Conrad: Get with it and I'll start opening Boyd bolts. *(pause)*

Pete is going back to the Central Station, either to finish releasing the magnetometer or to start on his next task of releasing the Central Station's sunshade.

Bean: Yup. That's going to do okay, Pete.

Conrad: Thank you.

Bean: Yeah.

Conrad: All righty. *(long pause)* If I wasn't at 3.7 psi, I'd whistle while I work. I can't do it. *(four short, monotonic, breathy attempts) (garbled; long pause)*

Interviewer – "Could you whistle?"

Conrad – "No, you can't at 3.7. You can't do it at 5, either. Couldn't do it in the spacecraft; and couldn't do it in Skylab."

Bean: Okay. That seismometer's right in position now. Here's hoping ... here's hoping nothing happens to it.

Conrad: Looks real good, Al.

Bean: Yeah, it does. It looks like it's going to clear. Houston, looks like digging that little hole might work. Setting up there nice now. And it looks like maybe it's not going to get down in the hole. Maybe we did it. Let's just see.

Gibson: Roger, Al. We concur; it sounds like a good plan. *(long pause)*

Bean: Whoops. *(pause)* You know something? Look at that.

Conrad: What?

Bean: Well, I [should] put a little dirt on it, I guess. When I put out the skirt of this foil, it doesn't want to lie down. I guess because it's been folded so long. I guess I can just probably just spread it out. *(Pete chuckles; pause)* It acts almost like it gets a little static charge on it. It's resisting the lunar surface. I'm sure that isn't it.

Conrad: Hey, stop one second.

Bean: Watch out. Don't come across ...

Conrad: I know. That's why [what] I want you to do is come take it.

Bean: Okay.

This exchange suggests Pete has just finished releasing the magnetometer and wants Al to take a break from deploying the PSE thermal skirt to get the magnetometer out of the way.

Conrad: That [connecting] cable isn't even unreeling. See, it's perfect. Take it out there and set it. That a boy.

Bean: It's right behind there. You may ...

Conrad: There you go.

Bean: ... want to just take it. *(pause)* You might want to trip it off that thing, Pete.

The cable is draped on something.

Conrad: Trip it off what?

Bean: It's on that post.

Conrad: Oh. All right. *(pause)* [I see] what you're talking about. *(pause)*

Bean: *(aligning the PSE or setting the magnetometer down)* That shadow's going to be just right. Looking okay.

Conrad: Ah, fiddle dee diddle. Come on. *(pause)*

Bean – "This wasn't what I was used to hearing in training. When things like this happened, Pete usually said things like 'You son of a bitch!' or 'Goddamn Boyd bolts.' And all of a sudden we hear these moderate things. Who's this guy with me in that other suit?"

Conrad – "He came to me one day before the flight and he said, 'Aren't you worried about swearing during the flight?' And I said, 'No, I've been doing it all my life. I know when to and when not to. It's guys like you who are going to slip.'"

Bean – "You didn't slip."

Conrad – "No, but you came close. *(hearty laughter)* I had him worried!"

Bean – "That's right. You never said one. *(laughing)* 'Who is that guy in the other suit over there? With this 'fiddle-dee-dee.' I'm sure I'd never heard that one. He didn't even know how to say it right! He didn't know how to do a mild curse."

> **Conrad:** Hey, everything is going great, but that whistle [on the communication circuit] is driving me nuts. Bothering you?
>
> **Bean:** A little bit. *(pause)* Okay.
>
> **Conrad:** Oh, darn! I can't help but get dirt on this darn thing, no matter how careful I am.

Pete is probably releasing the Boyd bolts that are holding down the Central Station top and sunshade. In the process of moving from one to the next, he is kicking dirt onto what will be the top of the Central Station.

> **Bean:** I don't think there's a way, Pete. I don't think there's a way. *(pause)* Unless you got, you know, where you could put it out in a package and then your last step like the SIDE over there – it just deploys off the thermal wrapping – or your dirt protector – and you end up with a nice, clean experiment sitting there. *(pause)*
>
> **Conrad:** How we doing on the timeline, Houston?
>
> **Gibson:** Stand by, Pete. *(long pause)*
>
> **Conrad:** Wheee! *(hearty laugh)*
>
> **Bean:** What happened?
>
> **Conrad:** Oh, it's just the way things [one of the Boyd bolts] pop off down here in 1/6 g. *(pause)* Won't be long and I'll get to my favorite task, pointing the [Central Station] antenni [sic].
>
> **Gibson:** Pete, we show you're about 8 to 10 minutes behind but it's no sweat. You've got lots of oxygen and feedwater. We'll give you an update on the time remaining when you finish deploying ALSEP.
>
> **LM Crew:** Okay.

Bean: Good. Looks like it's going real good. Houston. [On] this seismometer, I'm having a little trouble making the skirt lie down; but, other than that, it looks good. It doesn't want to just lay flat like it does on Earth at 1 g. It sort of wants to slip up. I guess it's just because it's this memory in it from being folded so long, and ...

Gibson: Roger, Al. You can make a two-man task out of that and put a little dirt around the outside edge to hold it down if you like. *(pause)* You won't have any trouble getting a little dirt on it, will you?

Bean: Houston, I noticed that if I just ... *(pause, hearing the second part of Gibson's remark)* I'm pushing it down. *(pause)* That'll do it. *(pause)* Okay, Pete. Let me level it [the PSE] up. I think it's pretty near level. It's lined up exactly. *(pause)* See if I can give it a couple of pushes

Conrad: I got this feeling there's a couple of Boyd bolts that I haven't done here (garbled). *(long pause)*

Conrad – "You didn't really know, in fact, that you got a Boyd bolt completely undone with those alignment covers [the guide tubes]. You couldn't see in there."

Bean – "I remember on Earth we'd do that, sometimes. We'd try to get it out and then we'd have to fool with all of them to find the one that wasn't undone."

Bean (from the 1969 Technical Debriefing) – "My experience working the Boyd bolts is that you can do them a heck of a lot faster it they don't have those little alignment tubes on them. I don't know how Pete feels about this, but I recommend that you throw those off and just use the Boyd bolts. I could always stick my tool in there a lot faster when there were no tubes, and I can also see when the little bolts jump up a lot better."

Conrad (from the 1969 Technical Debriefing) – "The Boyd bolts, as Al pointed out, were no problem, [although] it would probably be easier if the cups were lower [that is, if the guide tubes were shorter]. The bolts should be kept covered with tape, though, because of the dust problem."

Pete's suggestion of putting tape over the guide tubes to keep dust out was implemented on the J missions in the form of breakable aluminum foil covers.

Bean: This thing looks level as can be to me. *(pause)* Looks good.

Gibson: Roger, Al. Copy. You have the bubble centered. *(pause)*

Bean: I'll tell you an interesting thing about this bubble, Houston. *(pause)* No. It's okay. It's okay.

On Apollo 17, the bubble on the Central Station stuck to the edge of its container and Jack Schmitt had some considerable trouble breaking it loose. It is possible that Al's bubble hung up momentarily as he made a final adjustment.

Conrad: *(slowly enunciating each word)* I am not happy here, Al. I'm afraid some of these Boyd bolts ... There's not enough of them, and I don't understand why.

Bean: Why, one of them didn't do or what?

Conrad: Well, I don't know. I haven't the foggiest idea. I thought all of them did.

Bean: Looked like you're well ahead of the time. *(pause)*

Bean: Hey, take your stick and kick a little dirt up on it right there [on the skirt].

Conrad: Huh?

Bean: Take your stick and just put a little dirt on the edge, Just on the edge, though. Not much. *(pause)*

Conrad: Uh-oh. *(garbled)* bad news.

Bean: That's okay.

Conrad: I won't do that anymore.

Bean: That's okay; that skirt'll stay down enough. *(pause)* Okay. That's complete. Let me take a couple of pictures. *(pause)* Okay. Houston. The passive seismic is down; the alignment is exactly 90 degrees. And I'm going to take a couple pictures of it here.

Bean (from the 1969 Technical Debriefing) – "The aluminum foil, the skirt, didn't want to lie down. It wasn't that it had memory. When I placed it near the ground, the many layers seemed to separate. The skirt seemed to have some kind of static charge to it that would not allow it to touch the ground. It took quite a little pushing to get it to lie down on the ground. The only way I could make it lie flat was to put a little dirt on it, which I tried. But that wasn't a very good idea, because it's difficult to put little clods of dirt on it. I later got some Boyd bolts and made the little alignment tubes sit on it. That worked real well; it held down the skirt pretty well."

The *Apollo 12 Mission Report* repeats Al's comments about separation of the layers and indicates that delamination had been anticipated. "For Apollo 13 and subsequent spacecraft, the shroud lamination will be spot sewed together at intervals around the periphery, a weight will be sewed to each of the six attach-pullout points on the shroud, and a 5-foot Teflon blanket will be added for thermal control to decrease solar degradation."

Conrad: Hey, Al.

Bean: Yes, sir?

Conrad: I don't know what's the matter, but something is ... There it is. *(pause)*

Bean: Hey; look and see if the lens on my camera looks clean?

Conrad: Huh? Wait a minute. *(pause)*

Both of them have been kicking up dirt as they move around and Al wonders if any has gotten on his camera lens.

Bean: There, you got it made.

Conrad: No, I don't either. I don't know what's the matter.

Bean: Want me to hold down this part while you get that other bolt?

Conrad: Push that center down.

Bean: All right. Oops.

Conrad: There it is.

Bean: You got it. That's beautiful.

Conrad: Easy. *(sustained laugh)*

Bean: Man alive. It sure wants to come up, doesn't it? Whoo!

Up to this point, the Central Station has been a flat package with the top held tight to the bottom by the Boyd bolts. Once Pete released the last of the bolts, springs popped the top upwards about 3 feet, carrying the curtain-like sunshade with it. The foil curtain shimmies and shakes for a few seconds.

> **Conrad** (from the 1969 Technical Debriefing) – "The sunshield deployment worked perfectly well and as advertised in 1/6 g; it popped up, [and] lifted off the ground, actually. It was a real thrill."

Conrad: That did it.

Bean: That's beautiful.

Conrad: That did it. *(pause)* Houston, Central Station's up.

Gibson: Roger. Copy. Central Station up and [confirming Al's report on the PSE alignment] 90 degrees on the PSE gnomon. *(pause)*

Conrad: Hey, Al.

Bean: Yes, sir. *(pause)*

Conrad: Never mind. I'll get you in a minute.

Bean: Uh-oh. Don't ever move backwards.

Conrad: Hey, we're going to have to do a lot of . . . *(pause)* What did you do, fall over?

Bean: No.

Conrad: No?

Bean: No, but I just . . . I didn't even come close, actually. It's just that I think it's [his warning about moving backward] something you want to follow as a rule. *(pause)* But don't kick dirt on this *(garbled)* right there.

Conrad: *(garbled under Al)* right there. *(long pause)*

Conrad: *(garbled)* One problem [is that] the Central Station is not really level. *(long pause)*

Conrad: Not too bad.

Bean: [I'll] get a good shot of it here, Pete.

Conrad: Okay. *(long pause)*

Bean: Okay. I'll take out the magnetometer. *(pause)* Central Station went up so nice.

Conrad: Al, those are my last two Boyd bolts.

Bean: Okay. *(long pause)* I'm down in a little crater now, Houston. And, sure enough, right in the bottom of the crater there is a lot softer dust than up on the rim. Not much; but it's noticeable. I don't think the sides are slippery at all. I don't think it's going to bother us going over to get our Surveyor.

Gibson: Roger, Al. We copy.

Interviewer – "What you're saying here doesn't agree with the impression I've gotten that the softest soil is generally on the rims of craters."

Bean – "In that particular crater. We can't categorize . . . Before missions, we always try to categorize all craters as being one particular way. But this one was softer at the bottom, and some others were harder at the bottom. Some of them had glass at the bottom, some of them had rocks at the bottom. I think we were worried more about the possibility that, on the slopes of Surveyor Crater, it would be slippery. You know, that the dust would be hanging there and you'd get on it and it would slide you down like snow. But it didn't seem to do it, at least not in Surveyor Crater. I would agree that the thickest dust, generally, was on the rim of craters. But, in this case here, I'm sure it wasn't. All these craters are different. And some of them are newer and had the raised rims. Those are the ones that are soft up there. This might have been an old one and it was level [that is, it lacked a raised rim] and the rim had been knocked off by small meteorites and thrown in the middle and you had a soft fill. There were all kinds. Some of them, you looked down in there and they were – like Pete said – they looked like they had glass. So they weren't soft down there, for sure. And the glass wasn't even covered up. We shouldn't always be looking for one answer in this stuff, but people do have a tendency to want to have a single answer to every question, when there aren't. Before we went, they were always arguing about whether the Moon was volcanic or just impact. Instead of taking a more reasonable approach of probably some of each. Just like the Earth. So, some of these craters are soft in the bottom, and some aren't. And maybe someday people will say, 'Well, if it's got a raised lip, generally something.' We were more worried about the sides being slippery at Surveyor."

Bean – "People reading this may notice a difference of opinion about how to do different things or about looking at craters. And there shouldn't be an effort to try to find out who's right but, rather, there should be a thought that there is a diversity of observations of things. The first position might be 'There is a difference, and there may be other differences' and that's just the way the world runs."

Gibson: Are you on your way out with the LSM?

Bean: That's right. Got it right in hand; I'm out at the end of the line [50 ft SSE of the Central Station], and I'm deploying the legs right now.

Gibson: Roger. *(pause)*

Conrad: One antenna mast in place, Houston. Going back for my favorite thing.

Pete is going to install the Central Station antenna aiming mechanism.

Gibson: Roger, Pete. Copy antenna mast in place and good luck.

Conrad: What'd you ...

Bean: Look at that, Pete.

Conrad: What?

Bean: Hey, one of the fun things here, Houston, is all these Styrofoam packing blocks that are put on there to protect it during shipment or launch. When you take them off and throw them, they really sail. These things stay airborne up for 10 seconds, maybe. *(pause)*

Styrofoam blocks were used as protective packing material on several flights and, on Apollos 16 and 17, after some of the foam blocks had been sitting out on the surface for a while, heating up in the sunlight, gas trapped in the hollows of the foam pressurized and exploded the foam. In fact, during an Apollo 17 EVA, Gene Cernan found one piece out about 100 meters from the spacecraft.

Conrad: Got any more thoughts on our TV camera, Houston?

Bean: Hey, Pete? Pete?

Conrad: What?

Bean: Watch this. *(they both laugh, probably as they watch another piece of Styrofoam sail off into the distance)*

Bean: Try that on ... Hey, I just threw something. It hasn't hit the ground yet; it might have gone up 300 feet. *(Pete laughs)* Boing!

Conrad: *(laughing)* Stop playing and get to work. *(laughing)*. Come on. Maybe they'll extend us until 4-1/2 hours. I feel like I could stay out here all day.

Bean: I know it.

Conrad: *(garbled)* holds up. *(pause)*

Bean: Right there. This is going to be a good place for the magnetometer.

Gibson: Pete, we've been thinking on that [TV] camera, and when you get back [to the LM], we'll have a test for you to run on it. It looks as though part of it's come back. And we're going to try and see what we can salvage.

Conrad: Okay. *(pause)*

Bean: What do you think happened to it, Houston? *(pause)*

Gibson: Al, we're not sure on that. Why don't you wait until you get back to the LM, and we'll work it out a bit and see if we can determine it?

Bean: All right.

Conrad: Hey, Al?

Bean: Yes.

Conrad: See, that nice.

Bean: Hey, that's weird *(they both laugh)* On Earth, they [magnetometer arms] just flop out in position. Here they don't even want to stay. There they go.

Conrad: *(chuckling)* That thing hit the ground and it's still bouncing. *(pause)*

Conrad – "I threw something. I must have thrown something and you were doing the arms and they don't stay down, and then I'm laughing again and said 'That thing hit the ground and it's still bouncing.' So I must have thrown something at the same time you got the arms."

Bean: *(pleased with how easily the magnetometer is deploying)* That's slick. That's really slick! *(pause)* Boy, I hope ... I hope ... *(pause)*

Gibson: Hello, Yankee Clipper. Houston.

Conrad: You hope, you hope?

Bean: Oh, I was just thinking about something. *(pause)* Pete, no way to keep these things clean; I'm really worried about this white coating. *(pause)*

Bean (from the 1969 Technical Debriefing) – "The magnetometer was a beautiful experiment and it was easy to deploy. It was easy to align and level and it took quite a bit less time than the passive seismic because it was sort of self-contained. You just screwed the legs to make it level. You could grab one of the magnetometer arms and move it around so that it would be in alignment."

Figure 2.7. Dirt. Dirt everywhere. High-velocity meteorite impacts – over billions of years and at all scales – as well as some explosive volcanism have combined to cover the moon with thick deposits of lunar regolith. When you kick it, the soil scatters far and wide in the moon's vacuum and reduced gravity. In this water-free vacuum environment, particles stick to everything. Dirt patterns on Al Bean's knees indicate that he has been kneeling at some point in the EVA. With the sampling scoop punched into the surface and the tool carrier at hand, Bean appears to be ready to collect a soil sample.

Bean: Well, I just don't like all that dirt on it; don't know what we can do, though. There's no way to work around it. The radiator's got a little dirt on it, and there's no way to dust it off, there's no way to . . . All you can do is tap it a little bit and hope some of it falls off and that's about it. Don't want to tap it too hard. *(pause)* That's the best I can do. Okay, [that's finished]. *(long pause while he takes the pictures.)*

Later crews will carry dust brushes to clean off radiators, thermal covers, and – especially – each other before they go back into the cabin. They will have two kinds of brushes: small lens brushes made of camel hair or some similar soft material for cleaning camera lens, visors, and other surfaces they don't want to scratch and a large brush (the size of a house-painting brush) for cleaning the suits and other less sensitive surfaces.

Bean: God damn. *(pause)* Okay; I'll deploy the SIDE, Pete.

Conrad: Okay.

Bean: It's my last item [on the ALSEP portion of the checklist].

Conrad: [I'm] on my last one.

Bean: Okay. *(long pause)*

Conrad: Does the antenna look like it's pointed at Earth?

Bean: It looks close.

> **Bean** – "If Earth was near what the stowage position was, then it didn't take you long to move it. But if Earth was up high, you'd be fooling with it [longer]. It was hard to do."

> **Conrad** – [Looking at the checklist] "We had rough pointing angles in here. This is probably a good example of something designed by somebody that wasn't ever going to see the lunar surface."

Conrad: That's it. *(pause)*

Bean: Okay. [I'll] do the SIDE, right now. *(pause)*

Al will use the UHT to carry the SIDE 55 feet southwest of the Central Station. The package actually contains two separate pieces of equipment: the SIDE and a smaller experiment called the Cold Cathode Ion Gauge or CCG (later CCIG). The latter is stowed in a compartment in the side of the SIDE. Al will remove the CCIG and set it down 3 to 5 feet from the SIDE. The experiments are connected by an electrical cable because data from the CCG is sent to the Central Station via the SIDE ribbon cable.

As it will turn out, the CCG/SIDE connecting cable will retain a great deal of memory and will make it very difficult for the crew to get the CCG properly pointed.

Conrad: *(still getting the antenna aligned)* Got it.

Bean: Okay. *(pause)* Okay. I hope I don't tug on the [Central] Station here [with the SIDE cable] as I come out, Pete.

Conrad: If that happens ...

Bean: It looks pretty level, by the way. Your station does.

Conrad: Yeah. *(long pause)*

Conrad (from the 1969 Technical Debriefing) – "The antenna alignment went as advertised. I had played with it enough that I knew how to align it correctly. Apparently it is aligned all right, because you are receiving good signals."

Bean: I'm nearing the end [of the SIDE cable], I'll bet.

Conrad: I'm done. How's that, ah, ... *(pause)*

Bean: Gosh, this ... *(laughing)* They must have doubled the cable length on this one.

Conrad: *(garbled)* *(pause)*

Bean: That's it.

Conrad: You're at the end. You're at the end.

Bean: Okay. Let me set it down and work on it right here.

Conrad: That does it, old buddy

Bean: Okay. *(long pause)*

EPILOGUE

After setting up the first multi-instrument geophysical observatory on the surface of the Moon, Conrad and Bean returned to the LM for dinner and a well-deserved rest. They hooked up their hammocks and, still in their suits, tried to rest. Later, they said they had slept for only about 4 hours, not because of lunar gravity or excitement, but because the bulky suits were very uncomfortable. In reality, they hardly slept at all. They would have been more comfortable if they had taken off the suits, but that was time-consuming.

After some breakfast, they headed out on their second hike across the lunar surface. It was another busy foray, during which they sampled rocks and soils from

Figure 2.8. This photo, taken at Sharp Crater, shows the team of Conrad and Bean at its best. Working together, they are filling the "long can," a vacuum-sealed sample container designed to protect the soil sample from the Earth's environment when they return to Houston. The nearly full can is sealed with the lid, which is hanging by a cord from the bottom. This photo gives us a good view of the top of Al's Hasselblad camera as well as the cuff checklist and watch on his left sleeve. The visor reflection shows Pete, the HTC carrier, and the remarkable contrast between the lunar surface and the darkness of space.

several craters and collected cores of the layered lunar regolith, which may have included material ejected from Copernicus. They visited Surveyor III, photographed it, and – using bolt cutters – cut off the earlier spacecraft's soil mechanics scoop and the TV camera to bring home to Earth. During the first EVA, they were anxious about going into the crater where Surveyor III had landed because the walls appeared to be steep. Fortunately it was an illusion, and they had no trouble entering or leaving the crater.

Conrad and Bean didn't miss much, as is clear from their consistently insightful observations about working in the lunar environment. Equipment designers were taken to task, in a gentle and positive way, for not understanding the realities of the

Moon's hostile environment. Carefully maintained pure white surfaces, twisty cables, small letters in light-colored print, small buttons, traditional bubble leveling devices, and dust, dust, and more dust were all discussed in the context of a field environment that was a far cry from an Earth-bound clean room. Their comments about moving equipment around in 1/6 g showed that reality sometimes mirrored the simulators and sometimes did not. Their experiences during the longer EVAs and subsequent suggestions for overcoming space suit problems would be used in modifying the equipment for later missions. Later Apollo missions owed Conrad and Bean a great deal for being patient, observant, and practical explorers.

The seismometers left on the Moon as part of the ALSEP performed well. Their data were examined by an all-star team of geophysicists, including several National Academy of Science members (and one who would eventually be president of the Academy and a presidential science advisor). Their studies confirmed that, unlike Earth, the Moon was a cold, rigid body. The data provided evidence of some "moonquakes" that were probably related to tidal stresses, but studies indicated most of the activity occurred when a relatively large meteor impacted the surface, generating a signal that "rang like a bell." These data were not only scientifically interesting but supplied information on the flux of larger meteors that could be a hazard for a future lunar base.

For the engineers and scientists who supported the Apollo missions, it was Apollo 12 that truly highlighted the hazards of dust on the Moon. Lunar soils are similar to silty sands on Earth in some ways – with mean grain sizes ranging from 45 to 100 micrometers (the approximate equivalent of the diameters of two human hairs). Many of the grains are sharp and glassy. Lunar dust has very low electrical conductivity and dielectric losses, which permits the accumulation of an electric charge under ultraviolet radiation. Consequently, when the astronauts kicked up dust with every step, it settled on their suits and equipment. Dust coatings reduced the reflective capability of thermally sensitive surfaces, obscured visibility through helmet visors and camera lenses, caked and clogged moving parts, and created abrasion hazards. It's clear that even with filtration systems in the spacecraft cabin, individuals working on the Moon in the future will experience intermittent exposure to lunar dust as they suit up and when they remove their suits after lunar surface activities. For those who plan lunar bases in the next stages of exploration, dust management will be a major issue.

For those of us working in the Apollo Program, Apollo 12 was a very satisfying mission. Certainly we made giant strides scientifically, but we also learned to value the joys of exploration. The public may have forgotten some of this early elation during the dramatic rescue and return of the crew of Apollo 13, but it was revived by Apollo 14's successful exploration of Fra Mauro's rugged terrain.

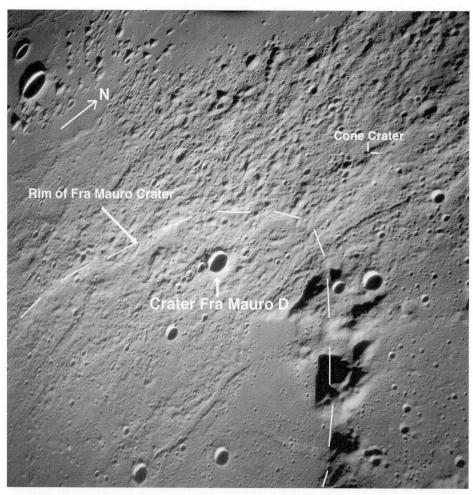

The Fra Mauro highlands were the primary objective of the Apollo 14 mission. The hummocky ridges here are interpreted as ejecta deposits from an impact that formed the 1200-kilometer-diameter Imbrium Basin, which is located over the northern horizon. These deposits overlie and partly mask the 94-kilometer-diameter Fra Mauro crater (seen as flat plains with a linear rift, located below center-left).

3

A Damned Hard Walk Followed by a Little Golf

G. K. Gilbert, a well-known American geologist, suggested in 1893 that among the last major events in the Moon's history was one that formed the 1500-kilometer-(930-mile)-diameter Imbrium impact crater and its extensive ejecta blanket. This impact debris layer is considered a major boundary (a geologic "timeline") used to subdivide the history of the lunar surface and therefore was particularly interesting to the scientists of the Apollo Program. The widespread debris layer was named the Fra Mauro Formation after a region north of Fra Mauro crater, where the deposit appears to protrude above the surrounding mare lava flows. [Fra Mauro was a 15th century Italian monk, who was a renowned mapmaker of his time.]

On Earth, such large and distinctive features are normally studied both in the field and with aerial or satellite imagery; in this case, the preliminary studies used orbital images of the Moon.

The Fra Mauro Formation is rugged terrain, marked with 1- to 4-kilometer-wide linear ridges that are oriented radially around Mare Imbrium. Samples from this area would include rocks ejected from deep within the highland crust of the Moon that could be used to determine radiometric ages for the Imbrium event. The Apollo 14 mission would be the first to leave the relatively smooth maria surfaces and to venture into hilly terrain.

Apollo 14 astronauts Alan Shepard and Ed Mitchell landed February 5, 1971 on the Fra Mauro Formation at a site 1230 kilometers (764 miles) south of the center of the Imbrium basin. To guide them in, they had a distinctive landmark, the 300-meter-diameter Cone Crater. During the final approach, they had no trouble recognizing Cone Crater when, at pitch over, they got their first complete view of the landing site. Very quickly, Shepard picked up the familiar pattern of smaller craters near their aim-point (targeted landing spot), which was another mile or so to the west. There wasn't, Shepard said later, any really flat ground close at hand – there were either craters or sloping ground wherever he looked – but he soon found a crater-free, LM-size patch only 30 meters from his target. The LM landed 191 kilometers (118 miles) east of Apollo 12's landing site on the much flatter Oceanus Procellarum.

The only problem with the particular spot where they landed was that it was on an 8-degree slope; for 24 of the next 33 hours, the astronauts had to contend with a tilting floor that threatened to dump Shepard over onto Mitchell's side of the spacecraft. In fact, the tilt contributed to a sleepless night between their two EVAs on the lunar surface.

Officially, the Apollo 14 crew had four primary objectives:

APOLLO 14 FACTLIST	
Crew	Commander, Alan Shepard; CMP, Stu Roosa; LMP, Ed Mitchell
EVA CapCom	Bruce McCandless (EVA 1), Fred Haise (EVA 2)
Command Module	Kitty Hawk
Lunar Module	Antares
Launch from Earth	21:03 GMT 31 January 1971
Lunar landing	09:18 GMT 5 February 1971
Popular site designation	Frau Mauro Highlands
Lunar stay	33 hours 31 minutes
Splashdown	21:05 GMT 9 February 1971
Command Module lunar orbits	34
Time outside the LM	2 EVAs, 9 hours 23 minutes
Maximum distance from the LM	1.4 kilometers
Distance walked or driven	3.0 kilometers
Lunar samples	42 kilograms
Lunar surface photographs	401
Surface equipment	209 kilograms
Highlights	First extended trip away from the LM, first non-mare landing, oldest moonwalker (47 years)

- Perform selenological inspection, survey, and sampling of materials in a pre-selected region of the Fra Mauro formation (a geological field excursion).
- Deploy and activate the ALSEP. (This was one of the very successful Apollo geophysical stations, which returned data to Earth until the network was turned off at the end of NASA's 1977 fiscal year to save an annual monitoring cost of $200,000! The network cannot be turned on again from Earth.)
- Develop man's capability to work in the lunar environment (what worked and what didn't).
- Obtain photographs of candidate exploration sites (to improve our understanding of the landing sites for Apollos 15, 16, and 17).

In the 3 hours following the landing, Shepard and Mitchell reconfigured the LM for takeoff, took time out to have lunch, and passed along some excellent descriptions of the view out the window. They finally started to don their backpacks, getting ready, as Shepard said, "to go out and play in the snow." That mental picture came easily. Outside the window, the soft, rounded forms of the craters gave the look of snow-covered ground (although the color wasn't quite right … it was "mouse-brown or mouse gray" depending on where they looked in relation to the Sun), and the black lunar sky made it easy to think of a snowy winter's night back home. And to top off the

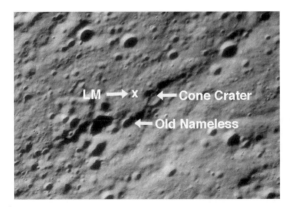

Figure 3.1. Cone Crater is situated on one of the ridges of hummocky ejecta above the more level swale where the crew landed. Cone Crater, surrounded by angular boulders, is younger than some of the more subdued craters in the area. One goal of the mission was to sample rocks that had been excavated from deep in the lunar crust by the Imbrium impact, a major event in lunar history (about 3.85 billion years ago).

Figure 3.2. In this photograph of the lunar module (looking east-southeast), Al Shepard is approaching the spacecraft late in the second EVA. In response to a request from Houston, he is aiming the TV camera at the equipment-storage assembly on the LM. The boulders at the lower right are likely to have been ejected from Cone Crater, their target for the geology traverse.

childhood image, Shepard and Mitchell only had to look at each other, standing there in their cumbersome spacesuits, donning the thick gloves, eager to get outside – play in the snow, indeed! Impressions were sometimes contradictory; Shepard shifted from the snow analogy to comments about how stark was their view of the lunar surface.

Because the lunar regolith completely blankets the surface, the subdued terrain makes it difficult to estimate distances – in part because there are no familiar objects that can be used for comparison. Antarctic explorers have much the same problem when all they see is snow. Apollo 11 astronauts Armstrong and Aldrin were the first to comment on the deceptive nature of the lunar surface:

"Distances on the lunar surface are deceiving. A large boulder field located north of the LM did not appear to be too far away when viewed from the LM cockpit. However, on the surface we did not come close to this field, although we traversed about 100 feet toward it. The flag, the television camera, and the experiments, although deployed a reasonable distance away from the LM and deployed according to plan, appeared to be immediately outside the window when viewed from the LM cockpit. Because distance judgment is related to the accuracy of size estimation, it is evident that these skills may require refinement in the lunar environment."

The purpose of the first EVA was to quickly grab some samples and set up the ALSEP on a spot about 180 meters (600 feet) west of the LM. Four hours into the EVA, they made their way across the rolling, cratered terrain to the LM to get some rest and then prepare for their trip to Cone Crater.

Although Shepard and Mitchell could have used a good sleep before making the climb up to Cone Crater, both of them were too keyed up to make full use of the scheduled rest period. In the interest of having time for a complete second EVA, they talked Houston Mission Control into cutting the rest period by an hour and, naturally, spent a good deal of the time lying awake. Anticipation was a major contributor to their sleeplessness, but there were other factors. As was the case with Apollo 12, the mission simply wasn't going to be long enough for them to get out of their EVA suits during the rest period. Because they were unpressurized while the crew was inside the LM, the suits were uncomfortable and irritating; among other things, the neck rings made it difficult for the astronauts to find comfortable support for their heads. And finally, after they were stretched out in the hammocks, the tilt of the spacecraft became more and more obvious. Rationally, they knew they were okay, but at intervals through the night they awoke from a half sleep and, in the light lunar gravity, had the uncomfortable feeling that the LM was tipping over. Several times one or the other of them felt that he had to push aside the window covers and, just to be certain, take a look outside.

In the morning, it was they who called Houston – half an hour before the agreed upon wake-up time. They reported getting about 4 hours of sleep, but that was certainly a generous estimate. The flight surgeons concluded that neither of them had slept much at all and, like the crew of Apollo 12, Shepard and Mitchell had to draw on mental and physical reserves to get them through the second day of their exploration at Fra Mauro.

In the portion of the Apollo 14 transcript covering the Climb to Cone, the CapCom is Fred Haise. On this first significant trip by a crew away from LM and toward a specific destination, the EVA CapCom was much more of an advisor to the crew and a more active participant than Apollo 11's and 12's EVA CapComs had been. Haise was particularly well qualified for the role. He was the LMP for Apollo 13 and, with Jim Lovell, had trained for the mission to Fra Mauro and the climb to Cone. Because of the serious accident on the way to the Moon, Apollo 13's landing was cancelled in favor of an emergency return to Earth, and Haise was very lucky to have returned safely. From the geologist's point-of-view, it was particularly disappointing that he did not have a chance to explore the lunar surface; he was one of the program's best observers, with a complete grasp of the geological purpose of the missions. Having trained for Fra Mauro, he knew better than anyone what Shepard and Mitchell were planning to do and what some of the challenges might be.

DOWN THE LADDER

We pick up the transcript just as Al and Ed have opened the forward hatch and Al is ready to go out.

Mitchell: Okay, Watch the hatch cover. Kick it closed with your knee ... I mean the handle cover. *(pause)*

This is probably a cover on the exterior handle.

Mitchell – "The handle wasn't a 'T'. It was a regular handle, and I think there was a little cover that came up and hooked over the end of it to hold it in place. And you had to release that cover to operate it. That would prevent you from hooking anything on that handle. I'm not sure that's true, but that picture comes to mind."

Shepard: Okay. *(pause)*

Mitchell: Okay, You're going to have to lean toward me.

Shepard: All right.

Mitchell: You're hung up on the purse. There you go.

Shepard: Coming over your way.

Mitchell: Okay. Okay. Now hold it *[that is, "stay still"]* while I get your hatch ... *[correcting himself]* get your antenna. Okay. You're Go. Go right on out. *(pause)*

Mitchell: Back straight on out. Now you're in good shape.

Shepard: Okay, Houston. Al is on the porch.

Haise: Roger, Al. *(long pause)*

Shepard: Okay. [as per his cuff checklist at 0 + 10] I'm ready for jettison bag, Ed.

Mitchell: Okay. Let me get my [cuff] checklist open here. *(long pause)*

Shepard: Okay. Got it, *(pause)* And it's clear.

Mitchell: Okay. *(long pause)*

Al has dropped the jett bag over the side porch rail and reports that it will not be in their way.

Mitchell: Wait a minute, I'll come down and get it [meaning the LEC, which Al will hand in to Ed.]

Shepard: That's all right.

Mitchell: Just hand it to me. I'm right here.

Shepard: Okay.

For about 2 minutes, while Al climbs down the ladder and Ed double-checks the contents of the ETB and gets it ready to go down on the LEC, the astronauts go about their business without saying anything.

Al appears to hop down the ladder rather than stepping down, letting his hands slide along the rail as he goes from rung to rung. It appears to be an effortless, confident performance.

Haise: Okay. We got ...

Mitchell: Okay. Houston, Al's on the surface.

Haise: Roger, Al. And we got a good picture here, and ...

Mitchell: Okay. Al the LEC *(garbled)*.

Haise: ... we just saw you hop off.

Shepard: Okay.

Mitchell: And the ETB's ready to come down. *(pause)*

Al goes to the left side of the ladder to get the dangling end of the LEC and then gets into position directly west of the hatch.

Mitchell: Wait a minute. Let me get a little tension on it [the LEC]. Okay.

Shepard: Okay. Coming down. *(long pause)* Okay. Slow it just a sec.

Mitchell: Okay.

The ETB is now at about helmet height and Al walks toward the MESA to grab it.

Shepard: Okay, I've got it.

Mitchell: *(garbled)*.

Shepard: Let it go.

Mitchell: I'm checking the circuit breakers. *(long pause)*

Mitchell: Okay, Houston. I'm ... Ed's ready to egress. *(long pause)*

Al positions the MET near the MESA. The Modular Equipment Transporter (MET, NASA-ese for a kind of lunar rickshaw) held tools, sample bags, core tubes, and the lunar close-up stereo camera. It was an awkward means of carrying a lot of equipment – but better than a heavy tool belt or a sled – and was used on only this one mission.

Mitchell: Houston, Ed ...

Haise: Roger, Ed. You're cleared to come out.

Mitchell: Okay. *(long pause)*

> **Interviewer** – "Was it nice to get out of that confined cabin?"
>
> **Mitchell** – "Oh, yes, We were looking forward to getting out, I remember feeling the tension of wanting to get on with it and wanting to get out and explore, but hating to see the clock tick away. Because it was closer to having to go home. There was a lot of ambivalence there. We couldn't stop the departure, but we would have liked to have slowed time, and yet, at the same time, enjoy being out there."

Mitchell: And Houston, Ed's on the porch.

Haise: Roger, Ed.

Mitchell: Starting down the ladder. *(long pause)*

Ed hops down the ladder, albeit more slowly than Al does.

Mitchell: Well, it's nice to be out in the sunny day again.

Shepard: *(as Ed jumps off the bottom rung)* Yeah, it's a beautiful day here in Fra Mauro Base.

Haise: Sun ought to be a little higher today.

It has been about 16 hours since the start of the first EVA and the Sun is now about 8 degrees higher in the sky.

> **Shepard:** Yup, going on oxygen today.
>
> **Mitchell:** Beautiful day for a game of golf. *(pause)* Okay. *(long pause as Ed reads his checklist)*

> **Interviewer** – "'Going on oxygen today'? What does that mean?"
>
> **Mitchell** – "It's a kind of obtuse reference, a play on words. Fred says that the Sun ought to be a little higher in the sky, but Al's interpreting it as a little higher emotionally. And pilots go on 100 percent oxygen to get a high."
>
> **Interviewer** – "A pilot's version of glue sniffing?"
>
> **Mitchell** – "Yeah."
>
> **Interviewer** – "And you were in on the golf conspiracy?"
>
> **Mitchell** – "Yeah, and that's the reference."

> **Shepard:** Ed, I started to get a picture of home sweet home [meaning Earth] right straight up there.
>
> **Mitchell:** Yup.
>
> **Mitchell:** Could you undo my EVA antenna, please?
>
> **Shepard:** Okay. *(long pause as Ed moves in closer to the spacecraft, perhaps to grab the edge of the MESA or one of the ladder support struts so that he can bend down to give Al access to the antenna)* Okay, you're now undone.
>
> **Mitchell:** Okay. I've been undone before.
>
> **Shepard:** Never like this! *(pause)*

Al seems more playful than he did during much of the first EVA.

> **Mitchell:** Al?
>
> **Shepard:** Yeah.
>
> **Mitchell:** One more problem here. [Is] my gold visor's caught? I can't seem to ... *(pause)*
>
> **Shepard:** Okay.
>
> **Mitchell:** ... pull it down. *(pause)*

The LEVA's gold-plated visor provides protection against ultraviolet light and can be raised or lowered. Ed's visor is stuck, possibly because of dust in the track or because of interference with the fabric covering on the LEVA. Al is at the MET to install some weigh bags and Ed joins him so that Al can fix the visor.

Shepard: Okay.

Mitchell: Good enough, Thank you. Want some help?

Shepard: No, it's okay. All righty.

Mitchell: Okay, we're all set. *(long pause as Ed goes to the MESA)*

For the next several minutes, they transfer equipment from the ETB and the MESA to the MET.

Mitchell: *(back at the MET)* [I'll] put the two spare [film mag]s right up here.

Shepard: Okay.

Mitchell: They'll fit up here okay. *(long pause as Al goes to the MESA and returns)* Watch your foot! Back up. *(pause)* Okay.

What just happened was that Al caught his right foot on one of the cables and, feeling it pull, hopped several times on his left foot, with his right foot raised behind him, to keep from pulling the cable any more than he already had. Al then lowered his right foot and turned clockwise toward the spacecraft to free his foot and then completed the turn.

Haise: That's a pretty neat jig there.

Shepard: Yeah. *(long pause)*

Shepard: Okay, while you're down there, pick up the [extension] handle, *(pause)* Okay, very good. *(long pause)*

> **Interviewer** – "You were actually down on your knees, but you had the MET to hold on to."
>
> **Mitchell** – "Yes; but you had to be very careful of that, though. You'd turn the damn thing over if you tried to use it to get up."

After Al got free of the cable, Ed got down on his knees to pick up some dropped gear. Although Al was blocking the camera's view of Ed's hands, it is clear that he was leaning at least a little forward as he grabbed the dropped objects. Then, just before he rose, he rotated his torso back until his PLSS was over his lower legs. After he was up, he kicked his feet forward to get them under his center of mass. He may have used the extension handle to push himself back. Members of the later crews sometimes rose from their hands and knees by pushing hard with their arms to rotate their center of mass up and back so that they could then use their legs to rise. Here, it does not appear that Ed was leaning far enough forward to use his arms to push himself back.

* * *

We are skipping the period in the transcript when the crew is unpacking equipment and taking stock. We pick up the action again as they begin their hike toward Cone Crater.

Shepard: Yeah, I just wanted to get a good direction. Actually. Our [line of] sight to [Station] A [is] directly toward the center of the [Cone] crater ...

Mitchell: Yeah, that's right over that way.

Shepard: And it's ... *(counting map squares)* two ... six. About 350 meters, a thousand feet.

Mitchell: Okay. We'll start off that direction and take a look around.

Shepard: Okay, and I'll aim the camera towards Cone.

Mitchell: Okay.

Interviewer – "On the video, the two of you were looking up toward Cone, which must have been interesting with the Sun that low."

Mitchell – "Of course, the Sun's moved up a little bit."

Interviewer – "You've got the MET. And you're going to pull it out to Station A."

Shepard: Okay, Houston. We're going to try to put the TV camera in the shade [of the LM], and aim it up toward Cone. I'm not sure we're going to be successful in doing that.

Haise: Okay, Al. We don't want to tarry too long on that one ...

Shepard: And I'll ...

Haise: ... We're about 2 minutes behind starting out. And, the settings, you can leave them just as they are right now.

Shepard: Say again.

Haise: Okay, The settings that are on the TV right now are good.

Shepard: You don't want to aim it toward Cone Crater?

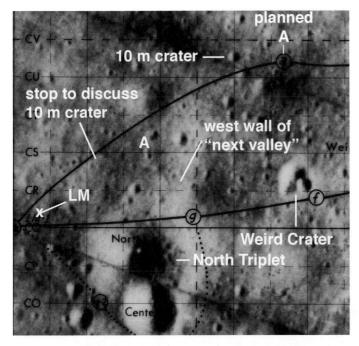

Map detail A demonstrates how difficult it was for the Apollo 14 crew to navigate on the lunar surface during the traverse to Cone Crater. When the crew stopped at what they thought was the planned Station A, they had only covered half the distance. Each of the squares in the grid is 50 meters on a side. (*map by Lennie Waugh, ALSJ contributor*)

Haise: That's affirm, Al. You can do that task, but we won't worry too much about fineness on aiming it. The settings on the camera right now should be good.

Shepard: Okay. We'll aim it up toward Cone. It's going to be fairly close to the Sun.

Haise: Roger, Al.

Shepard: We'll see what happens. (*pause as Al re-aims the TV, pointing at the ground to avoid a repeat of the Apollo 12 loss*) Do you have the image of the Sun, yet? (*pause*) Do you have the image of the Sun, yet?

Haise: Okay. We have a little bit of a glare there, but we have a picture, Al.

Shepard: I'm going to bring you a little further to the right.

Haise: Roger, Al. I think we can see the slopes ...

Shepard: How's that?

Haise: Left [north] flank of Cone coming in.

Shepard: Okay. Okay, you're looking at Cone.

Haise: Roger, Al. We have little bit of a glare across the center; but in the background, we can see the crest of Cone.

* * *

LOOKING FOR STATION A

Shepard: Okay. *(to Ed)* Head on out, man.

Mitchell: Yeah, let's go.

Shepard: You got it?

Mitchell: I don't know exactly where we are.

Shepard: Well, keep the map in your hand ...

Mitchell: Right.

Shepard: ... and keep going, I got this [perhaps the MET].

Mitchell: Okay. If I can locate a familiar crater ...

Interviewer – "Is it right that you have the map in hand and Al has grabbed the MET? Commander's prerogative?"

Mitchell – "It was kind of my job to do the navigation, I was kind of the budget-keeper. Keep us on course, make sure we are where we're supposed to be and so that's just kind of the way we worked together. Al concentrated on doing; I concentrated on getting us there, keeping us on procedure and on the timeline as much as possible."

Interviewer – "By the time the J-mission crews came along, the Rover was a respectable vehicle. And there wasn't any way a Commander was going to let an LMP drive." [Ed agreed.] "The MET does not fall in that category?" [It didn't.]

Shepard: Okay, Houston, We're headed just about toward the center of Cone Crater.

Mitchell: Okay, Al, Is this North Triplet right here to our right? It is, isn't it?

Shepard: Yes, sir.

Mitchell: Okay. This nice big depression over here. *(pause)*

Shepard: Houston, we're again proceeding directly toward the center of the crater [that is, toward the center of Cone, toward] point A. As Ed pointed out, we're passing north of North Triplet. The area over which we are passing again, of course, is pockmarked by craters. However, the land is generally flat right here. We have sort of a ... *(pause)* I was going to say "mesa" but I really don't think it's a mesa. It's more of a ridge, which extends to the southeast, almost normal to our path of travel. *(to Ed)* I think point A is probably down in that valley.

Mitchell – "It's a ridge and then it's just a slope down and then it starts back up toward Cone Crater again. It's kind of like sand dunes. And I'd say it's 20 feet from the top down into a gently sloping valley and then back up again toward the shoulder of Cone Crater."

Interviewer – "What kind of crest-to-crest distances? 100 meters?"

Mitchell – "Since they were caused by meteor impact, it's more random than that, I don't recall anything systematic."

Mitchell: Yeah. Look, Al, I've spotted it. See the crater almost directly up-Sun from us, in the valley? Right in the middle valley?

Shepard: Right.

Mitchell: That's Weird [Crater]

Weird Crater is actually a set of three overlapping craters, which means it has a distinctly different or weird shape when compared with most craters.

Shepard: Okay.

Mitchell: ... and if we head to the north of that, we're in business.

Shepard: Okay. That means that point A is, in fact, down in the valley.

Haise: Roger, Al. *(pause)*

Shepard: There seem to be quite a few large rocks as we progress along here. I see rocks of up to 2 or 3 feet in size, and one would fairly easily postulate these came directly from Cone Crater. Of course, we'll get samples of these a little further along.

Mitchell: A little further to the left. Okay. Point A, Al, is not quite in the valley. It's right beyond over here. *(pause)*

Ed is probably stopping every once in a while to check his bearings while Al goes ahead with the MET.

Shepard: Okay. *(garbled)* fairly subdued craters now.

Mitchell: Yeah.

Haise: Okay. Is there any basic change in the ...

Mitchell: Okay, this is *(garbled)* we're going through right here.

Haise: Any basic change in the surface texture as you're heading out across toward A, there?

Mitchell: No, It looks all the same, Fredo.

Haise: That's what I was afraid of.

Mitchell: We're ... Fredo, see the crater 60 meters to the west of point A? *(pause)*

Ed is referring to a crater on the map – a fresh, 10-meter-diameter crater. The crater is actually about 200 meters northeast of their current position and, clearly, Ed thinks that they are moving a lot more quickly than they actually are. The squares on the map are 50 meters on a side.

Haise: Roger, Ed.

Mitchell: The rather sharp one?

Haise: Okay. I think I have it on the chart.

Mitchell: Okay. We're coming up on that one right ... *(stops talking to listen)*. Okay, we're coming up on that one right now. It's the sharper one in the east, north-south line of about three craters. And our traverse, supposedly, passes right between them. Got it?

Haise: Okay. We got you right on the map, Ed.

Mitchell: Okay. The kind of doublet crater, supposedly just south of our track at 71 and CT ... and CT 0.3. We're passing exactly on the south rim of those two now.

Haise: Roger, Ed.

Public Affairs reports that at this time Al's heart rate is 84 beats per minute and Ed's is 90.

Shepard: Probably A right here, is it not?

Mitchell: It's right over here to our left a little bit, Al, I believe. *(pause)* Well, let me see.

Haise: And one other ...

Mitchell: *(garbled)*

Haise: ... question from here. Did the blocks you described as you moved across there, do they appear to be in the form of rays from Cone or are they pretty widely spread?

Interviewer – "When I was going through this the first time, it seemed to me a little unusual that the Backroom would be peppering you with questions this early in the traverse."

Among the many backrooms full of experts that supported various aspects of the missions, the Science Backroom – or here simply, "the Backroom" – was literally a nearby room full of scientific experts, mostly geologists, who were listening to the voice feed, watching the TV monitors, and providing assistance to the Flight Director, known as "Flight." The Backroom had a spokesperson, who was the only one who talked directly to Flight. Through the spokesperson, the Backroom answered questions from Flight, CapCom, or the crew, and, on occasion, had a question or request for the crew. If Flight thought the question or request was appropriate in the particular circumstances of the moment, he would ask CapCom to pass it on to the crew. Generally, CapCom was the only person who spoke directly to the crew. The Backroom during the EVAs was awash with maps, photos, reference books and journals, and – sometimes – conflicting opinions.

> **Mitchell** – "They obviously did that every time. They wanted to know answers we weren't quite ready to give them. And they did use that in kind of the form of cue, 'Hey, guys, we want you to tell us about this when you see it.'"
>
> **Interviewer** – "Would there have been the equivalent of a Backroom on any of the field exercises?"
>
> **Mitchell** – "Oh, yeah. Yeah, we practiced it that way. When we worked with radios and had geologists in the field in the later stages of exercises. We had a little harness on with a radio sticking up on it and they'd go play Backroom and work it from the van or from some vantage point. And we practiced doing that."
>
> **Interviewer** – "Were there any people you worked with particularly?"
>
> **Mitchell** – "Gordon Swann [U.S. Geological Survey Astrogeology Branch, Flagstaff, Arizona, who was part of the astronaut training program] [and] Bill Muehlberger [University of Texas]. [On the other hand] Gene Shoemaker [US Geological Survey Astrogeology Branch, Flagstaff Arizona] seldom came around. [I asked about Lee Silver, from California Institute of Technology, Pasadena]. Lee went along when we were doing geologic studies, but I don't recall that Lee did anything with regard to our training exercises, as far as getting ready for the lunar surface activities."
>
> **Interviewer** – "Now, I understand that Jack Schmitt [Apollo 17 astronaut] was heavily involved in, at least, developing the training program for you guys. Would he have been involved in the field exercises?"
>
> **Mitchell** – "As I recall, by the time we were doing that sort of exercises, Jack was involved in his own crew training [as a member of the Apollo 15 backup crew]. And he wasn't really that involved in our training."

Shepard: No. We don't see any ray pattern, I would say. They're fairly generally scattered.

Mitchell: They may form a pattern when we get to the top and can look at them in plan view, Fredo.

Haise: Very good.

Shepard: How about right here in the center of these three are A.

Mitchell: *(tentatively)* Okay.

Shepard: Buy that? *(pause)*

Mitchell: Well, it's pretty close, I don't think it's exactly at A, but it's close.

Haise: Okay, I'll ...

Mitchell: *(garbled)* I guess.

Haise: I'll clock you at A, right now.

Mitchell: Okay. That large crater to your right, Al, just doesn't show up. Ah ha! It does, too. That's the one. Just beyond that is A.

Shepard: That's what I thought. About 20 feet ahead of me, right?

Mitchell: Yup, yup.

Shepard: Okay, babe. Fred, the surface, here – we spoke about that – is textured. It is, of course, a very fine-grain, dusty regolith, much the same as we have in the vicinity of the LM. But, there seems to be small pebbles – more small pebbles – here on the surface than we had back around the LM area. And the population of larger rocks, perhaps small boulder size, is more prevalent here. Okay, this is probably pretty good.

Mitchell: Yeah, this a good place for A and I might also comment, Fredo, that we have an appearance here, quite often, like raindrops; [like] a very few raindrops have splattered the surface. It gives you that appearance. Obviously, they haven't; but it's that sort of texture. In places.

Shepard: Yeah, I think ... I was just about to say that there's a relationship between the texture and these small surface pebbles. Okay, point A.

Mitchell: Okay, at point A, *(pause)* we do a double core [and] LPM. I'll start with the LPM [Lunar Portable Magnetometer] and a pan [photographic panorama].

* * *

Shepard: And I see a fairly large rock here at the north of these three craters. It's embedded right at the rim. It's about 2 feet long. I can see some crystals in it. It has a good fillet pattern. I'm shooting a close-up of that [with the Gold Camera]. And the Sun angle again will be 9 o'clock.

Haise: Roger, Al. *(long pause)*

Haise: Okay, and Al, a word from the Backroom says go at least two crater diameters away from, I guess, the crater you're just describing, when you get ready to take the double core.

Figure 3.3. The LPM was flown on Apollos 14 and 16. The box (lower left) contained the three-component readout gauges, the on/off switch, and the high/low-gain switch; it was mounted on the MET for Apollo 14 and on the back of the Rover for Apollo 16. The tripod would be deployed 15 meters away from the MET/Rover, and as the astronaut carried the tripod to the site, the ribbon cable unwound off both sides of the centrally mounted cable reel. The astronaut returned to the MET/Rover and, after waiting a pre-determined amount of time for transients to damp out, read the X, Y, and Z values on each of the two gain settings. He then went back out to the tripod and changed the orientation of the sensor head at the top of the tripod, returned to the MET/Rover and made another measurement.

The geologists in the Science Backroom want to get a sampling profile of the top of the Cone Crater ejecta blanket and, if Al were closer than two diameters of a smaller crater, he would get an inverted section of the Cone blanket.

Shepard: Okay, we'll try to put it in the center of the three craters to get all three . . . Well, to get whatever stratigraphy we have here, and the last [Gold Camera] fillets picture – shadow 9 o'clock – was 18. *(pause)*

Haise: Rog, Al.

STATION A – THE DIFFICULT RIBBON CABLE AND ED'S TROUBLESOME GLOVE

Jumping forward about 7 minutes, we rejoin the mission as Ed is trying to wind up the LPM ribbon cable.

Mitchell: This is a can of worms, Agh!

Haise: You're having some problem reeling it in there, Ed?

Mitchell: Yeah, An awful lot of problem with it, Fred. The set in the cable is so much that if I ever let go of the handle, it winds down about three or four turns on me – at least – and then I have to take it back out. And the cable is all bunched up and curled out here. I'm not sure I'm going to get wound or not.

Haise: Roger, Ed. *(pause)*

Mitchell – "I had the same problem in training, But not nearly as severely as with the flight equipment on the Moon. And it's kind of like backlash. As I recall, in turning it, you have to turn loose of it and get another grip and turn it some more. But, when you turned loose of it, it'd spin back out on you, So it was a pain in the ass."

Interviewer – "The Apollo 16 LPM had a ribbon cable (about 2 inches wide). Was yours the same?"

Mitchell – "Yeah, Same thing. And it had a lot of set. And it was stiff. It was just a badly designed piece of equipment for lunar surface work."

Mitchell (from the 1971 Technical Debriefing) – "The LPM cable was very difficult to roll up. The spring in it and the backlash were such that I could roll four or five rolls and not crank up appreciably any of the tape. It was just tightening up inside the reel. When I loosened my grip to grab it again, it would unwind three or four turns. In addition, the tape was rolling up in bends, and it looked like a giant bow, very fluffy with lots of bends all balled up around the reel. It was really a mess to handle. It took about two or three times as much time as expected to get the cable reeled in the first time. I would have objected strenuously if I had to do it a second time, and we had planned three measurements. I was seriously considering just trailing that ribbon behind us and taking our chances with it. It was really difficult. It was complicated by the fact that my glove problem was making the mobility in my right hand difficult."

Interviewer – According to the *Apollo 14 Mission Report*, "Corrective action for Apollo 16 consists of adding a ratchet-and-pawl locking device for activation with the gloved hand, and providing a better grip for the reel and crank." Because of the design change, John Young had relatively little problem rewinding the LPM cable.

Mitchell (from the 1971 Technical Debriefing) – "I might say, in defense of the LPM, it gave good readings. We had to stay on high-scale most of the time, so obviously there was a magnetic field that we were measuring."

Mitchell: [I'll] try a different method of holding it. *(pause)*

Mitchell – "I hated to use that LPM, because it just tore your wrist up, winding it back up. I remember in training and on the surface, not only did you have this problem [of the ribbon snarling], which was frustrating, but it was terribly difficult on the wrist. And with my glove cable problem I had, it was doubly difficult. So it was a pretty frustrating thing."

After suit pressurization for EVA-2, Ed reported that his right glove was pulling his hand to the left and down, but he hadn't had any such problems during the first EVA. Although he was able to manipulate the glove, whenever he relaxed his hand and forearm, the glove went into this awkward left/down neutral position. Although Ed is left handed and wasn't as bothered as he would have been had there been a problem with his left glove, the errant right glove made his EVA-2 work more difficult – particularly in this case, when he needed to use his right hand because of the way the take-up reel was designed.

Interviewer – "The crank was on the right side?"

Mitchell – "Yeah, so I had to use the right hand."

Interviewer – "And you were abrading on the wrist ring?"

Mitchell – "Yup."

Shepard: Okay, Houston. A couple of quick stereos in the "locator" [documenting photograph] of the core tube as it's about to be driven; and, in the "locator," the LM is in the background.

Haise: Roger, Al. *(long pause)*

Al takes a stereopair from the north of the core tube, stepping to his right between frames. He then goes around to the east of the core tube and takes another picture across it toward the LM.

Mitchell: Okay, Fredo. I got the LPM reel reeled in just enough to keep it off the ground. I'm trailing a can of spaghetti here.

Haise: Okay, Ed. *(pause)*

Mitchell: Al, you haven't taken a pan, have you?

Shepard: Nope.

Figure 3.4. Al Shepard has just driven a double core tube into the surface, extracted it, and is carrying it to the MET. His left hand is on the joint between the top core section and an extension handle attached to the top of it; the bottom core section rests on the hammer, which he holds in his right hand. Al has a pair of tongs tethered to his waist at the left side, with the handle on the outside of his left hip. The large crater in the background at the left, called Old Nameless, is about 2.5 kilometers southeast of Shepard.

Mitchell: Okay, I'm starting with the pan.

Haise: Okay. Just in the way of bookkeeping, we need the double core, and the pan, and a sample.

Mitchell: Okay.

Shepard: Okay, Houston. We've got almost two complete tubes here, about 1 and 7/8 tubes, I would say.

Haise: Roger, Al. *(long pause)*

Shepard (from the 1971 Technical Debriefing) – "Point A is where I took the double core. It went down all the way. It went down relatively easy for the first section. The next half – of the second core – wasn't bad, requiring just general tapping. Then I had to bang it very hard to get the last half of the top core. But I did get it. I got 1-3/4 to 1-7/8. That's about the deepest penetration we had."

Al may have just finished driving the core tube in. If so, then, during this communication break, he extracts it and takes it over toward the MET to begin disassembly and capping.

Mitchell: Okay, Houston. The pan is completed. I took it from the rim of a old crater with a fresh crater right in the bottom of it, and several small ones around it.

Shepard: Yeah. That's a fairly blocky one, that new one. I think if we take samples from right along that rim there, you'd probably get some of that from the bottom.

Mitchell: Yeah. Okay.

Haise: Okay. We copied, Al and Ed.

Mitchell: *(garbled)* just a minute. *(pause)*

Shepard: Okay. And the core bit, just for the fun of it, is going in bag 2 November. *(pause)* If we can get it back.

Shepard (from the 1971 Technical Debriefing) – "I suggested that we use the tip of that thing [the bit from the double core] for the bug sample [that is, a sample to be examined for signs of native lunar organisms]. We brought the bit back in a separate bag."

Although the bit undoubtedly did not yield any lunar organisms, it may have yielded practical information on abrasion and wear.

Haise: And Al, they'd like a description of the surface where you drove the core tube. *(long pause)*

Shepard: Okay, Fred. The surface was the same textured pattern of which we spoke coming up in this traverse.

Mitchell: Uh oh.

Shepard: What?

Mitchell: Where's our color chart? *(pause)*

The color chart/gray scale in question is supposed to be attached to one of the legs of the gnomon. When taking documented samples, they position the gnomon with the chart facing the Sun to provide calibration during photo processing.

Shepard: Here you go. Got it? *(to Haise)* Did you read the core tip?

Haise: Roger, Al. We've got that, and for your information, that we're about 5 minutes behind, in the total timeline, for departing A.

Shepard: Okay. *(pause)* Continuing our description of the surface, it appears to be a scattered population of very small blocks, some of which Ed is going to photograph here in his documented sample. I believe they came from the crater to the north of the sampling sites. Other than that, the double-core sample site is not unique to the traverse, so far. The first core went in fairly easily. Had some increasing difficulty with the last core.

Haise: Roger, Al. *(long pause)*

Mitchell: *(to himself as he adjusts his camera settings)* Okay, Seven. *(pause)*

Shepard: Get that by yourself?

Mitchell: Yup. *(long pause)*

Mitchell: And, Houston, the rock I'm sampling seems to be a fairly typical one of this little crater – multiple crater – that we're working around right now near A. And it's going into the bag 3 November.

Haise: Roger, Ed. Copy, 3 November.

Mitchell: Oops. It's breaking apart on me as I pick it up. I'll try to get most of the pieces.

Haise: Roger, Ed. And we need to move on here to B; and before we depart A, we're going to need an EMU check.

Shepard: Okay. This is Al's EMU: reading 3.75; oxygen is reading 71; I have no flags; I'm on Min cooling; and I'm comfortable.

Mitchell: Al?

Shepard: Yeah.

Mitchell: Can you hand me another baggy?

At the time of Apollo 14, commercial plastic sandwich bags were marketed under the brand name Baggies.

Shepard: Okay.

Mitchell: Houston, I can't get all of this sample in 3-N. Consequently, it'll go into 3-N and the next one. *(pause)* It looked like it was fractured, and when I picked it up, it fractured into about four pieces.

Shepard: *(grunting)* Okay. *(pause)*

Haise: Roger, Ed. *(long pause)*

TOWARD STATION B – HILLS AND VALLEYS

Shepard: Okay. I'm going to head on up the hill to [Station] B.

Mitchell: Okay.

The planned Station B is on the southern rim of a 20-meter crater, about 230 meters east of the planned Station A. Because Al and Ed are not where they think they are, and because they are overestimating their rate of progress by as much as a factor of 2, they actually will make the next stop about 70 meters south of the planned Station A site. East of the actual Station A, they will drop down into another valley and then up the far side. This modest climb may be the "hill" to which Al is referring and is the 100-meter-wide dark swath that extends from the planned site of Station G to the planned site of A. (See the map on page 127.)

Haise: Okay. And we still need an EMU check from Ed.

Mitchell: *(either still bagging the sample or taking his "after" photos)* Okay, Fred. I'll give it to you in a minute.

Shepard: Can you catch up with me all right?

Mitchell: Yeah, I'll catch up. Go ahead.

Shepard: Okay. Al's heading up with the MET. From A, we go down into a valley. We drop down on a fairly consistent slope of approximately, oh, 8 to 10 degrees. The texture, here again, is pretty much the same on the surface. The basic regolith, of course, is the fine material which is now, at this particular Sun angle, kind of a grayish brown, with the small pebbles on the surface making the raindrop pattern.

Map detail B shows Apollo 14's planned route (black line) and the actual station locations on the climb to Cone Crater. (*map by Lennie Waugh, ALSJ contributor*)

Interviewer – "Conventional wisdom ascribes the raindrop pattern to micrometeorite impacts, with visibility depending on regolith properties, sun angle, and surface age. Association with pebbles suggests association with secondary impacts. Perhaps, as usual, it's both and, with practice, one could learn to detect differences."

Mitchell – "I'm not sure. Al made an assumption … He made a causal determination which I'm not sure is valid. I don't recall making that association [of the raindrop pattern] with pebbles from secondary impacts."

Interviewer – "I think Al's the only one that makes that association. I'm pretty sure that, when other people have seen it, they haven't made that association. And it's often been out in places that are just a lot of fine-grained stuff and nothing much over millimeter size."

Mitchell: And, Houston, I'm trailing along behind Al now. I'm starting to catch up with him. And – it hasn't been described for you before – the MET tracks make a very smooth pattern in the surface, reminiscent of driving a tractor through a plowed field. It smoothes it out and makes a very smooth, distinct pattern, and probably, oh, a quarter of an inch deep, no more.

Haise: Roger, Ed.

Mitchell: It leaves gaps every now and then as it bounces.

Shepard (from the 1971 Technical Debriefing) – "I thought the MET worked very well. We had been living with it for some time during the training cycle, and it had been modified a few times to take care of some of the problems. I thought it was generally worthwhile. It enabled us to operate more efficiently than we would have otherwise."

Mitchell (from the 1971 Technical Debriefing) – "We would have been in real trouble trying to move all that stuff out with just a hand tool carrier, and still get the same amount of work done. I think that the MET stability was good at reasonable speeds. It was not hard to pull. It did make you change your gait a little bit. I didn't feel like I balanced quite the same way with the MET as I could without it. You could pull it up to fairly good speeds without any stability problems. It did bounce and hop and try to turn over if you hit rocks with the wheel or if you hit a crater with the wheel. It was not too hard to stabilize it with that triangular handle. Al seemed to be able to move faster with it than I did. That's because I didn't feel comfortable with the stability of it. When we hit rocks and things, I was worried about it tipping over, and I really didn't want to see all that equipment spread out over the lunar landscape. So, I think I probably tended to be a bit more cautious. When Al was pulling it rapidly, he was controlling it well, and it didn't tend to turn over."

Shepard (from the 1971 Technical Debriefing) – "I really expected more dust to be collected by the tires and thrown up on the MET. That didn't turn out to be the case at all. We dragged it through some fine-grained stuff near the edges of the smaller

Figure 3.5. On their climb up to the rim of Cone Crater, Al and Ed made their first sampling stop about 200 meters northeast of the spacecraft. In this combination of two frames from Ed's panorama, we can see the MET tracks and Al's footprints and – in parallel – Ed's footprints. Ed tended to use a skip, and we can see in his footprints that his right foot landed flat and then made only a slight toe impression as he pushed off. His left footprints show a smaller, much deeper, toe-only landing. Al used a loping, foot-to-foot stride and we can see that his left and right footprints are similar. There are places where his right footprints seem deeper, but this may be a result of variations in bearing strength of the surface.

craters and, although the tires sunk in more in that fluffy, less-dense regolith, it still didn't throw up an awful lot of dust."

Mitchell (from the 1971 Technical Debriefing) – "Dust didn't adhere in any appreciable amount to the rolling surface of the tires. The MET seemed to mash it down, but it didn't adhere. It didn't throw out a rooster tail as we might have expected."

Mitchell: *(to Al)* Think you found B? Yeah. It's this big crater over here, isn't it?

Shepard: It [Station B]'s way up the hill.

Mitchell: Pardon?

Shepard: I think it's up the hill.

Interviewer – "When you and Al are talking about way up the hill, are you up above the break in slope of the Cone Crater Ridge?"

Mitchell – "No. I think we're coming out of that swale, that valley. As a matter of fact, it's on this map. This darkness in here [between Weird Crater and the planned location "A" in Figure 3.6] is what we were calling [the eastern wall of] that valley.

Figure 3.6. Ed Mitchell carried a copy of this photomap during the traverse to Cone Crater, and identical copies were used by mission support personnel in Houston. The planned landing point is at the lower left, at map coordinate CQ/65. Shepard and Mitchell more or less followed the planned traverse route (black line), but because of fatigue and some disorientation, came close to – but did not reach – the rim of the 400-meter-diameter Cone Crater. The map was prepared from images taken by the unmanned Lunar Orbiter spacecraft, which were flown between August 1966 and August 1967. The lines that cross the map diagonally from the lower right to the upper left are boundaries between filmstrips taken by the scanning camera; grid lines are 50 meters apart.

That was a very good depression. And that's where Al's saying it's up the hill. See, there's another one over here, and then we really started up the flank of Cone Crater."

Interviewer – "As indicated by the post-mission cross-section view at about B1."

Mitchell: Oh, that's right. B is the crater we go ... This is the crater we go by on the way to B. Gotcha. *(long pause)*

Shepard: Okay, Houston. I'm looking for a contact [the edge of the Cone ejecta blanket] somewhere in here, but it's not apparent at this point. Surface texture

Figure 3.7. Where are we? Ed Mitchell is studying the traverse map to determine their location on the way to Cone Crater. The rolling terrain, high-contrast lighting, and monochromatic nature of the lunar landscape make it difficult to spot landmarks. Later Apollo crews will have a navigation system and sun compass on the Rover and – with the added mobility the Rover provides – will have a much easier time navigating. Finding your way in new territory has always been a major challenge for explorers, and it is only very recently that, here on Earth, the job has become easier with good maps, global positioning systems, and some spare batteries.

seems to be very much the same; from the standpoint of soil-bearing properties, it's still about the same softness, and it still has the same raindrop pattern.

Haise: Roger, Al.

Mitchell: Oh, Fredo; you wanted a EMU check from me. I'm at 3.7 (psi), showing 67 percent [oxygen remaining]. I'm on Min cooling and no flags.

Haise: Roger, Ed.

Mitchell: And continuing the description a little bit, Houston. Trying to think of an adequate description or comparison to something we've already seen, but I don't think there is one. Incidentally, I see a string of craters down to the south – *(correcting himself)* a string of boulders – to the south of us that may prove to be a ray pattern from Cone. And I observe, as we get closer to Cone, the number of large boulders is increasing. We're going to go past some here in a couple of minutes near about a 20-foot-wide, fairly fresh crater. The boulders – a dozen of them or so – are 4 or 5 feet in diameter.

> **Haise:** Roger, Ed.
>
> **Mitchell:** *(Garbled under Haise)* thing around them.
>
> **Shepard:** Okay. Let's see if we can find us ...

Ed has the map, but the dialogue suggests that they are looking at it together, trying to figure out where they are.

> **Mitchell:** This crater is the one, I think. Al, it's halfway between A and B, isn't it? *(pause)*
>
> **Shepard:** Yeah. I think so. This little ...
>
> **Mitchell:** Can you see the boulders off to the side there on the map?
>
> **Shepard:** Well, they don't show very well. *(pause)* I think ...

It isn't clear just what boulders Ed is thinking about. In the Lunar Orbiter photographs, on the north side of their planned track, three boulders are tentatively identified just beyond the planned Station B location. It is possible that these are the boulders Ed is looking for.

> **Mitchell:** Ah! You should be able to spot that little chain of craters just to the south of us on the map, if that's where we think we are. *(pause)*

There are a variety of candidates for a "chain of craters" south of their present location.

> **Shepard:** Boy, that little chain of craters right there ...
>
> **Mitchell:** Kind of small.

Thinking that they are near the planned Station B location, they may be looking on the traverse map at a line of three craters.

> **Shepard:** That will make us right here, huh?
>
> **Mitchell:** Pardon?
>
> **Shepard:** There's no big one to go with it. No sharp one to go with it. *(pause)* *(garbled)* one right up there. How about that?
>
> **Mitchell:** Well, let's take a look.

Interviewer – "I gather from this that you two were standing together, consulting the map."

Mitchell – "Yeah, and as you can tell, we had a disagreement ... We have different interpretations ... We were really having trouble, on that terrain, figuring out where the heck we were. We knew where we were within a hundred meters, but not the micronavigation that those characters [the geologists] wanted us to do. It was frustrating. We wasted time. And that continued. That's what slowed us down the whole rest of this thing, trying to be a little more precise about where the heck we were. In retrospect, it would have been much better, frankly, had we worked basically on time and let us select our own sites, looking for interesting things. Plan on two or three stops on the way up there. Let us select them. And take pans that let them, afterwards, find out exactly where we stopped. But they tried to prejudge and send us to where they thought they were going to find something. And us trying to find it. We ended up wasting a lot of time."

Interviewer – "In thinking about the J missions, those crews had a lot more flexibility. That sounds like that might have been a result of your experiences."

Mitchell – "It could very well have been. But there's no question, we tried much too hard to satisfy getting to where the geologists said they wanted us to be. And it proved to be an inefficient and a very difficult and odious thing to try to do. We could have done a lot more work had we had the flexibility where we could just pick and choose. Because we were darn well trained. We knew what we were looking for. With that flexibility, I think we would have done a lot better."

Shepard: *(probably on the move again)* That's probably Weird right up there. We're probably about even with Weird right now, although you can't see it on the ridge.

Mitchell: That's Weird, that big one right over there, Al.

Shepard: Yeah. That's what I say. I think B is that deep crater right directly ahead of us, Ed.

Al may be indicating the fresh crater that is about midway between their location and the planned Station B site.

Mitchell: No, I disagree. I think ... See that crater right over there that we came by? To the south, the big one?

Shepard: Yep.

Mitchell: I think this is the crater that ... That's B. I think this boulder field, we can see it here if we look.

Shepard: This crater right here?

Mitchell: Yep. *(pause)* We have to be considerably past Weird.

Shepard: Not even halfway to the rim of Cone yet.

They believe they have completed a little under half their journey and, in reality, have done only about a third.

Haise: And, Al and Ed ...

Mitchell: *(garbled under Haise).*

Shepard: Let's go on ahead.

Haise: Yeah. I don't think you have to worry too much about the exact position of site B. If it appears you're getting close to the general area, and that should be good enough on B.

The following took place at the Technical Debriefing, December 17, 1971.

Shepard – "I felt that we had a navigation problem on EVA-2. I don't know why we didn't worry a little bit more about that pre-flight. We did discuss the fact that points A and B were not very well defined. They said, 'Well, it wasn't too important to get exactly to those two points from a geological point of view.' This may be true, because we're supposed to be in contact with nondescript material. But it sure made it tough to figure out exactly where we were as far as the progress of the EVA was concerned."

Mitchell – "Yes."

Shepard – "Until we really get a feel for navigation on the surface, there should be some strong [that is, well-defined] checkpoints to follow. First of all, it gives you a feeling of security to know where you are. You know where you are distance-wise and what you have left to cover. Second, there's no question in my mind that it's easy to misjudge distances, not only high above the surface [that is, during the landing or from the LM windows] – that we discussed before – but also distances along the surface. It's crystal clear up there [on the Moon] – there's no closeness that you try to associate with it in Earth terms – it just looks a lot closer than it is."

The fact that the Moon has no atmosphere means that there was no haze, no obscuration, which, on Earth, helps us distinguish distant objects from close objects.

Mitchell – "I certainly agree with that. I think there are two problems that affect your distance measurements. One, as Al described, and the second is there has to be a little bit of distortion in the bubble [helmet]. I don't know how much that contributed to it, but I think it contributed some. I believe that our primary problem in navigation was the surprise brought about by the roughness and the undulation of the terrain. We couldn't see one set of landmarks, the prominent landmarks – our next set of landmarks – from our present positions. Large craters, which we expected to be able to see standing out on a reasonably flat plain, were not on a flat plain. They were hidden behind other craters, ridges, and old, worn-down mounds. You'd say 'Well, this next big crater ought to be a couple of hundred meters away, or 100 or 150.' It just wasn't anywhere in sight. So you'd press on to another ridge, and you still didn't see it. All you would see would be another ridge. Finally, you'd get over to it, and there it was. You could not get enough perspective from any one spot to pin down precisely where you were. The undulations over the neighborhood were probably 10 to 15 feet [high]. Some of the big craters up to the north and to the south looked 50 to 100 feet below our level. It looked like we were in a large group of sand dunes. The wavelength of the sand dunes would be much greater here [on Earth], but that was kind of the feeling I had. I never knew what to expect when I went over the ridge on the sand dune, or what I was going to see on the other side of it."

Shepard – "I think that complicated our problem. I don't know what to suggest on that. I think that we talked about navigation problems before. We always felt that you'd see these craters out there. We planned for them and they're very well defined and we ought to be able to locate them easily, but that just isn't the case. There has to be more thought given to some better way of positioning oneself on the chart."

Mitchell – "Maybe this thought would help. We could put some work into a manual method of distance estimation better than your thumb up against the LM. We need a better manual method of estimating distance."

Mitchell: Okay. I think we're very close to it. I think this crater we just went by is probably it, but it's very hard to tell, Fredo. I don't see anything else that might be it, unless it's the next crater up. Ah; Al, I've spotted it. That next crater up is this one right here.

Shepard: Where are you pointing?

Mitchell: Pardon.

It is all but impossible to figure out which craters are being discussed in this section.

Shepard: Where are you?

Mitchell: Right behind you. That crater is that crater right up there. That crater is the crater over to the left of it.

Shepard: Where do you think B is?

Mitchell: I think B's the one we just passed; back there where we were talking.

Shepard: *(sounding a little impatient)* All right.

Mitchell: And here's the little … Ah hah, it is! Here's the little double crater right beside it. Look here. See, there's that crater; see, there's the little double crater; it's right there in front of you.

THE CLIMB TO CONE CRATER

Al and Ed only spent about 5 minutes at Station B, collecting a quick sample and taking a pan. We rejoin them as they are about to leave.

Mitchell: Okay, the next stop is the top of Cone. Let's get everything secured for that trip.

Haise: Okay. And we'd like the frame count before you depart.

Shepard: Okay, Houston – *(listens)* Yeah; yeah. You've got a frame count of 34 from Al. *(pause)*

* * *

Mitchell: Okay. I've got the MET.

Shepard: Okay. You want to go first and I'll follow.

Mitchell: Okay. To the top of Cone Crater.

Shepard: Yeah. Now let's ...

Haise: Okay, we're starting the clock.

Shepard: *(garbled)*

Mitchell: *(responding to Haise)* Okay.

> **Interviewer** – "Was Fred's 'We're starting the clock' a spur-of-the-moment comment, or had there been some discussion about how long it was going to take you to get up there?"
>
> **Mitchell** – "No, there hadn't been any discussion. He was merely keeping elapsed time from point-to-point. We hadn't discussed that a lot. But I think they, in the Control Center, with the Backroom and Fred, they were playing games with watches, just to keep a projection of some sort going, to help us manage time better."

Shepard: We'll have to go almost to the east here, and then on up by Flank.

Flank Crater is about 675 meters ENE of Station B, which they have just left. In frames from the 16-millimeter camera, shot out of Ed's window during the landing, it appears that Flank Crater may not be visible from their current location, being hidden by an intervening ridge.

Mitchell: Yeah, East and a little to the ...

Shepard: See, there's Flank up there.

Mitchell: Uh ... Yeah, right up ... I can just barely see the rim of it on the far side of it.

Shepard: Right, so we probably ought to head directly for Flank and on up from there.

Mitchell: Okay. *(pause)*

Shepard: Okay, and ... *(pause)*

Mitchell: Houston, as we go across here, this ground is – Al's probably previously described it – but it's very undulating. I would suspect that there is not 10 yards at the most between what were once old craters. They are most of them worn down, but the surface is continuously undulating. There's hardly a level spot anywhere.

Haise: Roger, Ed.

Mitchell: Lots of ... As we come on up toward Cone, we're getting to see lots more buried rocks, bigger rocks. *(pause)*

Shepard: We're keeping our eyes open for a contact here. But I guess the Sun angle makes it very difficult to see. However, I expect that by the time we get a little closer up to Flank ... Let me pull it for a while.

Mitchell: I have to shift hands. I'm good.

Interviewer – "Sounds to me like you two are breathing a little harder than you were previously. You're up on the break of the slope now."

Mitchell – "Well, what are we looking at here [on the traverse cross-section]? We're looking at 700 meters [horizontal] and we're going up about 80 meters. So that's 10 percent slope. That's a pretty tough climb."

Interviewer – "And with the soft surface ..."

Mitchell – "And pulling the MET. No question that we started breathing harder. We were laboring. And trying to make time, too."

Interviewer – "Would you have been using the skip-step?"

Mitchell – "Yeah, except it was very difficult to do with the MET. I'm pulling it at this stage. And, if you shake the MET too much, you're afraid you're going to shake things out. So, trying to do that loping gait that I liked was very difficult to do with the MET behind you. You had to more use the [foot-to-foot] stride like Al used and, even then, the MET was rolling from side to side."

Interviewer – "And you would, of course, have swapped map and MET, because he mentions something here about the crater opposite E. Was it any particular difficulty reading the map and moving along?"

Mitchell – "Yeah, you had to stop to read it and get in the right Sun-angle."

Shepard: Okay. By the time we get a little closer up the Flank, we might find some kind of a contact. *(pause)* The ridge of Cone Crater to the north is very apparent, as we expected that it would be. It stretches off into the distance and meets with the far horizon.

Haise: Roger, Al. *(pause)*

Mitchell: Fredo, I'm trying to find something distinctive to say about some of these craters we're going by, and it's very hard to do so. They're all smooth-walled except the very freshest ones; and we're coming by a very fresh one now, which is rubblely on the in [side] . . . Hey! It may even . . . That has some pretty good chunks of rubble on the inside. This is about the freshest crater this size we've seen, Al.

Shepard: That's correct. This is a very fresh crater. It's about *(garbled)* – it's about opposite to the crater at stop E. It's a crater about 20 meters in diameter and about 2 meters deep, and I'll get a quick *(pause)* rock from the side.

The crater where Al is getting a 'grab' sample is certainly not the one north of planned Station E but is probably the prominent crater midway between planned stations A and B.

Haise: Roger.

Mitchell: *(garbled)*. Al just dropped down on a knee to pick up a rock, and he went in 3 or 4 inches. Need some help [getting up], Al?

Shepard: Yeah. I think so. I can't get any . . .

Mitchell: Okay. Come on, give me your hand.

Shepard: Wait a minute. I got it now. *(pause)* Okay.

Mitchell: Okay. Come on up.

Shepard: *(grunts)* Okay. Thank you.

Mitchell: You're on your feet.

The J-mission crews had suits with a waist joint so that they could sit on the Rover with relative ease. An additional benefit of the waist joint was that, even though they still had some trouble kneeling, they were able to do it with some regularity and to get up without any particular trouble.

Shepard: Okay. That's just a quick hand sample from the side of that crater.

Mitchell: Do you think you're following us and know about where we are, Fredo?

Haise: Well, the board . . .

Shepard: *(garbled)*

Haise: ... I think, has reading for you just past the position abeam of [Station] E. *(pause)* Looking about halfway between D and B ...

In this case, Haise is using *abeam* in the sense of *left of their direction of travel* or, roughly, north.

Shepard: *(responding to "abeam of E")* Yeah, that's good ... *(listens to the second half of Haise's remark)* Rog. And the crater ...

Mitchell: Yeah. And we're starting uphill now. Climb's fairly gentle at this point but it's definitely uphill.

Shepard: Okay, Baby! Okay, I got it.

Mitchell: Almost turned, didn't it.

The MET almost turned over.

Interviewer – "It was bouncing pretty regularly from the small stuff. Was it a tipping hazard?"

Mitchell – "Yeah, there was a tipping hazard. We couldn't move too fast because the wheels hit rocks or dropped into a crater. It was a pretty unstable beast in 1/6 g. Much more so than we would have thought. And we were moving faster. On Earth, we didn't try to cover this much territory, so we moved at more of an Earth pace. And we trained in the suit, so we had to move slower. The MET trailed along pretty nicely in 1 g. But if you tried to move at the speeds we moved in 1/6 g, and the MET as light as it was, it was bouncing all over the place. Every time we'd hit a rock, it come up in the air and one wheel would hit and it would kind of flop and bounce. *(chuckles)* I'm glad we had it, so that we could carry all the equipment; but we could have done better."

Interviewer – "And Al's 'Okay, Baby' ..."

Mitchell – "I was on the handle and could feel it turning my wrist but he caught it in the back."

Shepard: Yeah. Okay, that sample from the west rim of the crater, which we described as blocky is in bag 6.

Haise: Roger, Al. Bag 6. *(pause)*

Shepard: Okay. The going is still very smooth as far as the area that we're able to pick out. Of course, we're tracing a kind of sinuous course here, staying out of the craters. *(pause)*

Mitchell: And, Fredo, to help further locate us, if you can, we're going by two very ... Well, (two) fairly fresh craters. I don't think quite as fresh as the one we were just talking about. The east-most one is fresher than ... *(correcting himself)* The west-most one is the freshest. They're separated about 75 to 100 feet, and they're about 25 to 30 feet across and 5 or 6 feet deep; 5 feet deep, I guess. The west-most one has got small blocks in it. The east-most one is very smooth.

One or both of them is breathing heavily. Unfortunately, Ed did not mention whether they passed north or south of this pair of craters. There are a number of possible candidates in the area. The undulating terrain makes it difficult for Al and Ed to give us sufficient information to pick out specific pairs of 9-meter craters.

Haise: Roger, Ed. And, you described the blocks there a couple of times. Now, I think you used the term "rubble." By that I assume you implied they were just lying loose [with] nothing really in place.

Mitchell: I'm not sure that's quite true, Fred. Some of it looked like stuff that belonged there, that had not fallen there.

Shepard: There's a lot of glass in that rock, Ed.

Mitchell: Yeah. Ooh, there sure is. It looked like some of that so-called rubble looked like it might be the residual of an impact just lying in the bottom. And, Houston, we're passing a rock much too big to pick up. There's a whale of a lot of glass in it.

Interviewer – "'Residual lying in the bottom' is presumably pieces of a secondary projectile?"

Mitchell – Yes. As opposed to something that rolled down the side of the crater into it."

Interviewer – "Had you done much work during training on secondaries [craters formed by ejecta from larger impacts]?"

Mitchell – "Not a lot."

Haise: Roger. About how big is it?

Shepard: *(garbled)* like it was splattered with glass.

Mitchell: Yeah. It looks ... *(listens)* It's about foot and a half, 2-footer ... Yeah, about a foot and a half across.

Haise: Roger, Ed. And we copy the glass ...

Shepard: That was a glass splatter, Fred. *(long pause)*

Mitchell: And, I'm going on Medium cooling for a minute.

Fairly heavy breathing is occasionally audible.

> **Interviewer** – "I gather from Gene, I think it was, that the sound of one's breathing often depended on whether your chin was forward."

> **Mitchell** – "It depended on exactly where the mike was positioned or how you were turned or looked up, because the mike did not stay precisely with your lips at all times. If you turned a little bit, the tension in the cord might pull it. It was only moving a little way, but it was enough to change the quality of the sound. But you're right, we were beginning to puff a little bit here."

According to the heart-rate records reproduced in the Mission Report, neither of them is much above 100 beats per minute at this point.

> **Haise:** Okay. And, Al and Ed, why don't we take a little rest here for a minute, and we'd like another camera count, too.
>
> **Mitchell:** Like a what? We haven't taken any pictures since the last ones, I don't think.
>
> **Haise:** Okay, Ed.
>
> **Shepard:** Okay. We'll slow down the traverse here. *(pause)* Okay. Should be Flank right here, Ed.
>
> **Mitchell:** Pardon?
>
> **Shepard:** Should be Flank right over here.
>
> **Mitchell:** Just out of sight, you mean.
>
> **Shepard:** Right … Yeah, right there.
>
> **Mitchell:** Let's go. Let's go over and see. *(long pause)*

Ed has been pulling the MET and sounds a little winded; Al does not sound winded. Author Jones asked Ed if he agreed with that assessment.

> **Mitchell** – "Oh, I think so. I've been pulling the MET. I think I was fairly consistently running 5 to 6 beats higher heart rate than he was. Baseline and everything else. Now, I wasn't really out of breath, nor was I really tired, but I was starting to perspire in Minimum cooling. My metabolism was burning at a little higher rate."

> **Mitchell** – "I didn't have a cold or anything, didn't have the sniffles or anything. But I always had a feeling of fullness in the sinuses. You can tell that in the sound of my voice. And Al didn't seem to be affected quite as badly."

> **Interviewer** – "Which reminds me, did you have any kind of a reaction to the dust in the spacecraft after the 1st EVA? Jack Schmitt clogged up pretty badly."

Mitchell – "We were pretty darn clean. Really surprising. There was a little bit in there, but we were surprised at how clean we were."

Interviewer – "You spent a pretty good period cleaning each other, although I don't think you spent a lot more time dusting each other than the other guys."

Haise: Okay, Al and Ed, I assume you're on the move now and heading toward Flank. Is that correct?

Mitchell: That's correct. Heading toward where we think Flank is.

Mitchell: I'll pull for a while, Al.

Shepard: *(taking the handcart, again)* That's okay. I got it for a while. *(long pause)*

Mitchell: Why don't we pull up beside this big crater.

Shepard: Okay.

Mitchell: Take a break, get the map, and see if we can find out exactly where we are. Press on from there. This one should be distinctive enough.

Haise: And, Al and Ed. While you're stopped here, we could use a photo pan.

Mitchell: Yeah, going to suggest that. *(pause)* If you'll take the pan, Al, I'll grab the map and get over here and see if we can find ...

Shepard: Let me pull it up on a little more level ground.

Mitchell: Okay. Give you a push.

Shepard: Okay, there we are. Level?

Mitchell: That looks good. *(pause)* Okay. *(long pause)*

REST STOPS AND MAP READING

They have stopped at Station B1 and are only about 50 meters southeast of the planned Station B location.

Mitchell: That old LM looks like it's got a flat over there, the way it's leaning.

Haise: Say that last again, Ed.

Mitchell: Just talking. Never mind. *(long pause)*

Haise: And, Ed, now we're going to have a site handover here [from NASA's Goldstone Tracking Station in California to the Honeysuckle Creek Tracking Station near Canberra, Australia]. *(long pause; light static)*

* * *

Mitchell: Yeah. Just one second, though. I think I got us [located].

Shepard: Okay. I'll head on out. *(pause)*

> **Interviewer** – "It sounds to me like stopping for a couple of minutes while Al took the pan did wonders for you. A nice little rest. Your breathing's back to normal."
>
> **Mitchell** – "It did help."

Mitchell: Fredo, can you read?

Haise: Go ahead, Ed.

Mitchell: *(to Al)* All I'm getting is the feedback on my own voice. *(pause)*

Shepard: Okay, Ed, I'm coming through.

Mitchell: Okay. Do you want me to pull awhile, Al?

Shepard: No, that's all right.

Mitchell: *(garbled)* hand with it. *(pause)* I can't really spot this crater [on the map], but I think I know where we are. We're pretty close to where you said we were. *(pause)*

Shepard: Houston, your transmissions are still unreadable.

Haise: Roger, Al, I hadn't been talking. How do you read me now?

Shepard: Is that Flank over there?

Mitchell: I think it's dead ahead of you, Al. Oh, wait a minute. This is probably it, right here. Yeah.

Shepard: To my right?

Mitchell: Yeah. Let's just double check and see.

Shepard: It's got about a 4-meter-radius crater in the south wall. *(pause)*

Mitchell: That has to be it.

Shepard: Okay, Houston. We're going by Flank on the way up. We're passing to the north side of it.

Haise: Roger, Al. Copy.

They are actually passing the large crater immediately west of planned Station E.

Shepard: Fred, you're still unreadable.

Mitchell: Let me pull awhile, Al. *(pause)* You're having all the fun!

The Flight Surgeon reported that, while they are moving, their heart rates are about 120 beats per minute.

Shepard: Well, we still have a little way to go.

Mitchell: Yes. We sure do. Putting the map away.

Shepard: Huh?

Mitchell: I'm just putting the map away.

Shepard: All right. *(long pause)*

They may have switched positions here, with Ed taking a turn pulling the MET.

Mitchell: Okay. *(pause)* Fredo, you back with us. *(pause)*

Haise: Okay, I'll try again. How do you read, Ed?

Mitchell: Okay. That's much better. You got a background squeal.

Haise: Okay. Evidently, that station switch gave us some problem. I've been copying both of you all the way though. We have you now just passing Flank. *(pause)*

Haise: Okay. We've been copying both of you all the way most of the time, and I have you by Flank now.

Mitchell: That's affirmative. And the grade is getting pretty steep.

Haise: Have you got any estimate? *(pause)*

Mitchell: *(breathing more heavily)* And the soil here is a bit firmer, I think, than we've been on before. Except around the mounds in between craters where it's been thrown out. But, by and large, it seems to have a little firmer footing. We're not sinking in as deep.

Haise: That should help you with the climb there.

Mitchell: Yeah. It helps a little bit. Al's picked up the ... Al's got the back of the MET now, and we're carrying it up. I think it seems easier.

Shepard (from the 1971 Technical Debriefing) – "Ed was in the front of it [the MET] one time, and I was in the rear. We lifted it up and carried it."

Mitchell (from the 1971 Technical Debriefing) – "One time when Al was pulling it, I picked up the back, and we carried it. We could move at a fairly rapid clip that way. It was not as free a pace, as fast a pace, or as relaxed a movement as you could make without it."

Shepard: Left, right, left, right.

Haise: There are two guys here [Cernan and Engle] that figured you'd carry it up.

Mitchell: Say again.

Haise: Said there's two guys sitting next to me here that kind of figured you'd end up carrying it up.

Mitchell: Well, it'll roll along here, except we just move faster carrying it.

Shepard: Okay. You want to rest here by this rock?

Mitchell: Okay.

Shepard: This is the first big boulder we've seen, Houston. I think it's worthwhile taking a picture of it with the close-up. Go ahead and keep going.

Mitchell: I'll pull on up. We probably ought to take a pan to locate everything here, while you're taking a close-up [with the stereo close-up camera].

Ed is breathing heavily and Al sounds a little winded. This stop is known as Station B2.

Haise: Okay, I understand, Al. You're shooting a close-up shot of a big boulder. *(pause)* About what's the size of this one, Al?

Shepard: Okay. The shot's been taken on the close-up counter number 317. Sun angle was 8 o'clock. This particular one is only about 12 feet long by about 4 feet wide. It's about one-third buried. It's old, very weathered. There are some evidences of some crystal shining through some of the fractures.

Mitchell: And I've taking a Hasselblad of the rock and will take a pan now from this location. Help document our course going to the top of Cone Crater.

Haise: Roger. Copy.

Ed's breathing is rapidly returning to normal.

Mitchell: *(probably taking a pan frame toward the south)* And I can look right across into the breach in the north rim of Old Nameless. We're about even with it now.

Interviewer – "Is this large crater off to the south of you Old Nameless?"

Mitchell – "I think that's it. I don't know how far that is away. It looks like that could be 200 or 300 yards and yet 200 or 300 yards would put it on the traverse map."

Mitchell – "The Sun's position is almost due east. It's not any significant variation from that. So, it's the Sun line that's giving us, really, our directions. And that [meaning Old Nameless]'s a lot more than 200 or 300 yards away. That's clear off of the [traverse] map. That's a mile away."

Actually, Old Nameless was about 2 kilometers SSE of them.

Mitchell – "And that may be part of the ambiguity here; your angle [meaning the bearing to something like Old Nameless] is changing very slowly, but if you look at this as we were doing a moment ago, you could easily say that's a couple of hundred yards away. Just enough to drive you crazy trying to figure out where you are. Your distance perception was just totally off."

Haise: Okay, copied, Ed. And was there any noticeable …

Mitchell: That was frame … *(listens)*

Haise: … dust on the large boulder?

Shepard: Not where I took the picture [with the close-up stereo camera], but some fillets around the bottom.

Haise: Okay; copy, Al.

Mitchell: Okay. And 44, Fred, was my frame count.

Haise: Roger, Ed.

By the time Ed finished his pan, Al had already headed off with the MET.

Mitchell: I believe that was; if I remember it. And I'm going to move on out. Al's ahead of me here [with the MET]. *(long pause)*

Interviewer – "Your standard procedure seems to be that whoever has the MET goes on ahead so that the other guy can watch for things falling off. Had you worked that out in training?"

Mitchell – "I don't remember whether we worked that out in training; but, if we didn't, we sure worked it out in a hurry here. In training we weren't terribly concerned about it, because we're moving slowly and the MET wasn't bouncing. But as soon as we saw how it was just flopping around. it was pretty evident that somebody had to be trailing along behind, and I think that was a real-time decision that we made."

The following exchange is taken from the 1971 Technical Debriefing.

Figure 3.8. High on the slopes of Cone Crater in a field of coarse blocks, Al and Ed were very close to the crater rim; great frustration came from not knowing how close they were and having to return to the LM as time ran out. This photo (toward the southwest) gives us a good view of the MET, including the triangular handle brace, the 16-mm camera, and the gnomon. The dust brush that the astronauts use to clean the suits before climbing back into the LM hangs below the 16-mm camera on the right side of the MET.

Shepard – "Didn't something come off the MET one time? It was that little SESC (Special Environmental Sample Container) can. One of those popped off."

The SESC, manufactured by Union Carbide's Nuclear Division, was created to return pristine samples in a vacuum canister. The SESCs were complex and difficult to use with spacesuit gloves. A knife-edged seal ensured that the sample was not exposed to the Earth's atmosphere after it was returned. After filling the can, the astronaut was supposed to remove a seal protector, then press the lid into the knife edge of the can lip with a torque handle. The experience was somewhat like replacing a small jar lid while wearing boxing gloves.

Mitchell – "Yes, that was on the way back [from Cone Crater to the LM]. I stopped and picked it up. Other than that, things did not bounce off of it. Anything that was tied down well stayed in place. The 16-mm camera started oscillating. It came out of its hold-down and was just sitting in those two retaining rings. Coming down Cone Crater, it was swinging around very wildly. The magazine that we had on there didn't have any film worth looking at, although it would have been darn interesting to see it."

Ed is making a reference to the fact that this magazine was accidentally left on the Moon.

Mitchell – "The camera was whipping around from side to side making 360-degree pivots. It would swing halfway and swing back very rapidly, and it had come out of the tension fitting that held it in position."

Shepard – "That was really held in only by gravity."

Mitchell – "It kind of brushed up against one side of the MET."

Shepard – "The close-up stereo camera had a slight flange fitting on the other side, and that baby never bounced up."

Mitchell – "It stayed in very well."

Shepard – "It stayed in there just as solid as a rock. In summary, we had very little trouble with things bouncing off."

Mitchell – "I think that's due in large to the fact that we insisted that everything have a good retaining clip on it. All the bags had covers to help hold them in."

Shepard: Okay. We're starting up the last flank of the crater now, Houston. The slope is probably about, oh, 18 percent. The surface texture is still pretty much the same as far as the raindrop pattern is concerned. But we seem to find an increasing population of smaller rocks.

Haise: Roger, Al.

Mitchell: *(breathing heavily, again)* The small rocks and smaller, fresher craters, as well. Well, wait a minute, maybe I'm being deceived. With this slope, the Sun angle is entirely different than it is on the flat land. The craters look sharper in these shadows.

Shepard: Okay. Let's make an EMU stop.

Mitchell: Okay. Let me pull a while.

Shepard: I'd like to stop and rest here for a minute, Ed.

Mitchell: Okay. *(pause)*

Shepard: Boy. I tell you, we're really going to get a panorama. We've got a tremendous one here, Houston, already. And we're not quite to the rim. Back towards out [sic] Old Nameless over there, right along our track, or just south of our track I should say. We made the right approach; we came up through the valley and over the ridge and down into the bowl. Couldn't have planned it better.

Mitchell: I thought we were in a low spot with the LM, but it turns out we're really not in the lowest spot around, I don't think.

Shepard: Well, I don't know, I'd say it's probably the lowest spot right ...

Mitchell: Oh, right in that particular local area.

Shepard: Right there; yeah.

Mitchell: But that's the lowest spot over to the right [north] that I was talking about. And there's a low spot ...

Shepard: Well, there's a crater over there, it's true ...

Mitchell: Yeah.

Shepard: Yeah. *(pause)*

Mitchell: Doggone it, you can sure be deceived by slopes here. The Sun angle is very deceiving.

Shepard: Yep.

Mitchell: Okay, let me pull a while. *(they trade off)* You ready to go?

Shepard: Yeah. All set.

Mitchell: Okay, let me go back to Min cool ... Minimum cool first. *(pause)*

Shepard: I guess right straight up is the best way to go.

Mitchell: Beg your pardon?

Shepard: Right straight up is the best way to go.

Mitchell: Yeah. I think so.

Shepard: Stay away from the rocks.

Mitchell: Okay. Get a little momentum going. *(pause)*

Shepard: Okay, Houston. We're proceeding onward now.

Haise: Roger, Al.

Mitchell: And the boulder fields that Al pointed out, the rocks and boulders are getting more numerous toward the top here. However, it's nothing like the rubble and the large boulders that we saw at the Nevada Test Site. Now, this is surprising to me, I expected it to be more like that. But it is not, at least not where we're looking now.

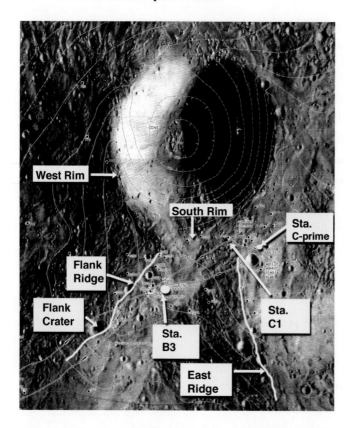

In Map detail C, physiographic map of Cone Crater, heavy white lines are breaks in the slope at the edge of ridges. The crew collected samples from Stations C-Prime and C1 (which was their closest approach to the rim). (*map by USGS*)

Ed is breathing heavily. During training they made a visit to the Nevada Test Site and, in particular, to Sedan Crater, which was dug with a shallowly buried nuclear explosive in 1966. Sedan and Cone are of similar size, but because Sedan is a very recent crater – even by terrestrial standards – it is surrounded by numerous blocks.

At this point, they have climbed high enough to see past Flank Ridge and get a view farther ahead.

Shepard: Well, we haven't reached the rim [of Cone], yet.

Mitchell: Oh boy, we got fooled on that one, I'm not sure that was Flank we were at a minute ago, either.

FLANK RIDGE – ED KNOWS WHERE THEY ARE

Mitchell: Wait a minute, Yes, it is. The rim's right here. That's the ... That's the east [ridge] ... [That's the] little shoulder running down from the Cone. That's

Flank over there. We're going to hit it [perhaps meaning Flank Crater] on the south side. We'll have to move on around of it. This looks like easy going right here. See, there's the boulder field that shows in the photograph; it's right up ahead of us.

Ed finally seems to have a good idea about where they are. He recognizes that they have not yet reached Flank Crater but are high enough on the ridge they have been climbing (labeled *Flank Ridge* on a small version of the USGS map and in the labeled frame from the landing film) to see another, higher ridge line (labeled *East Ridge*) beyond it. The boulder field Ed mentioned was undoubtedly the one on the south rim of Cone Crater.

The next extract (from the 1991 Mission Review) shows that Ed is beginning to believe at this point that Al thinks the west rim of Cone is ahead of them.

Mitchell – "Right. And I think he believes that what he's seeing is this [west] rim of Cone right here, where, on the normal [that is, the planned] traverse, we would be coming up on this [southern] edge right here, and he thinks this is it. And we see in here, shortly, I keep urging us to turn north. And we start to get a little confused as to exactly where the heck we are. This is where we start to get our ambiguity and we're deceiving ourselves or we're being deceived and it starts to get confusing in here."

Interviewer – "And part of the problem is that Cone is sitting up on that ridge of Imbrium ejecta, and so there are more things that look like rims than just the rim."

Shepard: There's the crater [meaning Cone] up there, Ed.

Mitchell: Yeah. Pardon?

Shepard: Crater up here. *(pause)*

Ed is breathing very heavily, although not enough to change his speech cadence. According to the heart-rate charts reproduced in the *Apollo 14 Mission Report*, during this period Al's heart rate is hitting peaks of 150 beats per minute; Ed's peaks are less than 115. In fact, in a few moments, the NASA Public Affairs commentator will report that Ed's peak heart rate is about 128 beats per minute. In his next transmission, Haise passes along a request from the Flight Surgeon that they take a rest.

Haise: Okay, Al and Ed. They'd like you to take another stop here.

Mitchell: Okay. We're really going up a pretty steep slope here.

Haise: Yeah. We kind of figured that from listening to you.

Since leaving the previous rest stop, they had been traveling for about 2 minutes 10 seconds and had covered another 85 meters.

Shepard: Okay. Well, now, that's apparently the rim of Cone over there. And we're about ... almost 2 hours [into the EVA] now. Is that right, Fred? *(pause)*

Haise: Okay. We're showing 1:57 and a half now, Al.

Shepard: Okay. *(pause)* That's at least 30 minutes up there.

Mitchell: Yup.

They will reach their farthest point east (and highest elevation) in the traverse at Station C-Prime – a point actually well past the southernmost point on the Cone Crater rim.

Shepard: And I would say we'd probably do better to go up to some of those boulders there; document that [and] use that as the turnaround point.

Mitchell: Yup, It's going to take longer than we expected, Our positions are all in doubt now, Fredo. What we were looking at was a flank, but it wasn't really ... The top of it wasn't the rim of Cone. We've got a ways to go yet.

Haise: Okay, Ed. And ...

Shepard: Well, perhaps you can think with us if you want. I'd say that the rim is at least 30 minutes away. We're approaching the edge of the boulder field here on the south flank.

Mitchell: Let's look at that [traverse] map.

Shepard: And what I'm proposing is perhaps we use that as the turnaround point. It seems to me that we spend a lot more time in traverse if we don't, and we don't get very many samples.

Haise: Roger, Al. And, just a couple of questions they have up now. They'd like you to note if you do see any dust, particularly on the top surfaces of boulders in the area. And, any comparisons between the boulders you see distributed around. Are they all the same or do some types appear different?

Mitchell: It's too early to make that sort of judgment, but we'll tell you when we get there. We're not really in that boulder territory yet.

Shepard: I think, Fredo, if you'll keep those questions in mind, the best thing for us to do is to get up here and document samples of what I feel is pretty sure Cone ejecta. And then, when we head down-Sun, we'll be able to see these subtle variations and rock types a lot better than we are right now.

Haise: Roger, Al.

Shepard: Well, let's head for these two babies up here. *(long pause)*

Mitchell: Hey, Al?

Shepard: Yeah.

Mitchell: I'd … *(pause)* No, let's keep going around this crater [probably Flank], but … *(pause)* Think that's right here.

Shepard: Well, maybe. I thought we'd get those boulders up there, Ed.

Mitchell: Yeah.

Shepard: They undoubtedly came from …

Mitchell: Yeah. Let's head right for that boulder field at the top. I think we'll be where we want to be.

Shepard: Right here.

Mitchell: Pardon.

Shepard: Right here.

Mitchell: Yeah, right … Clear on up at the top, you mean.

Shepard: No.

Mitchell: Huh?

Shepard: I don't think we'll have time to go up there.

Mitchell: Oh, let's give it a whirl. Gee whiz. We can't stop without looking into Cone Crater, *(garbled)* everything if we don't get there.

Shepard: I think we'll waste an awful lot of time traveling and not much documenting.

Mitchell: Well, the information we're going to find, I think, is going to be right on top.

Evidence gained in fieldwork on terrestrial craters formed by explosives and impacts suggests that the closer one gets to the rim, the greater the depth from which the ejecta was derived. Therefore, a climb to the very rim would offer the best chance of getting samples of the underlying bedrock – not to mention the view into the crater; in addition, the rim was a readily defined goal, akin to the summit of a mountain.

Mitchell – "We're not really in a pissing contest but what I'm trying to do is angle us … If this is where we turned out to be (a little short of B3), Al's trying to get us to this boulder field here (at C-Prime) and I'm trying to get us to go this way [north]. Because I'm pretty sure we're angling too far to the east here. I am now pretty confident of where the edge of the Cone is, and to get to the boulder field Al is talking about, which is kind of where we ended up, it would have been shorter to go

Map detail D shows that the south rim of Cone Crater was covered with clusters of boulders. One of these boulders is shown in Figure 3.9. (*map by Lennie Waugh, ALSJ contributor*)

right on straight up the hill, because the bigger boulders and everything are right there. So it's all of this confusion that got into where the heck we were, in here ... Al had one [mental] picture; I had another. So we got into a little disagreement here, but I'm becoming pretty certain that we should go this way, we end up going here and, of course, as it turns out, when we stop here at C1, we're only 50 feet from looking into that damn thing."

Interviewer – "Hindsight's a wonderful thing."

Haise: We establish ... *(pause)*

Shepard: Okay, Ed. Look at this; you're going through ... [We] just kicked up a layer of some very light gray fine underneath the ...

Mitchell: Yep, As you look back along your path, there's quite a bit of it.

Shepard: Get out of this crater ... *(long pause)*

Mitchell: Fredo, how far behind our timeline are we?

Haise: Okay. As best I can tell right now *(pause)* about 25 minutes down now.

Mitchell: Okay.

Shepard: We'll be an hour down by the time we get to the top of that thing. (pause) You got six samples.

Al is urging that they stop and sample.

Mitchell: Well, I think we're going to find what we're looking for up there.

Haise: Okay, Al and Ed. In view of your assay of where your location is and how long it's going to take to get to Cone, the word from the Backroom is they'd like you to consider where you are the edge of Cone Crater.

Mitchell: [I] think you're finks! *(pause)*

Haise: Okay. That decision, I guess, was based on Al's estimate of another, at least, 30 minutes and, of course, we cannot see that from here. It's kind of your judgment on that.

Mitchell: Well, we're three-quarters there. *(pause)* Why don't we lose our bet, Al, and leave the MET and get on up there? We could make it a lot faster without it.

Shepard: Well . . . I think what we're looking at right here in this boulder field, Ed, is the stuff that's ejected from Cone.

Mitchell: But not the lowermost part, which is what we're interested in.

Shepard: Okay, We'll press on a little farther, Houston. And keep your eye on the time.

Haise: Okay. And, as of right now, we have a 30-minute extension. *(pause)*

Haise: And Al, did you copy 30-minute extension?

Mitchell: We got it.

Shepard: Yeah. That's affirmative, Fred. Thank you.

Shepard: Okay. Get up at this little rise here and take a panorama.

Mitchell: Okay. *(long pause)*

Shepard: *(probably having stopped)* Okay, Al's going to Medium flow.

Mitchell: And I'll take a pan from here.

Haise: Roger, Ed. *(pause)*

They are now at Station B3 and will stop only long enough for Ed to take the panorama.

Shepard: Well, I'll tell you, it's a fantastic view from here. *(pause) (breathing heavily)* As this pan will show. *(pause)* We're approaching the edge of the rugged boulder field to the west rim. It appears as though the best for us to do will be go to the west rim and document from there even though the Sun angle may not be quite as good. Well, we're pushing on in that direction.

Mitchell – "He thinks we're further north than where we are and that we're going to the west rim."

Haise: Roger, Al ...

Shepard: Al's back to Min flow.

Haise: (confused by Al's statement) You're moving to the west then. (pause)

Shepard: Al is back to Min flow, and we're moving again. (pause)

Haise: And, Al and Ed, Deke [Slayton] says he'll cover the bet if you'll drop the MET.

Mitchell: Well, it's not that hard with the MET. We need those tools.

Shepard: No, the MET's not slowing us down, Houston. It's just a question of time. We'll get there.

Haise: Roger, Al. (pause)

Mitchell: Give you a hand, Al.

Shepard: It's all right.

Mitchell: You caught a boulder with your wheel when you went around that corner.

Mitchell: Al?

Shepard: Yeah.

Mitchell: Head left. It's right up there.

Ed knows they need to angle toward the north to get to the rim and is trying to get Al to head more in that direction. Al is headed for a cluster of boulders that are in the direction of Station C1 from the current position.

Shepard: Yeah. I'm going there. (long pause)

Mitchell: Just bear a little more left. Go right up through there. (pause) I'll give you a hand. (long pause)

Shepard: Okay. We're now right in middle of the boulder field on the west rim. We haven't quite reached the rim yet. (long pause) Okay. Want to rest here a minute?

Mitchell: Yeah. (breathing very heavily; speech labored) Let's take a look at the map. I think we're closer than that. (pause)

Shepard: I'll just go ahead slowly with this. (pause) Okay. Another crater.

They have been traveling for 4 minutes since leaving Station B3, at 40 meters/minute.

Mitchell: Yeah. The rim's right up here. *(pause)* Let's see if we can spot this one, Al.

* * *

Shepard: Okay. *(pause)* Want to pull for a while?

Mitchell: Yup.

A short rest works wonders.

Shepard: Okay. We're about the maximum elevation now, Houston. It's leveled out a little bit. And it looks like we'll be approaching the rim here very shortly.

Haise: Roger, Al. And you can leave the dial in Intermediate. We're fat on the feedwater [according to the readouts, they have plenty of cooling water remaining].

Shepard: Okay. Thank you.

Mitchell: Let me set mine. If we're in that good of shape, let me set mine, Houston, if I'm okay, too.

Haise: That's affirm, Ed. I guess the low item is the battery.

Mitchell: Okay. *(pause)* Oops! It [the MET]'s going over. No, got it. *(long pause)* Fantastic stabilization; Al, it's going to turn over. *(pause)*

Shepard: Okay. We better reconnoiter here. I don't see the crater yet.

Mitchell: I agree, Rock under my wheels. *(long pause)*

STATIONS C-PRIME AND C1 – FRUSTRATINGLY CLOSE TO THE RIM

Shepard: See this boulder pattern and all that we're in here right now? This boulder field and all?

Al is asking if the boulder field they are in is on the map. They have reached Station C-Prime, about 90 meters SE of the nearest point on the Cone Crater rim. Since leaving B3 they have traveled about 225 meters.

Mitchell: I thought it [meaning the boulder field] was on the south rim.

Ed is confident that they are in the area south of the Cone rim.

Haise: And, Al and Ed, do you have the rim in sight at this time?

Mitchell: Oh, yeah.

Shepard: It's affirmative. It's down in the valley.

Al and Ed both hear "LM" instead of "rim."

Haise: I'm sorry. You misunderstood the question. I meant the rim of Cone Crater.

Shepard: Oh, the rim. That is negative. We haven't found that yet. *(pause)*

Mitchell: This big boulder right here [on the traverse map], Al, which stands out bigger than anything else [undoubtedly Saddle Rock] ought ... We ought to be able to see it.

Because he has no familiar objects in sight to help him judge size and distance, Ed does not realize that the large boulder clearly visible on the map is actually in sight. Al will call attention to the "white boulder" in a few minutes. Later, they will go over to Saddle Rock – at what will become Station C1 – and collect samples.

Shepard: Well, I don't know what the rim is still way up here from the looks of things.

Haise: And, Ed and Al, we've already eaten in our 30-minute extension and we're past that now. I think we'd better proceed with the sampling and continue with the EVA.

Mitchell: Okay, Fredo.

Shepard: Okay. We'll start with a pan from here. I'll take that.

Mitchell: All right, I'll start sampling. *(long pause)*

Mitchell – "Right now, as I listen to this, I feel an enormous sense of frustration, just like I did then. It was terribly, terribly frustrating; coming up over that ridge that we were going up, and thinking, finally, that was it; and it wasn't – suddenly recognizing that, really, you just don't know where the hell you are. You know you're close. You can't be very far away. You know you got to quit and go back. It was probably one of the most frustrating periods I've ever experienced. There's no feeling of being lost. I mean, the LM is there; we can get back to the LM. It's not reaching and looking down into that bloody crater. It's terribly frustrating."

Figure 3.9. This 3-meter-long boulder is located near the summit of Cone Crater. The well-defined contact between the white rock on the bottom and the more normal gray rock on the top separates different components of a fragmental rock type known as a breccia, which consists of broken fragments excavated during large impact events and welded or consolidated by hot debris or molten rock. Samples collected from deep in the Fra Mauro formation provide valuable information on the violent history of the Moon.

Interviewer – "Still, 20 years later."

Mitchell – "Still, 20 years later. Well, I'm tapping back into those same feelings. The only thing that really makes it palatable is that 6 weeks later, after the flight, when we realized that we really were there and, from point C1, another 10 or 20 feet it would have been obvious. That really is distressing."

Shepard (from the 1971 Technical Debriefing) – "If we'd gotten to the point where we'd been willing to do away with the rest of the traverse [that is, do their work at the Cone rim and then proceed directly back to the LM without stopping], we could have made the rim all right. But I personally wasn't willing to do that. I felt that gathering more samples was the better of the two choices. We looked at the map again today and described two boulder fields that indicate that we were probably within 150 to 300 feet – depending on these two boulder fields – of the rim and still were not able to see it. That was a pretty good-sized lunar feature, to be that close to the top of the thing and not see it. That is just part of the navigation problem."

Mitchell (from the 1971 Technical Debriefing) – "At this point, in spite of personal frustration – and I know Al felt frustrated in the same way – to have us stop at that point and turn around and come back was the proper decision."

For the first time in the traverse, Al and Ed will stop to collect a number of samples. Because the boulders at Station C-Prime are large, the samples are certain to have come from deep down in Cone Crater.

Shepard: Okay, Houston. We are in the middle of a fairly large boulder field. It covers perhaps as much as a square mile. And, as the pan will show, I don't believe we have quite reached the rim yet. However, we can't be too far away and I think certainly we'll find that these samples [come from] pretty far down in Cone Crater.

Haise: Roger, Al. *(pause)*

Mitchell: *(to himself)* Okay. Come on! *(long pause)*

Shepard: Okay, you about to start taking documented samples?

Mitchell: Right here.

Shepard: All righty. Let me say, Houston, that most of these boulders are the same brownish gray that we've found. But we see one [Saddle Rock, which is numbered 1107 on the USGS map] that is definitely almost white in color. A very definite difference in color, which we'll document. We noticed that beneath this dark brown regolith, there is a very light-brown layer. And I think we'll get a core tube right here to show that. As a matter of fact, I think I'll do that right now.

Haise: Roger, Al. And for your information, we won't be doing the polarimetric experiment.

Shepard: I understand; you will not be.

Haise: That's affirm. You can delete that one. *(long pause)*

The list of tasks planned for the Cone rim were listed on two cuff checklist pages. The polarimetric experiment involved taking pictures at various orientations relative to the Sun with the Hasselblad lens covered with a polarizing filter.

Shepard (from the 1971 Technical Debriefing) – "We stopped and started taking samples. I feel pretty sure that we have some new and strange rocks. They looked – even to our eye, without looking at them through the magnifying glass – decidedly different from some of the rocks we had seen on the way up the slopes to the crater. I feel pretty sure there are going to be some types of rocks that they haven't seen before."

Almost all of the rocks collected at Fra Mauro are breccias (rocks containing numerous fragments of other rocks, bonded by pressure and/or rock melt from large impacts). Variations in breccia properties result from differences in the parent rock types. Although small breccia samples were collected at both the Apollo 11 and 12 sites, the Apollo 14 samples were the first large breccias returned from the Moon. Most of the Apollo 16 samples are also breccias, as are the Apollo 17 samples collected on the slopes of the North and South Massifs at Taurus Littrow.

* * *

Shepard: And can you give us a feel, Houston, about when you'd like us to leave the area.

Haise: Okay. Estimated time of departure is in about 8 minutes, 7 and a half minutes [from now].

Shepard: Okay.

Mitchell: Okay. Want the hammer? I'll grab it.

Shepard: Okay. I guess we just *(pause)* run down there this way, huh?

Mitchell: Yeah.

Shepard: Okay.

Mitchell: *(garbled)*. One of these boulders, Fredo, is broken open. They're really brown boulders on the outside, and the inner face that's broken is white, and then another one that most of it is white. They are right in the same area.

Haise: Okay, Ed. I assume you're going ...

Shepard: Okay. I believe that's probably a ...

Haise: ... to sample some of those.

Mitchell: That's where we're headed right now. It's about 50 yards away.

The distance from C-Prime to Saddle Rock was about 60 meters, so Ed's estimate is quite good.

Shepard: Why don't you go on down and start, and let me bring the MET down.

Mitchell: All right. Yeah. It's further than it looked.

Shepard: That's the order of the day. *(long pause)*

Mitchell: Okay, Fredo. I'm right in the midst of a whole pile of very large boulders here, Let's see what I can do to grab a meaningful sample.

Haise: Roger, Ed.

* * *

Haise: To get us back on the old timeline here, when you depart C here, we'd like to proceed directly to [Station] F, Weird [Crater]. And we'll pick back up from that point. En route you can make grab samples as you see fit.

Shepard: Okay.

Haise: And another note I'll remind you of ...

Mitchell: Hey, Fredo; I've chipped some ...

Haise: ... later on. Go ahead, I'm sorry.

Mitchell: I've chipped off one of the white rocks. I put it in bag 13-N. I'll photograph it. There don't seem to be any samples of the white rocks lying around that are small enough for me to sample and be sure they're what I'm looking for.

Haise: Roger, Ed. 13-N. *(long pause)*

Shepard: And Al is just going around picking up hand-size grab samples [possibly with the tongs] from the immediate vicinity of where Ed is operating. I have a couple that are going in bag 16.

Haise: Roger, Al. *(long pause)*

Mitchell: *(garbled)* help with that one?

Shepard: That's all right, I think I got it. There's a football-size rock, Houston, coming out of this area, which will not be bagged. It appears to be the prevalent rock of the boulders of the area. Got it?

Mitchell: Got it.

Haise: Roger, Al, we copy.

A football-size rock would have been too big to pick up with the tongs, so Al had to get down low enough to get it with his hand. There are several examples in TV films from later missions that show the gymnastics required to pick a large rock up off the ground. The most entertaining and instructive example shows Apollo 16 astronaut Charlie Duke collecting the 11.6-kilogram rock known as Big Muley at the rim of Plum Crater. The rock being collected here at Saddle Rock is sample 14321, a 9.0-kilogram breccia now known as Big Bertha. It is by far the largest rock collected during Apollo 14 and rivals not only Big Muley (the largest rock returned from the Moon, named for Bill Muehlberger, who headed the Apollo 16 geology team), but also the runner-up, a 9.6-kilogram Apollo 15 basalt known as Great Scott (after Dave

Scott, who collected it). In collecting Big Muley, Duke got down on one knee while using a long-handled tool in the opposite hand to help keep himself balanced. He then used his free hand to roll the rock against his knee, getting the rock far enough off the ground so that he had his fingertips on the lower curve of the rock. Then, holding the rock against his knee, he rose to his feet and worked the rock up higher on his leg until he could get back to the Rover, put his tool down, and grab the rock with both hands. Scott used a similar technique. Here, the dialog suggests that Ed helped Al pick up the rock and, apparently, took it from him once he was up.

The name Big Bertha derives from a massive World War I howitzer produced by the Krupp Armament Works in Essen, Germany, and named for Gustav Krupp's wife. The gun was capable of throwing a 1000-kilogram (2205-pound) shell over a distance of more than 14 kilometers (8.7 miles). Two of these guns were used to devastating effect in the battle for Liege in August 1914.

The Big Bertha breccia sample was very important to scientists' understanding of impact breccias because it was large enough that the textural and compositional variations one would find in an outcrop are visible. After the mission, scientists were debriefing the crew in the LRL quarantine facility; double-paned glass separated the scientists from the crew and the large samples that sat on a nearby table. Shepard was enthusiastically pointing out features on the rock while holding a cup of coffee in his pointing hand. The scientists were alarmed, but the lunar sample curator quietly said, "Al, please step back from the boulder and put down your coffee." A great sigh of relief spread through the room.

Mitchell: Have to go in one of the Z-bags.

Shepard: Okay. Do you have a sample of that white rock?

Mitchell: Yeah, I got one batch of particles.

Shepard: Put it [that large rock] right in here.

Mitchell: I don't think it'll go.

Shepard: Yeah. Core tube's out of the way. Put her down.

Shepard: Okay, We'll just carry it back that way.

Haise: Okay, Al and Ed, We have . . .

Mitchell: (garbled under Haise).

Haise: . . . about 1 more minute here at C.

Shepard: Okay. We're moving on down the hill now. (to Ed) Okay; can you see Weird from here?

On the trip back down to the LM, their elevation will allow them to pick out landmarks much more easily than they could during the uphill climb.

FAST TIME DOWNHILL

Shepard: Let's go on. Are you on the thing back there?

Mitchell: No.

Shepard: Okay.

Mitchell: Want me to hold you back?

Shepard: No, that's all right.

Mitchell: Let me grab it, you're going to go over here in a minute. *(long pause; garbled)* I can't get you. Okay. *(long pause)*

Interviewer – "I certainly don't hear the labored breathing on this run downhill."

Mitchell – "The run downhill was a piece of cake. It was easy to go downhill, and the MET was the only thing that really slowed us down, because it couldn't go as fast as we could go. It was bouncing around. And, having to navigate around craters and rocks and boulders, it was flipping/flopping all around. That was really the only thing that slowed us down. We could have been at a dead run and it's almost no effort."

Interviewer – "Do you remember feeling any instability, or was it a real stable run?"

Mitchell – "That nice skipping gait that I liked was very easy to do going down the hill. You could take longer steps because of that bound and, with the slope, I'd float further down before I hit again. So, it was easy coming down the hill."

Shepard (from the 1971 Technical Debriefing) – "With respect to familiarization with the terrain: if we were coming at a relatively high rate of speed, as we were coming back down from Cone Crater, where we were kind of running in a down-Sun direction, there were times when we had to be careful. If you were running directly down-Sun, there's an area where [in] two or three more steps you're going to be in a crater. So, you're going to have to zigzag a little bit, but I didn't have any problems avoiding those things. As long as you can see two or three steps ahead of you, it's sort of like broken-field running. The whole process is so totally much slower than you are used to on the Earth [because of the lower gravity]. Even when you get going in leaps and long strides, you can change direction and get around the craters fairly well. You can do this even with a fairly high-speed lope."

* * *

Al and Ed made a brief stop at the edge of Flank Ridge to get their bearings. It took Ed a few moments to spot Weird Crater and as soon as he had it, Al spied a prominent boulder just east of the crater – a feature now known as Weird Rock.

Interviewer – "I remember from reading in the technical debriefing that you used the boulder next to Weird as a tracking point."

Mitchell – "Yeah, Because, as I recall now … The reason it got called Weird is because of the triple *(garbled under the voice of Ed's rambunctious young son, Adam)* inside of it. The triplet inside of Weird was, I think, what gave it the name. And it [the boulder] was a good landmark for me to see it. So, once we were sure we could see that, and coming downhill, we lost Weird again when we came down through this valley and I think we come on to that pretty quick."

Shepard (from the 1971 Technical Debriefing) – "We started back down the hill. From the elevation where we stopped, the view down in the valley was just fantastic. But, outside of that, we could see exactly where we were going at this time. We said we were going back to Weird and we could see Weird. There was no question about it."

Mitchell (from the 1971 Technical Debriefing) – "It was just like a map. Fortunately, there was a boulder between us and Weird. We used it as a reference; and, if that baby hadn't been there, I'm not sure we would have found Weird when we got down on the flat."

Shepard (from the 1971 Technical Debriefing) – "It wasn't there. That's another remark about navigation."

Shepard: Okay. We're now out of the boulder field, Houston, and proceeding on down the flank.

Haise: Roger, Al.

They are slightly east of the top of Flank Ridge and will run along parallel to the ridge crest as it slopes down toward the southwest.

Shepard: And, I believe I'll just get a shot … Let's get a sample of that baby right there. Let's grab some from that one.

Mitchell: All right.

* * *

Haise: Roger. And one other question that's up here is to check for the stratigraphy reported earlier of the light gray-white layer below the top, if you see that exposed anywhere.

Shepard: Okay. Now, we did not see that until we started approaching the edge of the boulder field. There's no evidence of that at all that we noticed.

Mitchell: Not down this far. One thing I did notice further outside of where we saw the white underneath; it looked like an impact had either been of the white rock or it was a splatter of white. And it was just outside where Al was reporting that the underlying layer was white. As a matter of fact ... No, that's just ... The Sun angle was causing it. Right now, some of the spray that we're kicking up looks white underneath, but I'm convinced it's just the [Sun] angle. I looked back the other way, and it's not substantiated.

Haise: Roger, Ed. *(long pause)*

Mitchell: Hold it!

The MET almost turned over.

Shepard: That's what I'm trying to do! *(long pause)* Okay, we're moving along pretty well, Fred, at this point. And I'd say we're still probably about 10 minutes away from Weird.

Haise: Very good, Al. Looks kind of like you're making a little better time going down than up.

Mitchell: Yeah, the slope's a different way, Fredo. *(pause)* In this case, the MET helps.

Shepard: Okay, don't let me lose that baby [meaning Weird Crater]. That's it right there with the three ...

Mitchell: Yeah.

Shepard: ... with the three rocks beyond it.

Mitchell: Yup.

Shepard: We're getting down to the place where we won't be able to see it. *(pause)* This is probably Flank right here, isn't it?

Mitchell: I'm not going to say until I get down and look at the exact pattern. It probably is, Al. But if this is really Flank, we should have been at the top of Cone Crater where we were.

Shepard: Yeah, I know.

Mitchell: I think we've already passed Flank.

Haise: Okay, It maybe looks down here, Ed, that may be what you're looking at there, if you've got another Flank-size crater, is the one by [Station] E.

This is the crater they mistook for Flank on the way up.

Mitchell: No, this is a big crater. It's 40, 50 meters across. It has a fairly sharp crater in the south edge of it, which is ...

Haise: Okay, that looks like it may be the one by E.

Mitchell: ... 20, 30 feet across. *(listens)* Yeah, I think that's it, Fredo. And it's ... Oh, it's at least 50 or 60 feet deep. *(pause)*

Shepard: Why don't we just grab a couple from right here. Yup.

Mitchell: Okay. *(long pause)*

Shepard: That baby came apart. *(pause)* Very soft [rock].

Mitchell: Yeah, it's falling apart as you pick it up; very crumbly, isn't it?

Shepard: Okay. You got a bag ready?

Mitchell: Yeah.

Shepard: Very, very soft rock from the rim of that crater, plus another one fairly close to it with crystal in it, that's flashing now; going into bag ...

Mitchell: 15-N.

Shepard: Okay.

This "very, very soft" rock may be a regolith breccia (a clump of soil consolidated by an impact). Regolith breccias are also known as *instant rock*.

Haise: Okay, copied 15-N.

Shepard: *(garbled)*.

Mitchell: Not quite; let me get it in there.

Shepard: Stay behind me; we don't want to lose anything now.

Mitchell: Okay.

Shepard: Okay; that's where we're going, right there.

Mitchell: Yeah, we're going right for Weird. Head right for the big boulder. Then Weird's right beyond it.

Shepard: All right. *(long pause)* Easy!

Mitchell: Okay, keep going. *(long pause)* This is Ed, I'm going back to Medium cooling.

Haise: Roger, Ed. *(long pause)*

Mitchell: One of the problems of going downhill here is that you have essentially diffraction, I guess, around your body, and it creates a halo effect in your shadow, and you just can't see a darn thing right in front of you.

Interviewer – "Is there an example of the halo effect in one of the pictures?"

Mitchell – "Yeah, I think you can almost see it in the pictures. Most shadows are absolutely black and white, but when you're looking down-Sun, it's like it blossoms, just like it blossoms on the TV. Around the image. When you're looking down-Sun, everything seems to blossom around the shadow, right out in front of you. Around your own shadow. It's like you're getting a diffraction pattern – a reflection – in some way. Like the light's bending. It gives you a screwy effect, it really screws up your vision."

Interviewer – "That plus the zero-phase reflections that tend to wash things out."

Haise: That's okay, Ed, ...

Mitchell: It's either blacked out or washed out right immediately down-Sun of you. We're going predominantly down-Sun now. *(long pause)*

Shepard: Okay, Fred, we're still moving, and MET's about 3 minutes away now from Weird.

Haise: Roger, Al. *(pause)*

Mitchell: The crater we are going by now – we're just to the north of it, Fredo – is an old subdued crater.

Mitchell: Uh-oh.

Shepard: If you want to, run over behind that boulder over there, and I'll try and talk to you.

Mitchell: You're the one that has to get behind it and try to talk to Houston.

Shepard: Oh, that's right. Okay.

Mitchell: I'll pull the MET, Go ahead.

Shepard: Okay. *(pause)* On second thought, maybe it's not big enough. *(garbled)* ...

Mitchell: No, don't think it is.

Shepard: No, I guess not. Sure is a big old boulder! I'll take a picture of it anyway. *(pause)*

Haise: Okay, and this big boulder, Al, is ... You're just about at Weird now. Is that right?

Shepard: Oh, probably a couple of hundred meters short of Weird. *(long pause)*

Mitchell: This country is so rolling and undulating, Fred, with rises and dips everywhere, that you can be going by a fairly good-size crater and not even recognize it.

* * *

Al and Ed only spent a couple of minutes at Weird Crater – just long enough to take a pan and grab a quick sample.

Shepard: Okay; let's press on.

Mitchell: Okay. This darn rig, it's hard to fold up.

Mitchell – "Although I haven't mentioned it throughout here, I'm still having trouble with my dexterity because of that right glove. So, even the task of holding the samples – doing anything that requires right-hand movement – is a little difficult. So I'm fumbling around here with that right glove. That's part of the problem."

Interviewer – "And 'this darn rig' is one of the sample bags."

NORTH TRIPLET

Shepard: We've got a pan and a grab sample. What else do we want from here, Houston?

Haise: Okay, that's it, Al. We would like to proceed on to the North Triplet, and I'll give you the tasks when we get there.

Shepard: Okay, we'll try to get to North Triplet.

Mitchell: You ran out from under me just as I was picking it up.

Shepard: Sorry.

Mitchell: Okay. *(pause)*

Shepard: Okay. Oh, man.

Haise: Okay, Al and Ed …

Mitchell: There's some blocks over there. *(garbled)*

Haise: … for your stop for the G, we'd like that … take an estimated one-crater diameter short of the crater – North [Triplet] Crater.

Mitchell: You want us to stop one-crater diameter short.

Haise: That's affirm; because some of the items coming up are the core and the trench … triple core.

Mitchell: … Okay. I think we're seeing the rim of the Triplet series right ahead of us, aren't we, Al?

Shepard: I would say so, yes. We can say that's the rim of the North right there.

Mitchell: Yes. It's got boulders on it, and that's the only thing big enough to have boulders. We're probably about one diameter out right now.

Shepard: I'd say we are. Right here.

They have reached Station G.

Mitchell: The way we've been estimating distances today, that rim has to be at least 6 miles from here.

Shepard: Okay, Houston; we're about one diameter to the east of North Triplets [sic].

Mitchell: To the west of ... Yes, east of the Triplets. Excuse me.

Haise: Okay, copied, and ...

Shepard: Okay.

Haise: ... the number 1 item is the triple core.

Shepard: Okay.

Mitchell: Where's the third core tube?

Shepard: Well, why don't you use clean ones?

Interviewer – "What does Al mean by clean ones?"

Mitchell – "Remember, he's tried to make core samples in the past. One, I think, at [Station] A and another one up at Cone. And, in both instances ... Well, the ones at Cone, they all fell out, so those would be considered contaminated or dirty tubes. So he obviously put those in one bag and he had some clean ones over in another bag. Since that was his task, I didn't know where he had stowed those things."

Mitchell: I don't have clean ones.

Shepard: Yeah, you do. They're down in this pocket right there. Let me ...

Mitchell: This one?

Shepard: ... let me get my camera tightened up.

Apparently, Al's camera handle has loosened again.

Mitchell: This one's been used.

Shepard: No, no, no. In here, Ed.

Mitchell: Oh, okay.

Shepard: The three tabs [that is, the core tubes marked with tabs] should be clean.

Mitchell: All right.

Shepard: Okay, we'll pull it back together here.

Mitchell: *(garbled). (long pause)*

Mitchell: Now, they're clear at the bottom of that, I think.

Shepard: Okay; we've got the camera back together. Okay, Fredo, for your info, the CDR's Commander's [means "CDR's camera's"] reading 117.

* * *

Shepard: You're still planning on a trip [to the ALSEP] . . .

Haise: Okay. This gives us approximately . . .

Shepard: Go ahead.

Haise: . . . 25 minutes at stop G here.

Shepard: 25 minutes until what?

Mitchell: *(garbled).* Okay, I got this one. Go ahead; start your trench, if you like.

Shepard: Okay. *(pause)* I'll dig the trench in the far wall of this crater here, Ed. *(pause)*

Mitchell: Right. *(long pause)* Fredo, I've tried to push in the triple core tube. I get maybe, oh, 3 to 4 inches pushing in by hand. And it's just surface stuff; a very soft . . . It will not support the weight of the core tubes. *(pause)* Now, I've got it balanced, and I can take a picture of it, perhaps.

Haise: Okay. We're reading you, Ed. *(long pause)*

Mitchell: Okay, We'll try to drive it. *(long pause)*

Haise: And do I understand correctly, Ed, [that] you're taking care of the triple core on your own there?

Mitchell: That's affirm. Al's digging, busy with his trench.

Haise: Okay; very good, *(pause)*

Mitchell: I'll go over and help him photograph it in a while. And it's not going in easy, Fred.

Haise: Roger, Ed. *(long pause)*

Mitchell: I'll try driving a bit more, but I think I'm on solid rock; and I'm about one core tube down.

Haise: Roger, Ed. Solid rock, about one core tube down.

Mitchell: Yeah. *(pause)*

Haise: Okay. The recommendation, Ed, is to pull it up and move over a bit and try it again.

Mitchell: The way this one feels, it'll be the same thing. *(pause)*

Haise: Okay, Ed; and when you pull it out, they'd like to save the bottom core, and replace it with another one there before you try again.

Mitchell: Okay. *(long pause)*

Haise: How's the trench going, Al? Are you getting down there?

Shepard: I've got a trench here, It's going fairly easily, but I need the extension handle [which Ed is using at the moment] to get it deeper so, we'll wait 'til Ed's through with that. I'm cutting into the rim of a crater which is approximately, oh, say 6 meters in diameter, has a depth of about three-quarters of a meter. And we're back in about one diameter away from the north member Triplet. The trench is going through at least three layers that I can see. The fine-grain surface, dark browns; then, a layer of what appears to be quite a bit of glass; and then, a third layer of some very light material. And, we should be able to sample all three of these.

* * *

The Flight Surgeon reports that Al's heart rate is about 100 beats per minute, and Ed's is about 110.

Shepard: And a very interesting looking rock with fairly fine-grain crystals in it. It's a grab sample, Houston, from that same crater in which I'm digging. It's too large for a bag; its dark brown ... [its] dark part is fractured. The fracture face is very light gray with very small crystals.

Haise: Roger, Al; and if you can't get any with your samples down in the trench itself that have any rock fragments, you might include those as part of your sample.

Mitchell: *(garbled)*.

Shepard: What's that, Ed?

Mitchell: Put it in that side bag if you can; these are all full back here.

Shepard: Okay.

Mitchell: Let me help you. *(pause)*

Mitchell: Okay, baby.

Shepard: You about through with the extension handle, or are you going to ...

Mitchell: Go ahead and take it. I don't really need it to drive [the first section or two].

Shepard: I'll go over and cut that baby, and we'll be through here. Thank you. *(long pause)*

Shepard: Okay, Houston; I know that we did not mention this white layer down in this area before that was so obvious to us just below the surface up near the flank of Cone. But it appears as though it is quite a bit . . . Well, it's relatively deep, as far as visual observation is concerned. And certainly not any would be kicked up by footprints, or tracks or the like. But there appears to be some of that here in this trench.

Haise: Roger, Al.

* * *

Shepard: Damn. You know what's happening in this trench; is the surface fines are so loose that they're just falling down covering the layering that we want to get. *(pause)* I'll tell you, we're not going to get a classic vertical wall here, Houston, on this trench. *(pause)* Damn. *(long pause)*

Trenches dug at the Apollo 12, 15, 16, and 17 sites maintained nearly vertical walls without difficulty. In each case, the vertical wall was dug facing the Sun so that photographs could be analyzed for such things as variation of reflectivity and roughness with depth.

Shepard (from the 1971 Technical Debriefing) – "It looked as if Ed and I should have changed positions, because it was not soft enough for him and it was too soft for me. We practiced digging the trench in the edge of the crater, because it was mechanically and physically easier to dig the trench on the side of the crater. [However], by the side of the crater the dust just wasn't cohesive enough to get a good sample of soil mechanics. We probably did get a pretty good idea of what the composition [means "cohesion"] of the soil was, because it wouldn't hold more than a 60-degree angle on the side of the trench before it all started falling back down in. We did the best we could without [getting] a vertical wall. We were running out of time, again, and it was either do the best we could with that particular trench, or not do it at all."

Haise: And, Ed, Are you having any better luck on the triple core this time?

Mitchell: I've got it in about half a tube. But I'm getting ready to take a picture of it so you can locate it; and then, we'll go ahead and drive it the rest of the way in.

Haise: Roger, Ed. *(long pause)*

Mitchell: Okay, Fredo, There's three frames here, probably 69, 70, 71 [on the frame counter], that have core tubes. The first one's the aborted one that I couldn't get in. The second picture is this new attempt, a 15-foot shot, then I raised up and took a "locator" shot on the horizon on this. I think it might go.

Haise: Very good, Ed. *(long pause)*

Mitchell: Okay, I'm getting down low enough [that] I'm going to have to have an extension handle to finish driving it, I think.

Mitchell – "I didn't remember that I was able to get that much of a core tube sample."

Shepard: Okay. I'll give it back to you. I'm really kind of through with this trench.

Haise: Roger, Al.

Shepard: Here you are.

Mitchell: Thank you. *(long pause)*

Shepard: Okay, Fred. Bag 19 for the sample of the surface fines. That is, from the surface layer of the trench.

Haise: Roger, Al. Bag 19 is a sample of the surface fines.

Shepard: I am unable to take from the walls of the trench the type of material – the glassy type of material that I could see when I was digging – so, I'll just get a shovel full of that, and it will mix the surface with the second layer.

Haise: Roger, Al. How deep did you finally end up getting down?

Shepard: Well, the trench is about a foot and a half deep. I gave up actually not because it was hard digging, but because the walls kept falling in on it and it was covering all the evidence of stratigraphy.

Haise: Roger, Al.

Mitchell: And, Houston, I'm over 20 feet, no, 50 feet from where Al is; and on the east side of these craters, I have the triple core in about a tube and a quarter and it's tightening up again. I just don't think it's going to go the rest of the way.

Haise: Okay, Ed.

The actual distance from the second core site to the trench was about 8 meters (about 25 feet).

Mitchell: I'm maybe driving a millimeter a stroke. I'll hit it a few more licks, and I'll see if we can break through or move it a little more. *(pause)* No, that's as far as it is going, Houston; one and a quarter.

Haise: Okay, Ed. We'll just take your judgment on that. When you don't think you can get it in any further, you can stop there.

Mitchell: Okay. I think I could probably beat it for the next 10 minutes, Fred, and not get another inch out of it.

Interviewer – "During your first try on that core tube, you said that it felt like you were down to rock. Here, you just said that it felt like it was tightening up again. Was there a different feel, that you remember?"

Mitchell – "Yeah, I think on the first one, I probably did hit something. But I think what we're experiencing with both, as I recall, is that, the deeper we drive into, it was like we were driving into tighter and tighter and tighter material. But on the first one, I did hit a rock too, as well as getting into that layer of compacted material. On the second one, I didn't hit a rock, but I was getting into such firmly compacted material that, by the time I was done 1 1/4 tubes, it just wouldn't go in any further. And I was hitting hard, I mean, I used a lot of energy with that hammer, and it just wasn't going anywhere."

Interviewer – "I've seen Gene [Cernan] get the hammer up to helmet height and bring it down sharply."

Mitchell – "That's right. And using the hammer on the side. You weren't hitting on the end, you were hitting on the side because you didn't have enough control to really hit it on the head of the hammer."

The core tubes on Apollo 11 wouldn't go in very deep because of a poor design; a new, more successful design was flown on Apollo 12. *The Apollo 14 Preliminary Science Report* indicates that the first core bit was damaged when it hit a rock that stopped penetration. In this second case, no rock was involved and the investigators suggest that a larger-than-normal average grain size and low bulk density (compared with that found in the 11 and 12 soils) may have been responsible.

Mitchell – "I would say that's just the wrong answer. The bulk density's the criterion. I would say it was higher bulk density. If you go onto a wet, sandy beach and you try to drive a stake or something into that wet sand, it compacts on you. After a certain point, you got too much sticktion [sic] and it just won't go in when it's compacted. That's exactly the way it felt. It didn't feel like I was hitting something. It just felt like, the more I was driving it, it was getting tighter and tighter and tighter and tighter and tighter. Just like it was squeezing it down."

Interviewer – "Basically, the grains couldn't move out of the way."

Haise: Well, I don't think you need the exercise; you may as well extract it now.

Mitchell: I agree. I'll take a picture of it, a final picture of it, to show you how far we got with it. *(long pause)*

Shepard: Okay, Houston; this is Al. And bag 21 is kind of a collection of the – excuse me – a combination of the top two layers. Second layer is a thin layer of small glassy-like pebbles. I was unable to separate that by the trench method, so I gave it to you mixed up in that bag; and the last bag will be [a] sample from the bottom layer.

Haise: Okay, Al. And about what's the thickness of the intermediate layer there?

Shepard: Well, it's really ephemeral. *(chuckles)* It's almost … It's very thin; I would say no more than a quarter of an inch thick, and I just noticed it because of the difference of the grain structure as I was digging the trench.

Haise: Roger, Al.

Shepard: And in bag 20 – two-zero – will go a sample of the bottom material; also, mixed up with some of the surface material that has fallen down in on top of it. And that's about … Call it 18 inches below surface.

Haise: Roger, Al; and when you and Ed can work it in, we need another EMU check.

Shepard: Okay. *(pause)* This is Al, at 3.75 (psi) and reading about …

Mitchell: Oh, hell.

Shepard: … Reading 35; I have no flags; and I'm in Medium flow – now, going to Min flow – and feeling good.

Haise: Okay; and what kind of misery are you having now, Ed?

Mitchell: 3.75; 32 percent, Minimum … Am I at Minimum … Medium … Just a minute. *(pause)* I'm in Medium cooling and doing great.

Haise: Okay.

Mitchell: Now, my problem is I can't get the … Driving down to that rock, I couldn't get the core cap [means "bit"] off. I'll get some help from Al, soon as he puts his handful of samples down. *(pause)* Oops. Okay, that's great.

Shepard: Okay, let me get rid of this trencher.

Haise: Okay. On the agenda, here, we have remaining documented samples, and we need a pan.

Mitchell: Rog. We'll get it for you.

Shepard: Oh, God.

Mitchell: Get another one. Skip it; we've got plenty. *(pause)* I'll get it.

Shepard: I got it.

Mitchell (from the 1971 Technical Debriefing) – "We had a lot of trouble with sample bags. We threw a lot of them away because the little metal flags that were supposed to help you roll them up were getting entangled with each other. It was almost impossible to sort them out and pull one bag out of the dispenser. Generally, we pulled two or three, and one or two of those would get lost [meaning "dropped"]. It was too much effort to bend down and pick them up. It didn't look like we were going to use all of them anyway. That particular piece of equipment is going to have to be smoothed out. It was time-consuming and hard to use."

According to Judy Allton's *Catalog of Apollo Lunar Surface Geological Sampling Tools and Containers*, the Apollo 12/14 sample bag dispenser was a metal cylinder. Later crews had a flat-pack dispenser that worked quite well.

Haise: Okay. And, Al, one question, did you get the SESC sample out of the bottom of the trench?

Shepard: Well, I told you the trench was kind of a miserable thing, because the walls kept falling down. And I could get a sample from the bottom, but it wouldn't be the bottom, I'm afraid.

* * *

Shepard: See this white stuff on the rim area, Ed?

Mitchell: Beg your pardon?

Shepard: See this white stuff on the rim here?

Mitchell: Yeah.

Shepard: Document some of that. Here's a rock right here.

Mitchell: Yeah.

Haise: Okay, has Al moved over by the rim of North Crater now?

Mitchell: No, no, no. We're still at the same place.

Shepard: Negative.

Mitchell: That's pretty well disturbed, Al; I'll grab it without much documentation.

Shepard: Okay. Still digging the bottom of the trench for you, Fredo.

Haise: Okay, Al.

Shepard: I'm redigging the trench. *(pause)*

> **Mitchell:** I'm picking up one of the so-called whiter rocks, Fredo, near the area where Al is digging. Since it's already disturbed, I'm not going to waste time on much [actually, any] documentation. Kind of a *(garbled under Haise)*.

* * *

> **Shepard:** Oh, no!
>
> **Mitchell:** What's the matter? *(long pause)*
>
> **Shepard:** I can't believe it!
>
> **Mitchell:** What's the matter, Al?
>
> **Shepard:** Oh, that seal came off that thing.

Like the rock boxes, the SESC had a seal that consisted of a knife edge on the body of the can and a strip of soft, indium metal in the lid. During the flight out from Earth, the knife edge and the indium strip were kept apart by Teflon spacers, which also kept the seal clean while the can was being filled.

> **Shepard** (from the 1971 Technical Debriefing) – "In attempting to fill the container [with soil from the trench], I had a problem. Fortunately, we had two of those containers. I took the cover off. The top and bottom part of the container have protective Teflon seals. When I took the first can apart, both seals came off together. This left the knife edge unprotected. We discarded that one. We went after the other one and, fortunately, that came apart all right. I filled it with material. Even after we put it on the MET, that's the one that bounced out. Fortunately, Ed was behind there and saw it bounce out, so we didn't lose it. I don't know why those two pieces of Teflon came off together. The top came off and there they were. I was looking at the unprotected seal."

When the can came off of the MET, it was likely that it was Ed who was pulling and Al who noticed the can bounce off. What Al is probably remembering is an incident a few minutes later, when a core tube bounced off. In the interim, they had traded places and, when the core tube bounced off, it was Ed who was trailing behind.

> **Haise:** Okay, Ed and Al, we're going to have to be departing Triplet here. And that one brief stop at the North [Triplet] rim to pick up one documented sample, and get on back to the LM area if we're going to pick up the remaining tasks there.
>
> **Shepard:** Okay. *(pause)* Okay, you're right.
>
> **Mitchell:** Okay, I'll grab the gnomon. We're on our way. *(pause)*
>
> **Shepard:** *(subvocal)* The last I see of that son-of-a-bitch.
>
> **Mitchell:** They're miserable, aren't they? *(pause)*

Interviewer – "I take it that the SESC was not one of Al's favorite toys."

Mitchell – "It wasn't! And digging trenches wasn't his favorite task, either. *(laughing)* So the combination had him pretty upset. I think we're seeing here that the equipment we were using was just not optimally designed. With all the money and all the testing and everything we did and you go back and review it here, and you can see we had far too many equipment problems. And too much of the stuff just wasn't suitable for what we were trying to do with it. Now, it's nice, for example, those sample bags – the ['Dixie'] cup's about that big [about a 13-centimeter-diameter] … It's nice if you're picking up pebbles; but it turns out [that] that's damn little of what we were picking up. On this particular one I was having trouble getting in, I finally jammed it in because it just barely fit. And the sample bags were plastic. They broke. They cracked. They weren't big enough. And then the ring came off the SESC. You know, as careful as we were with quality control and all of that, it was just darn near criminal that we had to battle those sorts of problems. And battling them with inflated gloves that you couldn't hold on to anything. It was quite an experience. Now, I know we re-designed and corrected a lot of stuff before the J missions later on, and I didn't get a full debriefing on how those guys fared. But, you know, if they had as much trouble struggling with all of that equipment as we did, it's amazing, really, that we got done even half of what we really got done."

Interviewer – "The thing that puzzles me, a little bit, is that the 12 guys had a lot of the same problems and there was the delay in your flight after the 13 accident to get some of this re-design done. Or was everybody so focused on the recovery from 13 … ?"

Mitchell – "Maybe so. And, you know, those are small details and, if the science investigators didn't pick up on it, probably nobody picked up on it. There were just so many of those details to work with. It was a matter of priorities and money."

Interviewer – "Which was declining at this time."

The J-mission crews experienced relatively few problems with the geology tools and containers; however, they did have troubles with various scientific experiment packages and also – because of the far greater time they spent out on the surface – had mechanical and thermal problems because of the dust.

Shepard: Okay.

Mitchell: Oh, Move it up. Let me grab it for you.

Shepard: What? That thing? What do you mean … that can?

Mitchell: Yeah.

Shepard: Forget it.

Mitchell: Okay.

Shepard: I'm never going to use it again.

Evidently, Ed has noticed the SESC that Al had discarded after the seal protectors came off it. It may be under the MET and Ed is asking Al to move the MET so that he can retrieve the can.

Shepard: Okay, headed for the LM. And we're probably about 10 minutes away from the LM, Houston.

Haise: Roger, Al. *(pause)*

Shepard: Okay, everything's on so far. *(long pause)*

Shepard: Okay, we're ...

Mitchell: I think we're crossing the ... Al, here's the ... Triplet right up ahead of us.

Shepard: Should be.

Mitchell: We'll have to do a little bit to the north to get around it, I think. *(pause)* Yeah. We're approaching Triplet from the east; that's North Triplet from the east, *(pause)* There's a little rock field down here; a small boulder field, Al, to get a documented sample from?

Shepard: Okay. *(pause)* Looks good. Yeah, looks like they might have come from there [that is, from North Triplet]. Oops.

Mitchell: Did we lose something?

Shepard: We lost you know what.

Mitchell: Oh, no, What?

Shepard: *(chuckles)* This shiny can!

Mitchell: Damn SESC, huh?

Shepard: Okay, the shiny can is retrieved. Press on. *(pause)* We're going to have to mush, Ed, right down the middle and get a documented sample there.

Mitchell: Okay. *(pause)*

Shepard: Right in that pile of rocks; beautiful. Let's see, right to your left. Oh, just the right size.

Mitchell: Okay.

Shepard: Don't walk over them!

Mitchell: No, I'm trying to stay away from them.

Shepard: There you go.

Mitchell: Are these the ones or the ones up ahead?

Shepard: Yeah.

Mitchell: Oh, okay.

Shepard: Oops; God damn that thing!

Mitchell: Okay. *(pause)* Gnomon is in place. *(pause)*

They are at Station G1 on the north rim of North Triplet.

Shepard: Okay, why don't ...

Mitchell: I'll get the ... Go ahead. I'm on this side; I'll get the stereo.

Shepard: Okay. [I'll] get the locator. Can't really see the camera settings.

Mitchell: Yeah, it's got so much dirt on them. Okay, 7 foot [focus].

Haise: Okay, Ed and Al, as soon as you wrap this one up, you're going to have to press on back to the LM, or we're going to be really tight on the close-out.

Mitchell: Okay.

Shepard: Okay. *(long pause)*

Shepard: *(garbled)*.

Mitchell: Yeah. God damn, it's bigger than we thought. Al, grab-sample that one; I'll get you another one here.

Shepard: Okay. Listen, just put it in that thing. And let's press, because we don't have the time.

Mitchell: All right. I'll grab it, and let me take an "after" picture here.

Shepard: All right, I'll grab one right here in the foreground.

Mitchell: Okay.

Shepard: Okay, bag 27 Nancy.

Haise: Roger, Al. 27, Nancy.

Mitchell: And another documented sample ... Larger documented sample than we thought we were getting here, Fredo. Again, it was a buried rock; and it's too big for the sample bag; so, it'll go into a weigh bag.

Shepard: *(garbled)* that one right there. Can you get it?

Mitchell: Yeah. *(long pause)*

Shepard: Okay.

Mitchell: It has a very definite shape; I think you'll be able to sort it out.

Shepard: Okay.

In principle, post-mission comparisons between the rock samples and the in-situ photographs should allow identification of the samples without having to put each one in a separate sample bag. During the Apollo 15 mission review, Dave Scott suggested that, because crew time on the lunar surface was so limited, it might have been a better use of their time to take a "before" photo, collect a number of rocks from the scene, and put the samples all in one bag. This would certainly work with hard rocks that were not likely to break or alter in appearance during transport to Houston. What is absolutely necessary, however, is the pre-sampling photo. A number of Apollo 12 samples could not be unambiguously identified because they did not appear in pre-sampling photos.

AT THE LM – THE GOLF BALL AND THE JAVELIN

Mitchell: Okay, let's mush for the LM. *(pause)*

Haise: Okay, Al and Ed. I guess we can skip the rim of North Crater ...

Shepard: *(garbled)*

Haise: ... and proceed right on back to the LM area.

Mitchell: Okay. That's where we are. We're at the ... we are right at the rim of North Crater.

Haise: Okay.

> **Mitchell** – "They were misunderstanding. We were at the rim of North Triplet."
>
> **Interviewer** – "And they'd missed that you'd moved there. Having listened to some of the other missions where there's an awful lot of conversation going on back in the Control Room, the misunderstanding is easy to understand."

Mitchell: We're on the west ... *(listens)*

Haise: I think you misunderstood the message. We can proceed right on by the rim. We have the buried rock samples now, and head on back to the LM. That's to Antares.

Mitchell: That's right. That's where we're headed.

Shepard: Okay, that's where we're headed. *(pause)*

Mitchell: Hold it. I'll get it; keep going. We lost a core tube.

Shepard: Okay. Got it?

Mitchell: I'll have it in a minute. *(pause)* I got it. *(pause)*

Shepard: Okay. *(long pause)* Everything still hanging on?

Mitchell: Yeah. Everything is still there.

Shepard: Good. *(long pause)*

Shepard: Okay, we're approaching the LM now. Coming in at Fra Mauro Base.

Haise: Roger, Al, and I guess from here, we can split up; and Ed can take the MET and proceed to the cluster of boulders he had reported earlier to the north of the LM; and you can proceed out to the ALSEP.

LM Crew: Okay.

Shepard: I'd suggest . . . Well, you can do it the way you want to . . . I guess you can do without the LM.

Mitchell: Without the MET, yeah; I think so.

Shepard: Without the MET, because there's nobody to . . . If anything falls off, we've lost all those goodies.

Haise: Okay, that's it . . .

Mitchell: I'll just take a couple of rock bags, Fredo, and my tongs and camera, and go.

Haise: That's a good point, Ed, . . .

Shepard: Okay, Al's on the way.

Haise: . . . Yeah. That'll be fine.

Shepard: Okay. Al's on the way out to the ALSEP. *(pause)*

Mitchell: As a matter of fact, Fredo, I'm just going to take a weigh bag and no sample bags; that way I can get more. *(pause)* The size of these rocks . . . The sample bags are too small, anyhow.

Haise: Roger, Ed.

Interviewer – "A little real-time crew decision?"

Mitchell – "Yeah. Illustrating, again, that the sample bags are great for fines and little pebbles and things. But for the size rocks they were wanting, particularly buried rocks. They wanted buried rocks, but when you looked at a buried rock you'd think it was an interesting sized rock but it turned out to be like an iceberg. You've only seen the tip of it and you'd pull the damn thing up and you had this humongous thing, after you'd gone to all the trouble of documenting it and getting the right photos etcetera, you'd pull it up and it didn't fit in anything. So you'd end up throwing it in the weigh bag, and that contaminated it . . . You know, it just wasn't very satisfactory that way."

Interviewer – "What was their interest in buried rocks?"

Mitchell – "Frankly, I don't know, I guess buried rocks would indicate that they'd been in place a lot longer, so presumably would be from an older period than those just lying on the surface."

Interviewer – "I wonder, just speculating, if things just lying on the surface might be more likely to be exotics thrown in from elsewhere."

Mitchell – "That could be, too. But, remember, the main thing we were looking for here was ejecta from Cone Crater. And, so, presumably, unless it was completely beneath the ejecta blanket, it should have been part of that ejecta blanket. I guess I don't really know what their rationale was, but they wanted the in-place rocks and fillets and things that seemed to be a little bit buried and, invariably when we did that, we'd pull up a lot bigger rock than it looked like on the surface.

* * *

We pick up the transcript again after the crew has reached the LM. In the interim, they have done some housekeeping with the ALSEP and have loaded samples and cameras.

Mitchell: *(garbled)* [at the MET] Fredo, correct me, now; Mag Kilo-Kilo has never been used. Is that correct?

Haise: Stand by. *(pause)*

Shepard: *(facing the TV)* Houston, while you're looking that up, you might recognize what I have in my hand as the handle for the contingency sample return; it just so happens to have a genuine six iron on the bottom of it. In my left hand, I have a little white pellet that's familiar to millions of Americans. I'll drop it down. Unfortunately, the suit is so stiff, I can't do this with two hands, but I'm going to try a little sand-trap shot here. *(pause)*

Interviewer – "He topped and buried it on the first swing. I assume that the six-iron was snuck on board."

Mitchell – "In his suit pocket."

Mitchell: You got more dirt than ball that time.

Shepard: Got more dirt than ball. Here we go again.

Al's second swing pushed the ball about 2 or 3 feet – mostly along the line toward the TV camera, rather than along the line of the swing.

Haise: That looked like a slice to me, Al.

Shepard: Here we go. Straight as a die; one more. *(long pause)*

Al's third swing finally connected and sent the ball off-camera to the right, apparently on a fairly low trajectory. He dropped a second ball, which rolled left and toward the TV camera. Al got himself in position and connected again. The trajectory of this shot appeared to be similar to the previous one.

Shepard: Miles and miles and miles.

Haise: Very good, Al.

Haise: And *(to)* answer Ed's question earlier there; Kilo-Kilo was used for the window shots, Ed; so, you ought to bring it back.

Shepard: Yeah, that's right. We got some of that to start with, didn't we?

Mitchell: Yeah.

Shepard: *(garbled)*. *(long pause)*

The Moonshot ''Photo''

It's interesting to note that, while the golf-shot picture in Al's book *Moonshot* bears some resemblance to the TV images, it is actually a composite created from various Hasselblad images. The only actual record of the golf shot is the TV coverage (Al and Ed had already put their Hasselblads into the ETB). In the composite, the LM and LM shadow come from a left/right reversal of an earlier photo. In the composite, the LRRR (laser ranging retroreflector – also known as "LR-cubed") is sitting in the footpad of the ladder strut. In reality, the LR-Cubed was deployed at the ALSEP site during the first EVA. Both of the astronaut images in the composite come from a pan Al took at the beginning of EVA-1. Even the image of "Al" is actually a left/right reversal of a shot of Ed. In the real photograph, Ed is doing a TV pan; in the composite, the TV camera has been removed and the golf club has been added. The image of Ed in the composite is taken from another frame in Al's earlier pan. And, once again, the TV had been removed from a left-right reversal of the original images. Similarly, the image of the U.S. flag has been taken from one of the other tourist pictures Al and Ed took during the flag deployment. The precise images from which the MET and the S-Band were taken are not yet identified, but the MET image is very similar to the one in a photo that Al took at the ALSEP site at the end of the ALSEP deployment. Finally, the ball and the shadows of the S-Band legs – like the golf club – appear to have been drawn in.

Not long after I (author Jones) bought a copy of *Moonshot*, Andrew Chaikin and I had a long telephone conversation about the composite and worked out – at least in general terms – how it was put together. Journal Contributor David Harland tells us that the 1994 hardback UK edition published by Virgin Books contains the composite; whereas Brian Lawrence tells us that the 1995 edition does not.

Al removed the club head and brought it home; it is currently on display at the US Golf Association Hall of Fame in Far Hills, New Jersey.

Mitchell: How many films (means "frames") did we take with this [close-up camera] Eleven, Huh?

* * *

Shepard: Now, let's see what we got left. *(pause as Ed gets set)* There's the greatest javelin throw of the century!

Mitchell: We'll see if it is.

Shepard: Old Lefty, himself. *(Ed makes his throw)* Outstanding! Right in the middle of the crater.

Mitchell: Stayed up.

Shepard: Stabilized spin!

Mitchell: Wasn't bad at all.

Shepard: Beautiful. Beautiful! *(pause)*

Mitchell: Okay.

Mitchell – "Okay, stop a minute, Hand me that briefcase."

Interviewer – "That was actually a fairly credible looking throw. You took a big hop forward, right leg leading, and got your left arm moving forward reasonably well, at shoulder height."

Mitchell – *(handing Jones a photo)* "Now, see if you can spot the javelin and the golf ball."

Interviewer – "Yeah, I've got them both there in a crater off to the north. There's Turtle Rock beyond them. This picture was taken out your window after the EVA. I see some MET tracks. I see your footprints coming back from the boulder field, going around this little crater. And there's the javelin and that's the golf ball there."

Mitchell – "Right there. The javelin went just a hair further than the golf ball."

[It seems likely that the golf ball in this picture is the first one Al hit. In an interview with Ottawa Golf, Al claimed that the second ball landed in the general vicinity of the ALSEP. Note that the golf balls and the javelin were traveling from east to west – right to left.]

ASSESSING APOLLO 14

Astronauts Shepard and Mitchell returned to Earth with about 43 kilograms (95 pounds) of rock and soil samples. The rock samples represented a spectrum of breccias formed during meteorite impacts; they ranged from poorly consolidated, compacted regolith ("clods") to fragments of rock enclosed in glass formed by impact melting. Breccia clasts (angular fragments) within other breccia clasts displayed the complexity of multiple cycles of impact after impact that created these jigsaw-puzzle rocks. In addition to the breccias of the Fra Mauro Formation, there were samples from mare lava flows and crystalline rocks ejected by impacts in the lunar highlands.

Lunar soil samples from the Fra Mauro site – both scooped from the surface and collected in cores – were considerably more coarsely grained than soil samples collected at the Apollo 11 and 12 sites. Many represented slightly modified ejecta from relatively young features like Cone Crater. The trenches Al Shepard dug kept collapsing because coarser soils are less stable, like coarse dune sand. Although the collapsing trenches were frustrating for Shepard, they provided useful information for the soil mechanics scientists planning future missions and lunar colonies. The sample collected in the much-dreaded SESC was useful to Earth-bound scientists whose analyses depended on uncontaminated material.

The crew struggled to dig up the partly buried rocks that scientists were so eager to examine. Both for general technical interest and for much-needed analyses of hazards that future lunar explorers would encounter, scientists needed to study the frequency and energies of small, high-velocity particles (micrometeorites) that bombard the lunar surface. These data can be interpreted by studying any large angular rock that has been thrown out of a large impact crater, landed on the lunar surface, and been partly buried by lunar soil. By comparing the unaltered, buried surfaces with the exposed portions of the rock, which are rounded by micrometeorites that naturally "sandblasted" its surface, it is possible to calculate the rock erosion rate: studies indicate the influx of small, high-velocity particles has abraded about 1 millimeter every million years. Many of the rocks dug up by Apollo crews had been there for a period of time comparable to the time it took the Colorado River to excavate the Grand Canyon! It was difficult to pry these rocks out of the regolith, but worth the effort scientifically.

The advantages of manned spaceflight were (again) clearly demonstrated as the Apollo 14 crew demonstrated an ability to diagnose and work around hardware problems and malfunctions that would have otherwise resulted in an early termination of the mission – or worse. Navigation was the most difficult EVA task because of problems in finding and recognizing small features, reduced visibility in the up-sun and down-sun directions, and the inability to judge distances in that environment.

On previous lunar missions, lunar surface dust adhering to equipment and spacesuits had created a problem in both the lunar and command modules. The special dust-control procedures and equipment used by the crew of Apollo 14 were effective in lowering the overall dust level in the cabins.

The Apollo mission EVAs were so short – those on the first three missions in particular – that all the astronauts really had time for was to make an intelligent selection of samples that emphasized variety. In the case of Apollo 14, the crew was particularly pressed for time. During their climb up to Cone Crater, there hadn't been time to do much more than grab samples; however, in the process, they demonstrated that it was possible to make long walking traverses over moderately difficult terrain. In retrospect, it's logical to wonder if they had snaked their way up the hillside rather than making a frontal assault, the climb might not have been so arduous. In any event, Shepard and Mitchell succeeded in climbing the hill – a landmark accomplishment in itself. Although the Lunar Rover used in the next three missions would provide a far more efficient way of traversing the surface to study the geology, it was clear that, with some care taken in avoiding steep, uphill climbs, the later crews would be able to walk back to the LM if there were a Rover breakdown.

Apollo 14 was an important part of NASA's preparation for the more sophisticated Rover missions that concluded the Apollo series. It was a confidence builder.

Apollo Lunar Roving Vehicle (LRV)
This sporty runabout is the perfect addition to an astronaut's
exploration gear. Double Ackerman, font and rear independent steering allows the driver to nimbly avoid rocks and
craters. It is *the* vehicle for lunar exploration, able to reach
speeds of 13 kilometers per hour and climb slopes of 19 to 23
degrees. The driver must be alert – hard turns at speeds above
5 kilometers per hour will result in skidding. The vehicle will
also tend to spin its wheels in the loose regolith near fresh
craters. In spite of these few minor restrictions, the LRV is
great for exploration and it certainly beats walking!

Base Price: $13 million (1970 dollars)
Vehicle type: All-wheel drive, two-passenger, no-door, all-
terrain vehicle
Cargo Mass: Two suited astronauts (363 kilograms or 798
pounds; 133 pounds in the field) and experiments, tools, and
samples (127 kilograms or 279 pounds; 46.5 pounds in the
field)
Wheelbase: 229 centimeters (90 inches)
Turning circle: 610 centimeters (240 inches)
Curb weight: 218 kilograms (480 pounds; 80 pounds in the
field)
Cross-country "mileage": 35 to 56 Watts/hour per kilometer
Passive restraints: Driver and passenger seat belts
Power train: Independent electric motors on each wheel, powered by two parallel, non-rechargeable silver-zinc batteries
(36 V). The electric motor on each wheel is coupled to a
harmonic-drive gear-reduction unit.

4

The Lunar Dune Buggy

CONCEPTS, PROTOTYPES, AND A REAL MOONCAR

According to historian Mitchell Sharpe, the notion of a wheeled, electric-powered lunar exploration vehicle made its first literary appearance at the virtual dawn of the Automotive Age. In 1901, almost as soon as it became plausible to think about driving on the Moon, a Polish writer by the name of Jerzy Zulawski described an early ancestor of the Lunar Rover. Like all those who would follow in his footsteps, Zulawski realized that combustion engines wouldn't be practical on the airless Moon and, instead, chose to power his vehicle with an electric motor. And like most later concepts of lunar rovers, Zulawski's was a big vehicle with an enclosed cabin. His, he said, could carry a crew of five and a year's worth of supplies.

For about one-half century or so, lunar cars made episodic literary appearances. Then, in the years following World War II, they began to receive occasional attention from members of the spaceflight fraternity, who were beginning to think seriously about what might be done on the Moon after Wernher von Braun's V-2 weapon grew into a real space launcher. Most importantly, people started to think about some of the engineering details – about vehicle weight, power supplies, and traction on various types of surfaces.

Perhaps the largest pre-Apollo rover concept was a 10-ton tractor trailer that appeared in a 1952–53 series of articles in Collier's magazine; collectively, they were titled "*Man Will Conquer Space Soon.*" This series, and some related programs by the Walt Disney Company, did a great deal to fire public imagination in the years leading up to the Space Race of the 1960s. Two of the articles dealt with missions to the Moon; written by von Braun, Willy Ley, and Dr. Fred Whipple, they included not only the large tractor trailer but also all the components for the first lunar base. The lunar articles and the beautiful illustrations by Chesley Bonestall were republished in 1953 as "*Conquest of the Moon.*"

Aristillus Crater

Autolycus Crater

Mare

Imbrium

Apollo 15 site

Hadley rille →

Apennine Mountains

The Apollo 15 crew's goal was to explore a segment of the eastern margin of the vast Mare Imbrium. The Apennine Mountains, which form part of the Imbrium crater rim, are most evident at lower right of this photo. Regolith covers lava flows that had partly filled the crater floor. Hadley Rille (near the bottom of photo) may have been carved by lava flows pouring into the basin or may be the remnant of a collapsed lava tube. Toward the horizon are the craters Autolycus (39-kilometer diameter) and Aristillus (55-kilometer diameter), both of which were excavated by later impacts into the surface of Mare Imbrium.

It took real money to move the vehicles out of the conceptual stage, and it wasn't until the early, heady days of Apollo – when Congress gave NASA all but a blank check – that the first design contracts were let. There was no immediate need for a rover but, with an eye toward an eventual lunar base, it seemed prudent to make a small investment. Planners believed that the design process might take several years and, in the beginning, progress could be made by teams spending sums as small as a couple of hundred thousand dollars. (For comparison, the whole of Apollo cost about $24,000,000,000 in the dollars of the '60s and '70s.) A number of contractors (Boeing, Grumman, Bendix, and others) even invested some of their own research money in the interest of having a competitive advantage when NASA decided that it really did want to fly a rover.

The culmination of these early efforts was a concept called MOLAB (Mobile Laboratory): a two-man, 3-ton, closed-cabin vehicle with a range of about 100 kilometers. In 1964, the Marshall Space Flight Center awarded design contracts to both Boeing and Bendix, and in 1966 both companies built 1/6-weight prototypes that, early the following year, were put through both field trials in the desert near Yuma, Arizona, and scientific trials near Flagstaff, Arizona. However, it was becoming obvious that a lunar base was very far in the future and that there would be no need for so large a vehicle. After the trials, the MOLAB concept was abandoned. These early research efforts were not wasted

APOLLO 15 FACTLIST	
Crew	Commander, Dave Scott; CMP, Al Worden; LMP, Jim Irwin
EVA CapCom	Joe Allen
Command Module	Endeavour
Lunar Module	Falcon
Launch from Earth	13:34 GMT 26 July 1971
Lunar landing	22:16 GMT 30 July 1971
Popular site designation	Hadley-Apennine
Lunar stay	66 hours 55 minutes
Splashdown	20:46 GMT 7 August 1971
Command Module lunar orbits	74
Time outside the LM	3 EVAs, 18 hours 33 minutes
Maximum distance from the LM	4.8 kilometers
Distance walked or driven	27.9 kilometers
Lunar samples	77 kilograms
Lunar surface photographs	1151
Surface equipment	550 kilograms
Highlights	First Lunar Rover; first comprehensive site exploration; first crew to doff suits during rest periods

because NASA turned first to Boeing and Bendix, with General Motors as an additional partner, when it came time to define a lightweight rover for the J missions.

The decision to proceed with a lightweight rover wasn't made until May 23, 1969 – the day that the crew of Apollo 10 left lunar orbit for the trip home. With the first Moon landing now imminent, NASA had hopes of flying the first of the J missions within a couple of years and, clearly, the rover would have to be designed and built on a tight schedule. Preliminary design requests had been issued the previous November (1968); final design specifications were released on July 11; and, on October 28, 1969, the contract was awarded to Boeing.

This was a cost-plus-fixed-fee contract initially valued at $19 million, and NASA wanted the first of the flight-ready Lunar Roving Vehicles (LRV, or as the astronauts called it, simply the *Rover*) delivered by April 1971. Seventeen months wasn't much time and, as it turned out, meeting that deadline meant a lot of extra people working a lot of overtime. The total cost of the project eventually rose to nearly $40 million. In a time of shrinking NASA budgets and some fairly widespread disillusionment with Apollo, the cost overrun generated far more press coverage than was warranted by the relatively small size of the Rover program. Some members of the public – not to mention a few Congressmen – didn't understand how three golf carts could cost 40 million dollars. The answer, of course, is that major automobile manufacturers typically spent far greater sums developing new-model passenger cars, and if only three of these new cars were built, they would be very pricey, too. When the time came,

Figure 4.1. This conceptual drawing of a lunar rover by Polish author Jerszy Zulawski was published more than a century ago in Poland, first in serial form and later as the science fiction novel *Na Srebrnym Globie* (*On the Silver Globe*).

Figure 4.2. One of many proposed NASA mobile laboratories (MOLAB – as opposed to LESSLAB) was a closed-cabin, 3-ton vehicle with a proposed range of 100 kilometers. Grumman built a full-scale mockup for NASA in the mid-1960s, but the concept was dropped when it became clear that it would be many years before a vehicle that large could be transported to the Moon.

the Rovers performed beautifully and proved to be worth every dollar that NASA had spent.

The Boeing Rover was a far cry from the 10-ton, closed-cabin, 7-passenger tractor/trailer of the Collier's articles, but it was a nifty little machine nonetheless. Empty, it weighed in at a spare 218 kilograms (460 pounds) and could be folded up and stored (for the trip out from Earth) like an intricate toy. As Dave Scott (Apollo 15 CDR) points out, "The volume into which it could be folded was one of its most significant features," – rather like a square sandwich 1.5 meters (5 feet) on a side and 0.5 meter (20 inches) thick, in which the wheels are the filling. Deployed on the lunar surface, the Rover was 3 meters long and 1.5 meters wide; and the tops of the twin seatbacks were about 1.5 meters off the ground. When fully loaded with two astronauts and all their gear, it weighed a hefty 700 kilograms (1500 pounds) on Earth, but only 1/6 as much on the Moon. The Rover rode on four wire-mesh wheels and, when fully loaded, had a ground clearance of about 35 centimeters (14 inches). It had four-wheel, all-electric drive with a 1/4-horsepower motor turning each of the wheels. Steering could be done with the front pair of wheels, the back pair, or both pairs; and, at low speed, the Rover could be turned on a radius equal to its own 3-meter length. Between the seats and slightly forward of the astronauts, there was a small instrument panel containing displays of (among other things) the Rover's speed and also a range and bearing to the last place where the crew had initialized the navigation system – always a spot within sight of the LM.

DEPLOYING THE ROVER

Apollo 15 was the first of three missions in which the Rover was unloaded and deployed, and like most first-time operations, it had a few awkward moments. In a sense, the Rover deployment was semi-automatic: the astronauts pulled steadily on lanyards or tapes to drive the unloading/setting up process, which used a clever mechanism involving springs and intricately designed levers and hinges. The crew exercised extra care during the maneuver. Although there were some minor glitches, there was nothing that couldn't be overcome thanks to good training and advice from the Backroom's Rover experts, all of whom were watching TV coverage of the deployment with the intensity of first-time parents.

We pick up the transcript from EVA-1 as the Rover is being unloaded from the Apollo 15 LM at the Hadley-Apennine site. The voices are astronaut Joe Allen (CapCom during the Apollo 15 EVAs), Dave Scott (CDR), and Jim Irwin (LMP). Author Eric Jones interviewed Irwin during a 3-day visit to Colorado Springs in December 1989; he interviewed Scott in Santa Fe, New Mexico on several occasions during 1992/1993. Other quotations are taken from the Apollo 15 Technical Crew Debriefing done August 14, 1971 at MSC in Houston, Texas.

Scott: Okay. *(pause)* Ready? Here we go. *(pause as the Rover starts to rotate down)* Okay. *(pause)* Oh! Oh! That a boy. A little more. Little more. Looks like

Figure 4.3. Jim Irwin, Al Worden, and Dave Scott (left to right) pose with their color-coded Corvettes and (in the foreground) the 1-g LRV trainer. Coordinating with the stylized birds on the Apollo 15 patch, the cars are red, white, and dark blue. Dave Scott notes: "As I recall, this particular photo of the corvettes was taken out by the launch pad. The corvettes were stylized to essentially show the flag and set a bit of unit pride ... primarily to let the troops know that we were about and paying attention to all they were doing; e.g., at almost any level of the launch complex one could look down and recognize the crew's cars; we went to the pad often for various spacecraft activities as well as to just say hello to the folks putting the Saturn V and its payload together." At the front of the trainer, a mockup of the TV camera is on the left, the gold-foil-covered Lunar Communications Relay Unit (LCRU) is centered between the wheels, and the high-gain antenna is on the right. The white box, farther aft but forward of the seats, is the Rover console; the low-gain antenna for voice communications is pointed straight up on the right-hand side of the console.

you're going to have to do the bulk of the work today. More. Keep it taut. Atta boy. Okay, we're coming up here, 45. *(pause)* Up to about ... Easy, Jim! Easy! Oop. *(they both laugh)* Okay. Here, let me help you. Take it easy; take it easy. Give you a hand. *(pause)* Okay, come on up. *(pulling Jim up)* Up we go! Come on. Easy.

Irwin: *(garbled)*

Interviewer – "When the wheel sprung out, Jim took a tumble."

Scott – "Yeah, I think so. Too bad the TV wasn't a little bit to the right to see all these little things going on."

Interviewer – "Sounds like you gave him a hand getting up. I know that the 16 and 17 guys learned fairly quickly how to get up."

Scott – "Yeah, getting up was easy."

Irwin (from the 1971 Technical Debriefing) – "I was pulling on the lanyard with one hand and trying to take pictures with the other. And of course I fell down there because I tripped backing up in that soft soil."

Scott (from the 1971 Technical Debriefing) – "Yes, but you recovered gracefully."

Irwin (from the 1971 Technical Debriefing) – "Well, you helped me up."

Interviewer – "When you fell, you wound up lying on the ground and, what? ... Dave came over and gave you a hand to help you get up?"

Irwin – "Yup."

Interviewer – "Was it difficult to get up, by yourself, from a fall?"

Irwin – "No, it wasn't difficult, but you get awfully dirty if you have to turn over on to your belly and push up. It wasn't difficult. You'd give a push and you'd just float up. But it was just much easier, you know, to get your buddy to lift you up."

The usual technique for getting up is to get on your hands and knees and push back with your arms so that you rotate backwards through an upright, kneeling position. Then, once your center-of-gravity is over your heels, you can hop up into a standing position.

Scott: Okay, just pull. Just stand there a little easy. *(pause)* Forget the pictures. Just pull real easy, right there. Okay? Just go easy now.

Irwin: Go ahead.

Allen: Pretty sporty there, Jim.

Dave comes back into the field-of-view for a moment as he gets the tape that he dropped so that he could help Jim get up. Rather than try to pick the end of the tape up off the surface, he has come in close enough to the spacecraft to get a section that is hanging down within easy reach.

> **Scott:** *(to Jim)* Okay? *(pause as Dave lowers the Rover another 10 to 20 degrees)* Okay, we're ... Oh, shoot. The walking hinge again.

As Dave continues to lower the Rover, the lower left corner swings out from the LM a few inches in response to the spacecraft tilt. A walking hinge is an intricate mechanism that supports what might be called the base of the Rover sandwich. Multiple joints and levers allow the support point to move – walk – thereby unfolding and lowering the Rover to the lunar surface. The Rover chassis actually appears to be suspended by fore and aft cables. (For more information, see the Don McMillan animation of the walking hinge in action at *http://www.hq.nasa.gov/alsj/VirtualLRV.html*)

> **Irwin:** Did it come loose?
>
> **Scott:** Yeah. Let's see. Houston, the walking hinges are unlocked again. Is that right? *(pause)*
>
> **Irwin:** *(garbled under Joe)*, Dave, ...
>
> **Allen:** They're supposed to be unlocked now, Dave.
>
> **Irwin:** ... at that point.
>
> **Scott:** *(responding to Joe)* Oh, okay.

Dave pulls the tape taut once again and, as he pulls on it, the rear end of the Rover continues its downward rotation.

> **Scott:** Once you see those things unlocked up there in the stowed position, it doesn't give you too good a feeling. *(pause)* Looks like she's coming down okay. *(pause)*

The rear wheels are now nearly on the ground.

> **Scott:** Okay, can you pull it out a little bit, Jim?
>
> **Irwin:** How's that?

Apparently in response to Jim's pull, the Rover's forward chassis section snaps into place and, as it does, the entire Rover swings outward from the LM. In the TV image, we can see the control console sticking up perpendicular to the center chassis. The seats are folded down on the forward chassis.

Scott: That looks good. *(pause)* Okay, that's good. Outrigger cables are … Well, the one over there's not … *(pause as Dave releases his hold on the right-hand tape)* Okay, outrigger cables are loose.

Allen: Roger. *(long pause)*

Irwin: Watch … Watch the rope, and watch the glass ball.

Scott: Rog. I got it. *(pause)*

The glass ball is a 1-inch, spherical ball of impact melt that Dave spotted from his window before the EVA. He and Jim are being careful not to disturb it until they have a chance to collect it.

Scott: Okay, outrigger cable is loose and off. *(pause)* Okay. *(pause as Dave retrieves the left-hand tape from secondary ladder strut where he draped it earlier)* Okay, let's come down with the left tape. *(pause)* Easy does it. It's coming okay.

Irwin: Okay. *(long pause)*

Dave is standing facing the side of the Rover, with his left side next to the LM, and pulls the tape out with his right hand a couple of feet at a time. He is either pulling the tape through his left hand so that he doesn't lose it as he grabs the next section with his right or he is alternating pulls with his left and right arms.

Interviewer – "The difference between this Rover deployment and the later ones is that you're standing at the left side of the Rover pulling the left tape, bending your arm repeatedly at the elbow. Other people chose to walk away from the LM and use their legs rather than their arms."

Scott – "Maybe we suggested they do that. You got to remember, this is the first time around."

Interviewer – "Did you have any particular problems bending the elbow?"

Scott – "Nope. The elbows have convolutes in them and they're built to bend. In the Gemini suits, the neutral position [of the inflated suit arms] was always straight out. In the Apollo suits, they put the convolute in the elbow and any place you put your arm was a neutral position, a non-force position. So it wasn't hard to move the arm. I don't remember this being of any significance at all, except that unwinding a whole bunch of tape took a long time and maybe it is better to go walk out with it. This is the first time around and one of the things that should be done after we do it the first time around is improve it. So hopefully, somebody saw this and discussed it – or we discussed it – and said 'Instead of doing all that, next time why don't you walk out with it?' That's the purpose of the demonstration, the first time of a new procedure: to let everybody see it and then, hopefully, next time, it will be improved. If it isn't, then everybody's asleep. Right? You hope you get some improvement the second time. *(deadpan)* Unless, of course we were perfect the first time."

Interviewer – "Always a possibility."

Scott – *(chuckling)* "But unlikely."

Scott: Okay. It looks like it's loose to me!

Irwin: Okay.

Scott: That's good.

Irwin: Okay.

Scott: Why don't you go put the ... Come on over and we'll *(garbled)* *(chuckles)* Man, this thing's nice and light. *(pause)*

Dave drops the left-hand tape, then grabs a handhold at the side of the center chassis, and now is lifting and slightly re-positioning the Rover by pulling it toward the TV camera.

Scott: Check the old hinge pins. Oop! Out. Pin out. *(pause)* Let's see. Got a hinge pin out. I'm going to get you the tool. Maybe you can reach it, Jim. *(pause as Dave reaches over to the ETB and grabs the Rover tool with his left hand)* Maybe I can reach it. Hey, Jim.

Irwin: Yeh.

The hinge pins lock the fore and aft chassis sections together and must be depressed so they are flush with the floor.

Scott: Need you to get this hinge pin over here.

Irwin: Okay.

Scott: Wait. Let me get the ... *(pause)* Oh, shoot. See my hinge pin on my side?

Irwin: Yeah. It looks like it's almost all the way in.

Scott: Yeah, but not quite. How about putting the tip of the tool on it and pushing it. *(pause)*

Irwin: There you go.

Scott: Okay. Now, let's ... Let's line this up a little straighter. Let's pull the rear end back towards me.

Irwin: Okay.

Scott: There. Okay. *(pause)*

Irwin: Okay, chassis hinge pins are good on my side.

Scott: Okay. *(pause)* Now if I could get the telescoping rods off. *(pause as Dave lifts the front of the Rover from the left side)* Okay, let's ... Jim? Hold on a minute there. I'm not sure the telescoping rods are disconnected. Let's pick it up and move it back and turn it around. Okay?

The telescoping rods – and the related "saddle" – are the last parts of the deployment hardware still attached to the Rover and need to be released. (Details of the deployment hardware can be found in Scott Sullivan's excellent book *Virtual LM*.)

Irwin: Okay, turn it what, your way?

Scott: No, your way.

Irwin: Okay. *(pause)*

They have picked the entire Rover up using the outboard handholds on either side of the center chassis, but they can't free it from the telescoping rods.

Scott: *(chuckles)* Wait a minute. It's not disconnected. Let me ... Put it down right there.

Irwin: And maybe take it forward a little bit, huh?

Scott: *(at the front of the Rover, examining the attachment hardware)* Well, the pin's out. The rods ... The whole saddle up here is still on. Both pins are out. See what I mean?

Irwin: I think we can maybe lift the front end up, can't we?

Scott: We can try. *(pause)*

Irwin: Let me get in there and lift it up. Maybe ...

Scott: Here.

Irwin: Let me pull it this ... *(pause)*

Scott: Wait a minute. Let me twist it this way to give you a little more room. *(pause as Dave pulls the back of the Rover 3 or 4 feet farther to the left)* Okay. See that saddle. Oh, you'll never get in there with the PLSS, Jim.

Irwin: Am I too tight?

Scott: Yeah. Forget it.

Allen: Jim ...

Scott: Hey, Houston …

Allen: … verify you pulled the saddle pin, please.

Scott: … any suggestions? *(pause)*

Irwin: Yes, the saddle pin has been pulled. *(pause)*

Allen: Rog.

Irwin: We've got to somehow … *(pause)*

Scott: Okay. Joe, the situation is that both pins are out of the saddle, and it still seems to be connected to the frame of the LRV.

Allen: Roger. We copy, and we're working it.

Scott: Okay.

Irwin: Let's finish setting up the Rover, huh?

Scott: Yeah. *(long pause)*

Scott: I remember a guy who once said "dirt dirt" and it is [sic; intended to say "is it"] ever! Whew! *(long pause)*

Dave is quoting Pete Conrad, who used the expression "dirt dirt" early in the first Apollo 12 EVA. Dave was Pete's backup Commander.

Scott – "We watched Pete and Al [Bean] very carefully, and listened to them; and they made the point of how dirty it is. And it is. Good clean dirt."

Interviewer – "Did you have a LM mock-up to do Rover deployments in training?"

Scott – "Well, a Rover mock-up to do Rover deploys. I don't remember exactly how much of the LM was there. Maybe it was on the full LM mock-up. Certainly we'd worked on the procedures for the Rover deployment."

Interviewer – "A few times?"

Scott – "Yeah."

Interviewer – "Doesn't look like the sort of thing you'd have to do a lot."

Scott – "Yeah. But you've got to be familiar enough with it that you don't have to read the checklist while you're doing it. It was pretty straightforward. We spent a fair amount of time on it because the first demonstration we saw didn't work at all. It was all too automatic. So the flight-crew-support guys went through several iterations of what you should do manually and what you should do automatically. By taking automatic stuff off, you took weight off, too, which was always useful. We spent a fair amount of time on who-does-what and what's really necessary – manual and automatic – and how you can do it quickly. And, again, we're going slowly as we do it. It's not wham, bang, drop it out. It's the only Rover we have and the only" chance we'll ever have to use the Rover, so we're going slowly. And, if there's something hung up, you don't force it. You work it a little bit; but then we said 'Hey, Joe, it doesn't look right' and let the guys in the backrooms, who are more familiar, see if there's a simple solution, rather than trying to force something. And, as you notice, Jim says 'Let's go on and work it,' and we went about our business doing other things that were necessary to get it ready, and the one step that we didn't finish, didn't need to be finished right then. Don't force it, take it slowly, a step at a time."

Interviewer – "I also noticed, earlier on, after you had it down to like the 45 degree position and the rear wheels had popped down, you saw that the walking hinges were off again. You weren't sure that that was normal, so you stopped, asked Joe, Joe said 'Yes, that's normal,' and then you continued."

Scott – "Same idea. Use the experts in the [Rover] backroom. And another part of this is that it's a two-man job and I remember we spent a fair amount of time on the procedures as to who would do what, so that we could be as efficient as possible as a team. In fact, there was a lot of effort put into the whole timeline in terms of optimizing the time of the two crewmen so that there was no wasted time. And that took a long time. It's like a pro-basketball team with a no-look pass. Same thing. You've got to work together for hours and hours and hours to develop the methodology and procedures to let you use the time as efficiently as possible."

Allen: Dave and Jim, pull the Rover as far out as you can away from the LM, and then pull on the front end, if you could.

Scott: Okay. Standby.

Allen: And, by that, we mean lift up on the front end.

Irwin: Does that mean pull up . . . *(stops to listen to Joe)* Yeah. Lift up on the front end. Yeah. We copy, Joe.

Scott: Get this stowed, so I don't lose the tool. *(long pause)* Okay, let's try that, Jim. Okay?

Irwin: Okay, pull it out as far as we can?

Scott: Yeah.

Irwin: Back as far as we can? *(pause)* Okay, I'm ready. *(pause)* That's about as far back as we are going to be able to get it, Dave.

Scott: Yeah.

They have pulled the Rover back about 3 or 4 feet and the front-end is off the ground, still attached to the saddle and telescoping rods.

Irwin: If you want to hold it there, I'll get in front of it ...

Scott: Okay.

Irwin: ... and try to lift it up.

Scott: Okay, I'm holding it.

Irwin: See how I can clear this ... *(pause)*

Scott: Now, your PLSS is hung up, Jim.

Irwin: Well. *(pause)* It's coming!

Scott: Okay.

Irwin: There we go.

Scott: Good show. Okay, let's turn it ...

Irwin: Okay, Joe, it's off.

Allen: Outstanding.

Scott: Let's turn it around now, Jim.

Irwin: Okay. *(pause)* Okay, I've got my grip here, Dave. We'll turn it ...

Scott: Yeah, *(garbled)* way. *(pause)* Come to your left; don't walk back! Just swing [to your] left.

Irwin: Okay.

Scott: That a boy. *(pause)*

Irwin: You want to get a downhill run here.

They turn the Rover clockwise until Dave has his back to the LM and the nose is pointing more or less to the northeast.

> **Scott** (from the 1971 Technical Debriefing) – "We turned it around and pointed it away from the LM so I could drive off in forward rather than in reverse. We found it very easy to pick up and turn around. Subsequently, we moved it several times and it was easy to handle."

Scott: Yeah, back up a little bit now. Just back up a little bit. Go in reverse. You. *(pause)* That's good, right there.

Irwin: Okay.

Scott: Watch the [glass] ball behind you.

Irwin: *(laughing)* I've been watching that all morning. I just about fell on it.

Scott: I noticed. *(pause)* Have you got your side of the [instrument] console unlocked?

Irwin: Yeah, it's unlocked. *(pause)*

Scott: Lock it.

Irwin: Okay. *(pause)* Okay, my side is locked.

Scott: And my side is locked. *(long pause)*

Irwin: This side looks okay, Dave.

Scott: *(subvocal)* Okay. Man, they've really done it with the Velcro.

Irwin: Yeah, you almost have to pull against the shear-force of that to get the seat up. I had to really ... really tug at it.

The seats are held in the stowed position on the chassis with strips of Velcro. We can see Dave raise his seat.

Scott: Yeah, man! *(pause)* It's awfully bouncy too, isn't it? *(pause)*

Because of the weak lunar gravity field, the Rover is very lightweight and Dave's tugs on the seat to release the Velcro tend to lift the Rover. Jim has come around to the rear of the vehicle.

Scott: Okay. Get your seat belt out later, I reckon.

Irwin: Mine's in the ... Yeah, I might as well get it now. *(long pause as Jim goes back around to his side of the Rover, deploys his seatbelt, and then heads for the MESA)* Give a holler when you're ready to drive, Dave, I'll come out and take pictures.

Scott: Okay. *(pause)* Sticky fenders. You've got a [left-rear] fender, Jim. Get your fenders?

Irwin: No, I haven't.

Scott: Go ahead. I'll get them. *(pause)*

The fender on each wheel comes in two sections; the section farthest forward or aft from the center of the chassis is sandwiched on top of the other section. In order to deploy the left rear fender, Dave grabs the forward edge of the rear section with the fingers of his right hand and, after some effort, manages to slide it back on the deployment rails until it latches into place. To keep from falling forward, he has his left hand on the back of his seat with his elbow locked. Next, Dave hops around to the right side of the Rover (the side away from the LM) to get the right-rear fender.

Bill Kimsey, who worked on the Rover design, provided the following tale about the seats and fenders. "In 1969 I was sent to Huntsville to help with the proposal for the Lunar Rover. While in Huntsville I worked on the suspension design. This was before the GM [General Motors] guys got involved. After GM got involved, I did general arrangement drawings and fender design. I'm leading up to a couple stories about this preliminary design."

"We were working 14-hour days, 7 days a week. Because the Rover had to fit in to this little, strange-shaped bay on the descent stage, the team was having a problem fitting everything in. The people working on the seats were having a very hard time. One of the engineers came back to work after going home for dinner. He brought a lawn chair back with him. This was the answer to the problem of folding up the seats on the Rover. By the way, this was the night of the landing of Apollo 11."

"There were quite a few of us sent from New Orleans to Huntsville to work on the proposal. One of the other fellows, Waine Borne, and I had been working together from 1961. We had gotten to be good friends. We both had Model A Fords that we had restored. The Model A has beads [curved, raised edges that keep the metal from bending] around the fenders. I was joking with Waine and drew a bead on the fenders of the Rover. The project manager asked me why I put the bead on the fender and I told him it was to stiffen the fender. The bead remained on the fender. Now it's sitting on the Moon. It is the same width as a Model A Ford. No one but the two of us really knew the story."

Scott: Boy, is this dirt soft! Man!

Irwin: Like soft powder snow.

Scott: Really is.

Irwin: Except it's a little different. Different. *(pause)*

A SHORT TEST DRIVE

Scott: Okay, looks like the brake's on, so I'll see if I can't hop in it. *(pause)*

Dave is beside his seat, facing forward with his right hand on the inboard handhold next to the console. He jumps up slightly and pulls himself inward and settles into the seat with just a little shifting of weight once he's down. Because of the problem with the walking hinge, they have lost about 7 minutes to the timeline.

Scott: That's a reasonable fit.

Before the flight Rover was stowed on the LM, Dave and Jim participated in a number of tests to make sure that the Rover was properly prepared. Ernie Reyes, member of the crew operations team, provides a story from one such session.

"Getting ready for the first LRV integrated test in the O&C [Operations and Checkout Building], the test team had much anxiety about the upcoming tests with the astronauts in lunar EVA suits. The day before, I went to Dave Scott and presented him with the idea that we needed something to break up the tension and suggested hanging a raccoon tail somewhere on the LRV. I went to Carol O'Toole, who was the MSC resident office secretary, and had just come back from her vacation in Cherokee, North Carolina. I asked her if we could borrow a raccoon tail off of her recently acquired Indian headdress. She agreed, and went home and got it. I took it up to the astronaut quarters and gave it to Dave."

"The next day at the beginning of the LRV integrated test, the crew walked around the LRV hissing and said to the assembled test team, 'The LRV is not properly configured.' The team gasped, but instantly Dave said, 'Don't worry about it, we can fix it.' Having said that, Jim Irwin helped him to pull a double bag from his leg bag and proceeded to fluff it out. With a pair of scissors the first bag was cut open revealing a second bag ... Upon opening the second bag Dave pulled out the raccoon tail."

"At that point the entire assembled test team and other onlookers started clapping, cheering, and laughing. Dave and Jim then taped the raccoon tail to the right rear fender of the LRV and proceeded to get on to the LRV and said, 'Let's go test this baby!'"

Allen: Okay, Dave. And buckle up for safety here.

This was a well know automobile-safety slogan of the time, but had serious implications for the crew driving a newly designed vehicle on the rough, cratered surface.

Scott: Oh, yeah. *(pause as Dave attaches his seatbelt)* Okay, safety belt's on. *(pause)* Oh, you sit up a lot higher than in 1 g, but that makes sense, does it?

In lunar gravity, the suit doesn't compress nearly as much under the seated astronaut's weight as it does on Earth. It takes Dave several seconds to get the belt attached.

Scott (from the 1971 Technical Debriefing) – "The seatbelt was adjusted properly. I attached it, although it took a fair amount of effort."

Scott: Okay, hand controller is locked. Brake's on, reverse is down. Circuit breakers – all except the Aux[illiary] and the Nav[igation system] – are coming closed. Okay, I get readings on bus B. *(long pause)* All the switches are off, by the way. *(pause)* Okay, switches are all closed. Okay, Houston, are you ready to copy some numbers?

Allen: Go.

Interviewer – "On 17, Gene and Jack both speak noticeably louder to Houston than they do to each other. Pete and Al tended to do that as well and I notice that, here, Jim is a little louder to Houston but you aren't. Do you have any thoughts on that? Gene and Jack thought it might have been psychological distance or something like that."

Scott – "That's interesting. We spent a lot of time in the field, with Joe in the Backroom. So this exercise of talking to the CapCom was a very normal exercise that we'd done many times. I don't know how many times 16 and 17 went out in the field [answer: many] and do the CapCom thing. The comm was good and I was never conscious of the distance. Distance doesn't mean anything."

* * *

Scott: Okay. *(pause)* out of detent; we're moving.

Allen: Extraordinary. *(pause)*

The hand controller has a detent at the neutral position so that it will sit there when Dave isn't applying forward or reverse power. Dave moves off slowly and out of view to the right.

Scott – "Once it starts moving, the people in the backroom were breathing again."

Interviewer – "Not only the people who designed and built the Rover, but all the geologists, too. Let's talk about that a little. I assume that there were contingency plans for walking EVAs."

Scott – "Oh, yeah."

Interviewer – "Did you do any specific training for those?"

Scott – "You don't really have to train for them because, although you have different stations, procedurally it's the same thing. You're just walking instead of riding. You don't go as far but you do as many stations, you spend as much time, take advantage of everything you have."

Interviewer – "You wouldn't have had as much stuff. You probably couldn't have taken some of the tools."

Scott – "Yeah. And all that was in the checklists and it was all figured out before we went. In fact, the walking EVAs were designed to be accomplished from any point. If at any point the Rover stopped running, or we couldn't get it to start running, the walking EVAs were flexible enough that we could continue on doing the work or head back, as the case may be."

Interviewer – "I think your walking EVAs were more thoroughly developed than those on 16 and 17, because they knew from your experience that the Rover would probably work."

Scott – "We spent a fair bit of time looking at this, because, as you point out, it might not have deployed. Or it might not have worked. So I think we had some pretty good alternate plans to go do the work."

In a 1996 review of a draft of the Apollo 15 ALSJ, Dave also pointed out that, "during pre-mission planning, we were also concerned about the possibility of boulders that would preclude driving the LRV all the way – or at all!"

We now switch to the 1989 Mission Review interview between author Eric Jones and Jim Irwin.

Interviewer – "You and Dave had originally been scheduled for a walking mission with a handcart ..."

Irwin – "And then they canceled the last two and they decided they better try to have the car and get as much information as possible, and stay as long as possible."

Interviewer – "And was the Rover still in development when that decision was made? Were you and Dave the first ones to drive the prototype?"

Irwin – "We never really drove the real vehicle on the Earth because you couldn't. If you sat on it, it would collapse. Charlie Duke [Apollo 16 LMP] was the one who was following the development of the Rover."

Interviewer – "Because he and John Young [CDR] had originally been scheduled for the first of the Rover missions."

Irwin – "Yeah. So he was the astronaut who was following that operation. Charlie spent a lot more time at Boeing – where they were building and testing it – than anyone else."

Interviewer – "There are pictures of you and Dave driving the Rover around at the Rio Grande Gorge up at Taos [New Mexico, during a March 11–12, 1971 geology field trip]."

Irwin – "That was the 1-g version. But I think the Flagstaff guys built that one."

Interviewer – "Boeing didn't build that one?"

Irwin – "No. It was just a Rube Goldberg Dune Buggy. It was just a dune buggy that we used."

Interviewer – "Did the Rio Grande exercise have any particular relevance? I would think with all that low (0.5-meter) brush around there ..."

Irwin – "Well, it did, because it turns out that Hadley Rille was just about those characteristics. [A rille is a trench-like valley, generally with steep walls and a flat bottom; they are common features of lunar maria.] About the same depth and almost the same width. It just wasn't quite as abrupt at the edge. I thought was a

Figure 4.4. Scott (right) and Irwin (left) drive a 1-g LRV training vehicle along the rim of the Rio Grande Gorge, near Taos, New Mexico. At this location, the Rio Grande Gorge is about the same width as Hadley Rille (near the Apollo 15 landing site) and exposes a sequence of lava flows. Describing outcrops on the gorge walls provided some of the experience used later to describe the walls of Hadley Rille. This training Rover has only crude mockups of the communications equipment up front and is equipped with rugged, off-road rubber tires. Dave and Jim wear lightweight mock-ups of the PLSS. Both wear microphones and earpieces so they can communicate with members of the geology support team, who are practicing for their own roles in the Science Backroom during the mission. A large sample collection bag is mounted on the right-hand side of Jim's PLSS; Dave's is on his left. Jim holds a map and is serving as Dave's navigator. The mountains in the background indicate that they are on the west rim of the Rio Grande Gorge, driving south.

good exercise. Sometimes we used the trainer in Houston, and we used it also down at the Kennedy Space Center on a training surface there."

Interviewer – "Did it have stiffer springs?"

Irwin – "Just stiffer construction, 'cause the chassis was so light on the lunar version that it would just collapse."

Interviewer – "I would presume that the oscillation frequency of the terrestrial version was different. So there would have been a fair bit of difference in the handling characteristics."

In his 1996 ALSJ draft review, Dave Scott commented: "The Earth geology rovers were intended to exercise the tools, equipment, and procedures, rather than the handling qualities of a lunar vehicle or the response to the lunar surface – much like the mock-up PLSSs we used on field trips."

Irwin – "You know, we had good 1/6-gravity simulations with that Pogo device in Houston. In fact, we could even suspend the little car from this device and remove 5/6 of its weight, so it essentially was at 1/6 g. And we could drive that 1-g version on a track. In fact, we used the centrifuge on it. The centrifuge was no longer used as a centrifuge, so this thing was suspended from the centrifuge track, and it would remove 5/6 of the weight and we'd just drive it on a surface and have that [similar] bouncing sensation as we would on the Moon. So it was good driving simulations and, of course, we also used that to practice walking. The Pogo device was like big suspenders attached to the spacesuit and they'd just remove 5/6 of your weight and you'd bounce along under that. And we could either use it underneath the centrifuge arm, which would follow us, or they even had it mounted on the back of a truck so we could go out and walk or run behind a truck, to get the feel of operating on the Moon."

Scott: Hey, Jim, you can probably tell me if I've got any rear steering.

Irwin: Yeah, you have rear steering.

Scott: Okay.

Allen: Do you have ...

Scott: But I don't have any front steering.

Allen: ... amps on Batt 2, Dave?

Scott: *(under Joe's voice)* Joe, you sure about that battery bit? *(answering Joe)* Negative. But I don't have any front steering, Joe.

Irwin: Got just rear steering, Dave.

Scott: Yeah. *(long pause)*

Dave crosses the field-of-view from right to left – more or less in a direction perpendicular to the line-of-sight. He moves a Rover length in 3.75 seconds. That distance is 122 inches and the implied speed is 3.0 kilometers/hour. During the traverses, his top speed on reasonably level ground will be about 12 kilometers/hour.

Allen: And, Dave, while you're rolling there, requesting Forward Steering to Bus C, Bus Charlie.

Scott: Okay. Steering, Forward, to Bus Charlie. *(pause)* Still no forward steering, Joe.

Allen: Roger.

Scott: Okay, got another suggestion? *(pause)*

Jim crosses the field-of-view in Dave's wake. He moves a Rover length in about 8.8 seconds, doing some hopping but not moving with the speed and agility he and Dave will show later in the mission. His speed here is about 1.25 kilometers/hour. Peak running speeds – over distances of several tens of meters – of 5 to 6 kilometers/hour were achieved by several of the J-mission astronauts.

Irwin (from the 1971 Technical Debriefing) – "During this time, I was attempting to take sequence camera [DAC, Data Acquisition Camera] pictures of you as you drove around the back of the LM. Then I met you in front of the LM [at the MESA]. About this time, I looked at the mag[azine] and it had apparently not moved at all. This was the first indication that we were going to have problems with the sequence camera. Out of all the mags we tried on the surface, only one mag drove. I really don't know what the problem was. I suspect that it was a film loading problem, because we checked the film mags when we loaded the ETB and they seemed to be very tight. It was hard to manually advance the film in the mags. That's about all I can say."

Allen: Cycle the forward steering circuit breaker, please.

Scott: Okay. *(long pause)* Okay, I go to Bus Charlie and the circuit breaker is cycled. *(pause)* No forward steering, Joe.

Allen: Roger, Dave. Press on.

Scott: Okay. That's a good idea. Here, Jim, I'm going to bring her around here and let's get on with it.

Irwin: Okay. *(pause)*

Dave is parking the Rover near the MESA. The Rover can be operated without particular difficulty even when using only rear-wheel steering.

Interviewer – "Do you remember if, during Rover checkout prior to the flight, you played around with rear-only?"

Scott – "Oh, yeah. We drove it with only front and only rear. If one of them didn't work, it was not a big deal, because we knew we could drive with either. In fact, with the double Ackerman steering, it was interesting to experiment with it. 'Cause none of us had ever had any experience with both [front and rear]. So it was fun to play with the 1-g trainer and see how tight a circle you can do with the front and rear, and just the front and whatever. And that's why Joe says, 'Press on, let's go with it.'"

Figure 4.5. Early in the second EVA, Dave is probably testing the Rover steering before maneuvering the Rover into a down-Sun heading – a position that will allow him to re-initialize the navigation system using on the known position of the Sun. In this down-Sun photo, the ALSEP is visible in the background. On the Rover, the maps and 16-mm movie camera are mounted on the accessory staff forward of Jim's seat and the tool rack is at the back of the vehicle. The umbrella-shaped high-gain antenna at the front of the Rover is pointed straight up, the TV is in its stowed position, and the low-gain antenna, just forward of Dave, is pointed straight up. The low-gain will provide good voice communications if it is pointed within about 30 degrees of Earth, which is only about 22 degrees from the local zenith at this time.

Interviewer – "So you have a great deal of confidence. You've also got independent drive on all four wheels so that, if one of them goes you haven't lost much."

Scott: Boy, we're going to have a great time with all these hills and mounds. *(pause)* Okay, think you can handle it there?

Irwin: Yeah, that's good.

Scott: Okay, brake's on. Drive Power 4 coming Off. Off on the steering. Off on a 15 Volt DC. *(pause)* Okay, temps look about the same, Houston. *(pause)* Jim, soon as you get that dustbrush out, I want to brush you off so we don't get the old Rover too dirty.

Irwin: Okay. *(pause)*

Jim got some dust on his suit when he fell and Dave wants to clean him off. Jim comes into view at the MESA.

> **Scott:** You know, as I look back behind us, it almost looks like we landed in a ... Another, oh, 10 meters aft and we'd have been in Surveyor Crater. *(long pause)*

Dave was the backup Commander on Apollo 12 and, on that mission, Pete Conrad planned to land very close to a large crater that contained the Surveyor III spacecraft that had landed some 31 months earlier. After the landing, Pete wasn't sure exactly where he was but, once he got outside, he saw that he was on the northwest rim of Surveyor Crater with his minus-Y (south) and minus-Z (east) not more than 10 meters from the drop off.

Here, Dave has finally realized that he put his rear (minus-Z) footpad in a small crater and is saying that, if he'd landed 10 meters short of his actual location, the entire LM would be in the crater. Jim goes out of the field-of-view.

THE DRIVE TO ELBOW CRATER

We pick up the communications again after they have loaded the Rover with their field gear and are ready for their first ride across the lunar surface.

> **Scott:** Okay, Jim, here we go.
>
> **Irwin:** Okay, Dave. We want a heading of 203.
>
> **Scott:** Okay, 203.
>
> **Irwin:** [To get to] Checkpoint 1.

Jim is probably looking at Part A of the EVA-1 and EVA-2 traverse map. Checkpoint 1 is at the edge of the rille near Canyon Crater. Had they landed at the planned spot, a drive of 2.0 kilometers on a heading of 203 would take them to the checkpoint. They are about 600 meters north and about 175 west of the planned landing site but don't know that yet.

> **Scott:** Going to miss that double Ackerman [front and rear steering], I can see that now.
>
> **Irwin:** Okay, we're moving forward, Joe.
>
> **Allen:** Roger. *(pause)*

Irwin – "Each wheel [pair] is on an individual axis, separate axis. Each wheel [pair] steers on its own radius, so it's independent steering. And it's referred to as double Ackerman. Optimum steering. I don't know whether Honda has that in their four-wheel steering or not. Double Ackerman; it's been so long since I used that word."

Scott: Whew! Hang on.

Irwin: And we're coming around left. *(pause)*

Irwin: Heading directly south right now to miss some craters off to our right – very subdued craters.

Scott: Okay, I'm going to take a little zigzag here …

Irwin: On the right is …

Scott: Hang on. Get a feel for this thing. Nine kilometers an hour, Joe. *(to Jim)* Hold the geology; let's get the Rover squared away first. *(pause)* Okay; 8 kilometers [per hour] up a little rise. Okay, turning back. *(pause)* 203, huh? Okay.

Irwin – "It seemed like we were going much faster."

Interviewer – "Because of the roughness of the terrain?"

Irwin – "I guess so."

For comparison, as a "fitness" runner, author Jones once upon a time managed an unimpressive 10 kilometers/hour over 5 km. A world-class marathon runner maintains a speed of 20 kilometers/hour over more than 40 kilometers, albeit on a smooth surface.

Stephen Ambrose's marvelous biography of Meriwether Lewis, *Undaunted Courage*, mentions that during the spring of 1803, Lewis made an extended trip out of Washington to secure necessary stores for the proposed exploration of the Missouri and Columbia rivers and to be trained in celestial navigation and other essential skills. After spending nearly a month in Lancaster, Pennsylvania, with noted astronomer Andrew Ellicott, learning to use a sextant, chronometer, other instruments to determine latitude and – more importantly – longitude, Lewis headed for Philadelphia to continue his education/training. "The ride from Lancaster to Philadelphia took him over the most modern highway in America, completed in 1795, made of broken stone, the country's first gravel road. Stage wagons were able to average five to seven miles per hour on it. Going that fast in a stage was a new experience for Lewis."

Scott – "On the Moon, perhaps the measure should be hummocks/minute or craters/minute, rather than 10 kilometers/hour; 8 to 10 kilometers/hour may have seemed to be fast because of the many features going by each minute. It certainly did appear that we were going quite rapidly across the surface features."

Dean Eppler, who works on Advanced Planetary EMU (the spacesuit) development in the ISS (International Space Station) Payloads Office at JSC, writes in April 2003 about some field work he undertook in the Antarctic during the southern summer of 2002/3: "We got on the order of 10–15 direct line miles from camp at the farthest, with distances on the order of 20–30 miles for total traverse miles. The vehicles we were using technically could go faster than 12 kph, but in a lot of the terrain, the roughness kept our speeds down to the 12-kph range. I got my snowmobile occasionally up to 40 mph, but that was not an average speed. Glancing through my field notebook, I see moving averages on the order of 6–8 mph. I guess it doesn't matter what planet you're on, if you don't have roads, the going is slow."

Irwin: [Heading] 203 for 2 miles [means kilometers].

Scott: Okay. *(noticing a sharp rimmed crater)* That's a nice, young, fresh one.

Allen: Dave and Jim, Houston.

Irwin: Speed's varying between 8 and 10.

Scott: Go ahead, Houston.

Allen: Roger. Our TV pan suggests, you can go straight for St. George Crater, and you'll find Elbow [Crater] okay. And we're suggesting you omit Checkpoint 1. Rhysling Crater should be a good landmark along the way, and head 208. Over.

Scott: Okay. 208, Joe.

Scott – "We're skipping the checkpoint to make up time, because we're so far behind time. They have probably been making their tradeoffs – whether you want to stop at Checkpoint 1 or go straight to Elbow – and they probably concluded 'We're behind time, let's go straight to Elbow.'"

Interviewer – "So, by going 208, you're going a little more west than the planned 203. They think you're northeast of the planned landing site and you're actually north-northwest. And all of that combined is going to run you into the rille well north of where you're expecting to."

Scott – *(chuckling)* "It's just like a little TV drama, right?"

Interviewer – "Absolutely. A little detective story."

Irwin: Okay, we're doing 10 kilometers, now. Now we're heading uphill; when we head uphill, it drops down to about 8.

Scott: No dust, Joe, no dust at all.

Allen: Yes, sir. Sounds great. *(pause)*

Scott – "Somebody said we'd be driving in a cloud of dust, because of the dust being thrown up from behind. That would have been a big problem 'cause suits get dirty and you can't see, or whatever. So it was comforting to find out that there wasn't a problem. And that was a conscious observation, because of the [pre-flight] speculation."

Interviewer – "Of course, John [Young] and Gene [Cernan] both had to drive in dust, 'cause they both lost rear fenders; and they both spent a hell of a lot of time dusting their Rover batteries as a result."

Scott – "I think we did find out, later on, that we had the rooster tail, but it didn't throw it up on us."

Allen: And, Jimmy, we're standing by for amp readouts.

Irwin: About 9 kilometers, now.

Irwin: Okay, amp readout is ... Looks like 15 on 1. I can't quite see 2.

Scott: Okay, I guess ... Could this be Rhysling right here, Jim?

Rhysling Crater is named for the blind space poet in Robert Heinlein's classic science fiction story *The Green Hills of Earth*.

Interviewer – "Did you and Dave do much of the crater naming?"

Irwin – "Joe Allen did a great deal of it. We did a few of the craters right around our site. There was a Scott. There was an Irwin. There was a Matthew, Mark, and Luke. There was Index. But we thought that was a good thing for the Mission Scientist to do. And Joe's a very resourceful Ph.D. and he did a good job on that."

Irwin: Probably is ... This large depression off to our left?

Scott: Yeah. *(pause)*

They have only been driving for about 2 minutes and, at 10 kilometers/hour, have only gone about 300 meters. The large depression to their left may be Last Crater. Rhysling is approximately 1 kilometer south of their present location.

Scott: Well, I can see I'm going to have to keep my eye on the road. *(pause)* Boy, it's really rolling hills, Joe. Just like 14. Up and down we go. Oh, and this must be Earthlight, huh? Could that be? *(pause)*

Earthlight Crater commemorates an Arthur C. Clarke science fiction novel. In the opening chapter of *Earthlight*, Clarke described a monorail ride across the Apennines

near Hadley and then onward for about 900 kilometers to an astronomical observatory in the crater Plato on the northern edge of the Imbrium Basin.

Scott: Boy, look at that; we're going to have to do some fancy maneuvering here.

Irwin: There's an elongate depression here [at about 73.2 and stretching from BL.8
to BM.6] before you get to Rhysling. I don't think we're to Rhysling yet …
Rhysling ought to be about 1.4. We've only gone, [let's] see, 0.4 [kilometers].

Allen: Roger, Jim …

Irwin: Okay.

Allen: … we think you're short of Rhysling now.

Irwin: Do you think that's probably Rhysling out about 11:00 o'clock to us, Dave?

Scott: Okay.

Irwin: Out about, maybe, 1 kilometer.

Jim is using a Rover-oriented clock: 12 o'clock is straight ahead. Rhysling is more or
less due south of them and – if they are really on a heading of 208 – could be the crater
Jim is seeing at 11 o'clock. Rhysling is a relatively fresh, 150-meter-diameter crater
and its rim would be noticeable, provided it is not hidden by intervening terrain.

Scott: Yeah. Okay, Joe, the Rover handles quite well. We're moving at, I guess, an
average of about 8 kilometers an hour. It's got very low damping compared to the
1-g Rover, but the stability is about the same. It negotiates small craters quite well,
although there's a lot of roll. It feels like we need the seat belts, doesn't it, Jim?

Irwin: Yeah, really do. *(pause)*

Scott: The steering is quite responsive even with only the rear steering. It does quite
well. There doesn't seem to be too much slip. I can maneuver pretty well with the
thing. If I need to make a turn sharply, why, it responds quite well. There's no
accumulation of dirt in the wire wheels. *(pause)*

Allen: Just like in the owner's manual, Dave.

Scott: Okay, we're heading right … *(stopping to listen to Joe)* Yeah, man! Okay,
I've got it on the wall here (that is, at full throttle) for a minute, and we're up to 12
[kilometers/hour].

Irwin: As soon we head upslope, it drops off.

Scott: Yeah.

Irwin: Or, are you deliberately slowing down?

Scott: Yeah, I slowed down in order to get my feeling before we start sprinting.

Irwin: Oh.

Scott: Look at this little fresh one. Boy, look at that! Lots of very angular frags [fragments] all over the thing.

Irwin: Yeah, we passed several of those.

Scott: Okay; I'm going to cut down to the south here, Jim.

Irwin: Yeah, that'd probably be best, because I think that's probably ... Let's see, range (is) point 7. That's still not Rhysling. Shouldn't be.

If they have maintained the average heading of 219 that Dave reports, they have traveled south about 540 meters and west about 440 meters.

Scott: Whoa! Hang on.

Irwin: And we have a large subdued one at our 1 o'clock position, I'd estimate 50 kilometers [means "meters"] wide.

Allen: Roger, Jim.

Irwin: [Range is] 0.8.

Scott: Okay, how we doing on the heading, Jimmer? *(pause)*

> **Scott** – "I had to keep my eyes on the road. I couldn't look at the instrument panel. I had to be focused on where we were going, and that's why I asked Jim about the heading. Even though I can cross-check very quickly – from airplane experience, out the window and back – I remember that you had to focus, consciously, so much on where you were going, because it was so irregular and things happened so quick at 8 kilometers per hour [or '60 features per minute!' as Dave commented in 1996] that I couldn't take my eyes off of where we were going – even for a second – to look at the instrument panel and make an interpretation of the direction. I relied on Jim to do that for me, and that's why he's giving the distances and I'm asking him for the heading. Which is a reason why you want another guy along with you."

> **Scott** – "And that brings up something else. I've learned more about the Russian Rover. Lunokhod was built for a man. For the one guy who landed."

> **Interviewer** – "Really!?"

The Soviet Union landed two Lunokhod vehicles on the Moon: one in Mare Imbrium on November 17, 1970 and the other in eastern Mare Serenitatis on January 16, 1973. They were operated remotely from Earth and each carried a TV camera that transmitted one frame every few seconds. Lunokhod 1 was driven a total of about 10 kilometers during 11 months of operation; Lunokhod 2 covered 37 kilometers in the 5 months it remained operational.

Scott – "Yeah. We can talk more about that. Lunokhod was originally built for their one cosmonaut who was going to the Moon, and then they put a robot chassis on it when they decided not to send him. How about that, sports fans!?"

Interviewer – "Marvelous! The way I understand it is that, on their first landing, Leonov was going to go down and do a footprint; and it's my understanding that, on that initial mission, they were going to do an umbilical EVA. That is, not a backpack-supported, autonomous EVA."

Scott – "Yeah. I don't know the details of the initial mission, nor do I know when they were going to take the Rover. But, in the early 60s, they designed Lunokhod for a single person. And then, when they decided they weren't going to send people, they took that chassis and they put the robot on top of it."

Scott – "And that brings up an interesting point when you start thinking into the future. Do you want a Rover for one person or two people? In thinking about it now, without making a lot of trade-offs, gee, it was nice to have another person along. Because you could get more done. I'd have to think through it to make all the trades, but I guess the system was right."

Interviewer – "I've been transcribing my sessions with Pete and Al and they said they were involved, for a while, with the Lunar Flyer and that Pete had great plans of being the first guy to be a repeat Commander (on Apollo 19 or 20) and to use the Flyer on the Moon."

The Lunar Flying Vehicle – or Lunar Flyer – was a concept for what might be called an open-cockpit, mini-LM that was being considered in the mid-to-late 1960s, when it seemed possible that Apollo missions to the Moon would continue for many years and that exploration activities would become more and more ambitious. Bell Aerosystems did a preliminary design of a two-person vehicle with a range of 80 kilometers in 1965, and North American did a preliminary design of a one-person vehicle with a range of 8 kilometers in 1969. No hardware was ever built for either concept.

Scott – "I wasn't aware of that one! I remember the Lunar Flyer."

Interviewer – "The Flyer was a one-man vehicle, was it not?"

Scott – "Yeah. It was a one-man vehicle, while it lasted. Until Max Faget shot it down real well."

Max Faget, who contributed to the design of every US human spacecraft from Project Mercury to the Space Shuttle, became the Director of Engineering and Development at MSC in 1962. NASA's interest in Lunar Flyer concepts ended in late 1969 because of Faget's opposition, formal opposition from the Flight Crew Operations Directorate, and the increasing likelihood that the Apollo program would not continue beyond Apollo 20, if that.

Interviewer – "On operational grounds or engineering grounds?"

Scott – "Operational. Well, both. That would have been adding a hazard to a hazard, in my opinion. That's like taking a whole new set of circumstances and requirements and putting them in a remote situation, untested and you have a third dimension – that is, you can fall down. I think the Flyer would have been a very hazardous exercise. Very hazardous. And I'm not sure what you gain from it. But, nevertheless, I remember a lot of debate on the Flyer."

Interviewer – "One of the things that Pete and Al talked about was that, according to Max Faget, the Flyer was marginal as designed. It was marginal that it would work at all. And, if you actually started to build the thing, you would get into safety issues that would add weight to it . . ."

Scott – "And I think he's absolutely right. I don't think we were ready for that. And I'd have to give it a lot of thought, even at this point. Why do you want to do that? You want to range farther? Sure. But you have to consider the benefits versus the hazards versus the cost. You want to go flying around the Moon, but we've never flown a thing like that without an atmosphere before. Going to run into lot of unknowns."

Interviewer – "Maybe when you've got a lunar base that's been there for a while, you can do some experimental stuff and fly around in the immediate neighborhood."

Scott – "Sure. You're all set up with your own R&D [Research and Development] facility on the Moon to experiment with things. That's when you go do it. Goodness. It would probably be useful to do that kind of R&D. Probably not here."

Interviewer – "And probably not on Apollo 19 or whatever."

Scott – *(chuckling)* "I don't think so."

NAVIGATION

Irwin: Okay, if we're heading right for Elbow . . .

Scott: Pick a spot here, okay. *(pause)* You really have to pick your way.

Irwin: Yeah. You're only about half way to Checkpoint 1. We shouldn't . . . What I thought was Rhysling was probably not Rhysling. Rhysling is a larger crater, and it's out at about 1-point . . . should be about 1.4 from the LM.

Allen: That's affirm, Jim. Right on.

Scott: Okay. *(pause)* Whoa. Hang on.

Irwin: Bucking bronco.

Scott: Yeah, man. *(pause)* You back off on the power, it keeps right on going. *(pause)* The zero-phase lighting is pretty tough, Joe.

When looking directly away from the Sun (the zero-phase direction), three factors combine to completely wash out the zero-phase view: the generally uniform grey color of the rocks and soil, the lack of shadows (they are hidden behind the objects themselves); and most importantly, a general brightening due to a process called *coherent backscatter*.

Interviewer – "You're talking about zero-phase here but your average heading is about 210 [roughly southwest]. So you must be going west to get around something."

Scott – "[The washed out area around] zero-phase is not [just] due west, it's probably plus or minus 20 degrees [that is, from roughly 250 to 290]."

Interviewer – "But you're doing a significant turn to get around something."

Scott – "Apparently."

Interviewer Jones asked Jim about this in 1989, and he, too, was puzzled about zero-phase on what was a southerly traverse.

Interviewer – "Could he have been doing some wide swings around craters?"

Irwin – "I guess. He must have, 'cause that's the only way you get any zero-phase. That's probably what he meant. *(chuckling)* It was rough-riding regardless of how well you could see."

Scott: We're going to have to make sure we keep at an angle. Once I look into zero-phase, it all looks flat. *(pause)* There's a nice little round 1-meter crater with very angular frags all over the bottom and the rims, and glass in the very center. About 1 meter across.

All of the crews – except Armstrong and Aldrin – commented on these small, glass-bottomed craters. They are dug by small, high-velocity impactors, which turn a quantity of soil into glass when they hit the surface. The angular fragments on the crater rim are probably pieces of regolith breccia; soil closest to the impact point is turned to glass, but farther out, the soil is compacted into a weak, rock-like mass.

Allen: Roger, Dave. And, Jim, as you come up on the rille, you may want to turn your 16-millimeter camera on.

Irwin: Yeah, when we get to the rille, we will, Joe. Can't see the rille at all from here. Still looking for Rhysling.

Allen: Roger.

Irwin: 1.1

Scott: Okay, right now our bearing is 039 for 1.1.

Allen: Roger.

Dave is reading the Rover navigation system and giving Joe a bearing to the LM (39 degrees east of north) as well as a range to the LM (1.1 kilometers) from their present position. These figures indicate that they have traveled about 850 meters (3.4 grid units) south and 700 meters (2.8 grid units) west since leaving the LM.

> **Scott** (from a 1996 draft review) – "Before the mission, we were not even sure the LRV Nav system would work at all. And here we are, plotting our exact position 20 years later!"

Scott: Okay. *(laughs)* Click. *(long pause)* Hey, Jim, give me . . . Well, I just have to drive around these craters; that's all there is to it.

Irwin: Yeah. We have a large subdued one on our right about 60 meters wide with several small ones in the center. By small, I mean about 10 meters in diameter.

Allen: Roger, Jim.

> **Interviewer** – "Do you have any feeling for how well you were judging distances and sizes? I know it was difficult."
>
> **Irwin** – "It was just a guess. *(chuckling)* I'd be curious to know how accurate it was. It would be good to have a final plot of where we were, actually, and see what craters we were looking at."
>
> **Scott** (from a 1996 draft review) – "That is something to look forward to in our later years. Someday, I will use the Journal [ALSJ] to go back!"
>
> **Interviewer** – "Did your ability to judge sizes and distances get better after you'd been at Hadley at a while? Did it start to be less of a guess?"
>
> **Irwin** – "Well, I thought I could estimate probably as well as I could on the Earth. And I don't know how accurate I'd be on the Earth! 'Cause, you know, during our training exercises we practiced estimating sizes of craters, depths of craters. Saying it's 60 meters wide, I don't know why I picked 60. It'd be better to talk in increments of 50 meters, so things would be 50 meters or 100 meters. A geologist would like you to be as accurate as possible. We talk about small, medium, and large and put some numbers associated with those. And I don't know how accurate we were."
>
> **Interviewer** – "Were you taking still photos during this traverse, at all?"

Irwin – "No. You mean with the Hasselblad? We wouldn't have done that while we were moving. I don't think we ever took any still pictures as [when] we were on the Rover. It's a shame that the movie camera wouldn't work"

Interviewer – "There's a little short stretch of it coming back on this traverse, I think."

Irwin – "Yeah, I was wondering if it was on this traverse or on EVA-2. It's not clear which EVA it was. It had to be EVA-2. Because, on this EVA, the face of Mt. Hadley was in the shadow."

Interviewer – "And got sunlit by [EVA] 2."

Irwin – "Yeah. And you could see that [in the short sequence of successful 16-millimeter photography]."

Partly as a result of the 16-millimeter camera's failure on Apollo 15, both Charlie Duke (Apollo 16) and Jack Schmitt (Apollo 17) took Hasselblad photographs spaced at 50 to 100 meters during their traverses so they wouldn't run out of film before reaching a destination. They particularly wanted to ensure enough photo coverage to satisfy geologists who were interested in both the exact path they took and the distribution of rocks and small craters along the route.

Irwin: Boy, it really bounces, doesn't it?

Scott: Well, I think there's sort of a ... The rear end breaks out at about 10 to 12 clicks.

That is, the back end tends to swing around if he tries even moderately sharp turns at such speeds.

Allen: Roger, Dave. It sounds like it's like steering a boat, with the rear steering and the rolling motion.

Scott: And, gosh, every *(lost under Joe)* ... *(responding to Joe)* Yeah, that's right. It sure is. *(to Jim)* Hey, here's a good fresh one right there ...

Irwin: Yeah, I was looking at that one at 1 o'clock to us right now. Very fresh angular blocks of lighter albedo [more reflective] material on the south rim.

Scott: Okay, let's ...

Irwin: We kick up a little dust when we go through these craters.

Scott: Yeah.

Irwin: Seems like when we get to the bottom ... And I can see the trajectory of the fragments coming from the ... It looks like ... Yeah, they're coming from the front wheels and coming up kind of around my arm and then forward.

Scott: Yeah, but it's not dusty. I mean, they're ...

Irwin: No ... No, it looks like millimeter-type particles.

Scott: Yeah. *(pause)* Hang on.

Irwin: Okay, let's see, the distance 1.3. Okay, I think there's a large one coming up about 12:30 or 1 o'clock that could be Rhysling.

Scott: Okay.

Scott – *(chuckling)* "You go to any new place, you want to find something that's familiar. Right?"

Interviewer – "Although, in your case, you've got lovely things to guide you on the horizon: St. George and all that."

Scott – "But also remember, we've never checked the Rover navigation system, so we don't know it's right. It may say 1.3, but we may be at 1.9 or 0.7 because we'd never done this before and there was a question of wheel slippage. You know, if your wheels slip a lot, your odometer's going to turn a lot, and you're going to think you're a lot further than you are. So we don't really have confidence in the Nav system at this stage of the game."

Interviewer – "So you'd really have liked to have found Rhysling to be sure."

Scott – "Find me a place I know, then I know that the Nav system's okay."

Interviewer – "Grant Heiken [co-author of Jones] told me once that if you really blow up the pan camera images [taken from the Command Module in orbit], you can see the Rover tracks."

Scott – "Oh, yeah. The tracks are clear on the pan camera. I've seen pictures of that."

Interviewer – "Even though the Rover footprint is smaller than a resolution element, the fact that you're stringing them together makes the track visible. In those same pictures, the LM is pretty marginal, even though it's a lot bigger than the width of the tracks."

Scott – "Maybe what you're doing is changing the albedo so that you have a swath instead of two tracks. Throwing all that dust around changes the albedo. So, what you're seeing is not the tracks, but the albedo change due to throwing the dust around."

Immediately after the mission, one of the tasks of the field geology team was to locate the traverses and stops, which is important for locating samples and astronaut observations. Before the J-missions, this was accomplished by studying photos taken by the crews and correlating them with the crews' descriptions. The same method was being used after the return of Apollo 15, but there was a lot of uneasiness as to the accuracy of traverse locations. On the Apollo 15 mission, there were also two cameras on the CM, which were used for mapping the lunar surface. One, the panoramic camera, had remarkable resolution. Later, NASA geologists went to the mapping group at the MSC and asked them to enlarge panoramic photos of the site using the greatest resolution that they could achieve. The result was startling: there were visible rover tracks of slightly lighter albedo. As Dave Scott suggests, the visibility was probably a combination of albedo change and the fact that it is sometimes possible to see linear features that are technically below the resolution of the film.

Allen: Jim, that sounds good or it could be the large one to the northwest of Rhysling. Rhysling may be coming up on your left now.

Houston, thinking that the crew landed near November Crater, would be estimating that Dave and Jim are 0.9 grid units south and 2.0 grid units east of their actual location at any point in the traverse. When Dave and Jim get to a known location – Elbow Crater – the Rover Nav readouts will allow Houston to pin down the LM location and remove this navigational uncertainty.

Irwin: Well, there's a large one over there, too, Joe, I ...

Allen: Roger. But your heading is beautiful. Continue on.

Scott: Okay.

Irwin: Our heading's averaging about 200 [to] 210. *(pause)*

Scott: *(laughing)* Man, this is really a rocking-rolling ride, isn't it?

Irwin: Never been on a ride like this before.

Scott: Boy, oh, boy! I'm glad they've got this great suspension system on this thing. Boy. *(pause)*

Interviewer – "The suspension didn't damp very quickly, I gather. So it was pretty bouncy, 'pretty sporty,' I gather."

Irwin – "We had a spring leaf that gave us the support, the suspension. I'm sure Dave would have a lot to say about how it handled and how it felt to ride it. I know there were a couple of times when we'd go up on two wheels – in this time frame – and I thought surely we were just going to roll over. I was concerned that, if we rolled over, would we be able to get out from under it. I'm sure we would, eventually, but it was a concern. With the seatbelt so snug, would we be able to get to it to release it and get out from under the Rover?"

Interviewer – "So were you tipping sideways, occasionally?"

Irwin – "Yeah, 'cause sometimes you'd come up and there would be a huge crater ahead of us and Dave would have to throw the control over to a hard right and we'd just go up on two wheels. And I thought, 'Man, we're just going to be rolling over!'"

Interviewer – "Did you go up the other way with Dave on the downside?"

Irwin – "I don't ever remember that. *(laughing)* I remember I was on the downside. It seems to me, once we went through this indoctrination phase, we always had the throttle full forward. An amazing little vehicle."

That conversation with Irwin took place in 1989; in 1992, Scott and Jones also had a conversation about the ride.

Interviewer – "You said earlier in our discussions that the damping was a lot less at 1/6 g than 1 g. You said that, once it started bouncing, it kept bouncing for a while."

Scott – "But it also may have been caused by the frequency of impacts of the wheels, because there's so much out there. But, as we've discussed, the pre-mission photos didn't tell you how irregular the surface is. It's never straight. Choppy seas. Real choppy seas. But probably more than choppy seas, because they tend to be somewhat regular. It was a bouncy ride. The Rover, being a rigid body, has to damp out this motion, so the rigid movement over the surface is a relative smooth curve – it has to be because it's a big Rover with people in it – which means that, since you're on an irregular surface, the four wheels are all doing extra duty. Independently. So it would be interesting to see the oscillation curves of those four wheels, because I'll bet it's a wild thing."

Scott: Okay, here's a big one right here on our left, Jim.

Irwin: Yes, but I don't think it's big enough to be Rhysling.

Scott: No, I don't think it is either. We got a ridge up here in front of us, we'll …

Irwin: What did I say, Joe, about 1.6, 1.7, for Rhysling?

Allen: It's about 1.4, 1.5, Jim.

Irwin: Get on top of … *(stops to listen to Joe)* That could be Rhysling, Dave; we'll find out when we get up on top of this ridge.

All of the Apollo crews had trouble judging sizes and distances on the lunar surface. Dave and Jim are having a particularly tough job during this part of the traverse because they had no clear landmarks during the descent to help them figure out exactly where they'd landed. In addition, they have not encountered any identifiable landmarks through this first part of the traverse.

Scott: Yeah. *(pause)* By the way, Houston, your comm is crystal clear for us up here.

Allen: So is yours, Dave. Maybe we ought to take this gear to Flagstaff next time.

Joe Allen is referring to Flagstaff, Arizona, where the crews did extensive field training with geologists and engineers.

Scott: Yeah.

Interviewer – "This sounds like there were some comm problems with some of the Flagstaff field exercises."

Scott – "Yeah. I think over the hill sometimes is a problem."

Irwin: Off in the west now, I can see Bennett Ridge.

Scott: Oh, yeah. I've seen it all the way. You can see just the peak of it almost all the time.

Allen: And Rover, this is Houston. Your range to Rhysling is about 1.7; so you may be short of that still.

Scott: It just clicked off 1.7, and our relative bearing's 036.

Scott: And we're coming up on the right ...

DRIVING ALONG HADLEY RILLE

Irwin: Hey, you can see the rille! There's the rille.

Scott: There's the rille.

Irwin: Yeah. We're looking down and across the rille, we can see craters on the far side of the rille.

Allen: Roger. Like advertised.

Irwin: A lot of blocks. *(garbled)* to turn the camera on.

Scott: Yeah. *(pause)* Now we're getting into the blocky stuff, about 1 foot, quite angular, irregular surface.

Irwin: We're right at the edge of the rille, I bet you.

Scott (from the 1971 Technical Debriefing) – "I believe you mentioned that the block distribution – or fragment distribution – did increase somewhat as we got to the rille rim. There were more fragments."

Irwin (from the 1971 Technical Debriefing) – "I'm wondering, was that a function of the rille or a function of craters there? I know that was true when we got to Elbow. There were plenty of rock fragments there."

Scott: Yes, sir. We're on the edge of the rille, you'd better believe it. I think we're heading right …

Irwin: I don't see Elbow though. Oh, yeah, I see Elbow. Dave, we have to stay up on the high part of the rille, here.

Scott: Yeah.

Irwin: See; Elbow is not as prominent as we thought, but there's a definite crater there. I see Elbow.

Scott: Yeah, it's subtle though; subdued.

Irwin: I'd better turn on the [16-millimeter] camera.

Scott (from the 1971 Technical Debriefing) – "We did see Elbow Crater from the side of the rille, quite a ways away. And there, again, I think the distances were somewhat deceiving and that it looked closer than it really was. When we did see Elbow Crater, I felt like we were almost there. Then there was a fair amount of driving before we got there. Everything looked closer; and, as I look at our landing site, relative to Pluton [Crater], I would have thought Pluton was just right around the corner from the site. I think the distances, again, as everybody has said in the past, they're really deceiving up there with no other objects to measure and compare."

Scott: Hey, look there's a big block on the edge of the rille there that must be 10 meters [across]. There are lots of outcrops. But, on the far side, I don't see anything that would suggest really layering. There's a lot of debris, big angular blocks all the way down, but nothing that you'd really call exact layers.

Allen: Roger. We copy.

They are looking for horizontal layers in the far rille wall, which would be produced by successive flows of mare lava. During the EVA-3, they will make two stops on the edge of the rille about 1.7 kilometers northwest of their present position and will take some excellent 500-millimeter photographs of several areas of distinct layering.

Interviewer – "Did you stop at all to look around, or did you just keep going?"

Scott – "I don't remember. I spent what, about 20 seconds on this description. I don't think we stopped, but I don't remember."

Interviewer – "Do you have distinct memories of some of this?"

Scott – "Some of it. Depends on what part of it it was. Another challenge is, as we go through it, I start recalling more, but I haven't had time to review it and, if we could go through it twice, or I could sit and take a couple of days to read through the transcripts and think about it, then more would come back. 'Cause, once we get into a discussion, I start remembering things. But, I have to get into that discussion and get into that memory bank. If we had done this a year afterwards, it would be easy. But it does come back as we go through it. I do start remembering."

The cuff checklist pages have a list of features of interest to the geologists, including skirts of loose soil that build up at the base of rocks, perhaps during impacts. Another item was the "possible ray." It was generally assumed from morphological evidence (from pre-mission photography) that the craters of the South Cluster were formed by the impact of ejecta from one or both of two very large craters – Aristillus and Autolycus – hundreds of kilometers north of Hadley; there was also a suggestion of an associated ray, presumably created by finer grained ejecta from the same large craters.

Irwin (from the 1971 Technical Debriefing) – "We were supposed to be looking for a possible ray, and I saw no evidence on that leg [of the traverse] of any ray."

Scott: Let me get us back up on the ridge, it's smoother.

Irwin (from the 1971 Technical Debriefing) – "You know, when we first saw Elbow, I think we were kind of downslope, down on the rille side of the levee. We saw it, and we went back up on top of the slope. It was smoother driving up there."

Scott (from the 1971 Technical Debriefing) – "You're right. Matter of fact, I think we commented at the time it was better driving back up on the ridge line, or the raised point if you don't want to call it a levee – which, I guess, I agree wasn't really a very profound levee, if it was at all."

The current scientific thinking about Hadley Rille is that it is a collapsed lava tube or channel – a relic of the mare-forming period of lunar evolution. The point on the edge of the rille where the slopes become steep more or less marks the original edge of the tube or channel. Over the billions of years since the end of mare formation, the mare surface – and the rille walls – have been hit by countless impactors, most of them quite small. Away from the rille, these impacts have produced the regolith layer, which is now more than 5 meters thick. Near the rille, some of the larger impacts spray ejecta far enough that the rille-ward portion of the ejecta blanket goes into the rille; consequently, the ground slopes down to the rille, and the closer one gets to the rille, the thinner the regolith and the more exposed boulders and bedrock outcrops one finds. Dave and Jim will make a close examination of the edge of the rille at Station 9 during EVA-3.

Irwin: Yeah, I think that heading was . . . We were on a heading that was a little too far west. We're getting back up on the higher part of the rille rim. At this point, I'd estimate the slope is probably – what? – About 3 degrees?

Scott: Yeah, there's a definite ridge or rim that runs along the rille, maybe 70 [or] 80 meters from this [means] the inflection point that drops down into the rille, don't you think, Jim?

Irwin: Yeah. *(pause)* And, we might as well . . . We're heading right toward . . . We'll head toward the east side of Elbow.

Scott: Yeah, we're in good shape. We can see Elbow, and we can see the Front all the way down to the Spur [that is, Spur Crater]. And, there's not a big block on it. *(pause)*

During their interview, Dave and Jones tried to decide if this is a reference to Spur Crater, to Silver Spur, or to some other feature.

> **Scott** – "Yeah, it is Spur Crater. We're looking at the trafficability. Remember? We're worried about being able to drive along the Front. That's why we did the SEVA [Standup EVA, during which Dave stood on the ascent engine cover with his head and shoulder out the rendezvous hatch, described the scene, and took photos]. So, what I'm looking at is down [that is, eastward along] the Front, so it is Spur Crater."

Allen: Keep talking, keep talking. Beautiful description.

Scott: *(to Jim)* Hang on.

Irwin: I see one large block, about a quarter of the way up the Front, Dave.

Scott: Okay, hang on there.

Irwin: Yeah.

Scott: *(chuckling)* There's a big one [a rock close at hand] partially buried. Oh, there's some beautiful geology out here. Spectacular! *(pause)* Oop, watch out. Hold on. *(pause)*

Geologists have determined that a fresh 25-meter (82-foot)-diameter crater is about 5 meters (16.5 feet) deep and that the absence of blocks excavated from bedrock indicates that the typical regolith depth at Hadley is greater than 5 meters.

Irwin: Looking up at the Front now, Joe, I sure see the linear patterns that Dave commented on before. With the dip and everything.

Scott: Whew, whoopee!

They've hit a small crater, which must have been hidden beyond a small rise. They are driving SSE, so there should be plenty of visible terrain definition when they are driving on level ground or going down slopes.

Irwin: Boy, that was a good stroke.

Scott: It's a good stroke, all right.

Irwin: And, I sure get the impression it's a ... Almost looks like a slump feature, but we'll take some good pictures of that. *(pause)* You see the same linear-type pattern in the east side of the rille. And note the linear pattern there is parallel. Almost like layering in the rille. And, then as you look upslope – up the Front – that layering takes that dip to the northeast that Dave had mentioned earlier.

Allen: Roger, Jim. And can you actually see the east side of the rille, towards the south there?

Irwin: Oh, yeah. I can see ... Looking directly south I can see the exposure that faces northwest. I can look down and I think I can see Hadley C down there!

Allen: Remarkable.

Irwin: Yeah, I think I can see the south rim of Hadley C. *(pause)* Okay, let's see. Well, we can see Elbow. But anyway, when we get there ...

Scott: Hang on; got one coming.

Irwin: Okay.

Scott: *(subvocal)* Oh, my.

Irwin: It should be 2.7 [kilometers to Elbow Crater]. So we got another .7 to go.

Scott: Okay, we're doing all right.

Irwin: Speed's been generally about 10 clicks [kilometers per hour].

Scott: Yeah, but it takes attention paying. *(pause)*

Irwin: Yeah. And again, looking to the south along the edge of the rille that faces to the northwest, I can see several large blocks that have rolled down slope. Very large blocks that are about three-quarters of the way down the slope, into the rille. That's just at the base of St. George.

Allen: Roger, Jim. Copy. *(pause)*

Irwin: And ... Let's see, we're heading about 165 right now. Tried to stay on the fairly level and smooth part of the Rille rim. But looking over to the edge of the rille at this point, I see a large concentration of large boulders – large rocks. And I'd estimate the size ... They're angular, and they're all of the same color and texture as far as I can tell from here. *(to Dave)* See that con[centration]? Well, you'd better watch the road, Dave.

Scott: Yeah. No, I see what you're saying there, but you keep talking; let me drive.

Irwin: Yeah. It's the first good concentration of large rocks that I've seen. Very similar [in size and angularity] to the large rocks that 14 saw up at the top of Cone.

Allen: Roger, Jim, we copy. And your range should be coming up on 3.1, at Station 1.

Scott: Okay, right now, Joe, our bearing is 18 and range is 2.3.

Allen: Roger. *(pause)*

CHOOSING NAMES FOR LUNAR FEATURES

Irwin: Okay; now, Joe, I can see the bottom of the valley – Head Valley – that leads down toward Hadley C.

Interviewer – "Head Valley is [named] for Jim Head [Bellcom, now at Brown University] and it's nicely disguised so the International Astronomical Union (IAU) would never tumble to it."

Scott – "It wasn't intentional. All of the IAU exercise occurred after the mission. And they complained about everything. We had Head Valley, Silver Spur, the Swann Range, Schaber something. Bennett Hill. [These names are traceable to Jim Head, Lee Silver (California Institute of Technology), Gordy Swann, and Gerry Schaber (both from the USGS) all members of the astronauts' geology training team.] I told you that story, didn't I? If we overshot the rille, we would land on the west side and we called the mountain over there Bennett Hill, after the guy who designed the landing trajectory for us, Floyd Bennett [NASA MSC]. Our landing trajectory was twice as steep as the others, so we spent a lot of time with him, and he with us. He put so much effort into this – making sure that we didn't end up on the far side of the rille or that we didn't run into the mountains – we decided to name this prominent feature Bennett Peak. So there are a lot of scattered names around there."

The chairman of a nomenclature committee of the IAU, Harvard astronomer Donald Menzel, had a running battle with NASA over names for lunar features and frequently demanded that NASA submit lists of proposed names prior to the flights. Those requests were largely ignored.

Scott – "Although we were probably aware of this, we really did not give it any consideration in selecting names. Did you know that Hal Mazursky [USGS] did a long letter to the International Astronomical Union to get the names on Apollo 15 approved? And got turned down."

Interviewer – "I've heard a similar story on 17. Tell me the 15 tale."

Scott – "Mazursky was a member of the IAU, and he prepared a letter which he submitted formally, to get official recognition of all the names. And the head of the IAU wrote a response, which is marvelous! Because he turned down every name for a reason, and he wrote a paragraph or so on every name and why it was unacceptable. It's amusing, it's interesting, it's educational. And Mazursky's [response], as I recall, was excellent. He wrote down all the names and the rationales for each name. And then the one he got back was like … I mean we had one crater called Nameless Crater, and the guy wrote back, 'You can't have a Nameless Crater, 'cause that's no name. And if I were naming it, you should name it this. You should have a name.'"

Scott – "Hal passed away a number of years ago. Great guy. There's another one of those marvelous professors. Inspirational professors. I could listen to Hal Mazursky all day and all night. And he'd come to the Cape [Canaveral], and he'd come to the crew quarters and he'd tell us about the Moon. Spellbinding. Absolutely spellbinding. Marvelous guy."

Interviewer – "One of my purposes in preparing the Journal is to make sure that the names that you folks decided were appropriate get thoroughly recorded and used. My personal feeling is that it's always the privilege of the explorers and/or pioneers to do the naming. As long as astronomers were exploring the Moon with telescopes, then it was appropriate for them to do that. But as soon as somebody landed, the responsibility shifted to the people who landed and the IAU could go to hell."

Scott – "I've never heard that approach. That's a good approach."

Interviewer – "And once there are folks living in the area, they can name things that you didn't name. Dot Crater at Apollo 16 is named for Charlie's [Duke's] wife and everybody should know that. And Punk for Gene's [Cernan's] daughter Tracy. So as we come along to names and you remember why certain names were chosen, let's talk about it."

Scott – "And there was a reason for having names rather than numbers. Because I can remember a name with a character, how that character looks with the name. I get lost in numbers. If this is Crater 46 or 48 or 72, it doesn't mean anything to me. I need to know the name, and then the crater will take on a character, like a person. So you get to Dune Crater. That is Dune. Boy, you look at it and it's Dune. You know, we did a lot of ours [based] on science fiction, because that was fun. It brings it to life, because there's some relationship. The names are very helpful when you're trying to make your way around the Moon."

Interviewer – "Were you a science fiction reader? As a kid?"

Scott – "I frankly don't remember when I did it, and I probably wasn't that deep. But I did like it, and I did find that it was fun to bring all of these things out when we were doing geology and, as a by-product, naming craters. And why name craters? And the science fiction was a good reason why, because there were a lot of relationships that you could mentally tie in. And you're probably a science fiction reader."

Interviewer – "Yes. So Rhysling, the Blind Poet in [Robert] Heinlein's *Green Hills of Earth*, and *Earthlight*, and *Dune* and all those were immediately obvious to me."

Scott – "Heinlein is one of my favorites; *The Green Hills of Earth*, boy, that's a great story. And [Arthur C. Clarke's] *Earthlight*. That's still a great story. In fact, I still give copies to people once in a while, because that's [a story about] the first lunar base. And that's [located] at Hadley. And that's a great story, not only because it's entertaining, but because it could be a very valid forecast. Of course, Arthur Clarke does great work. And there's a real message in that story, which people should pay attention to, especially today with the things that are going on. When I talk to people in Washington about the lunar base, I tell them 'You're going to have to take care of it, protect it.' It's going to happen. You're going to have a lunar base. And all those things you don't pay any attention to now, if you pay attention to them now they'll help you out when you get to your lunar base."

Interviewer – "The crater naming was a cooperative venture between you and Jim and Joe ..."

Scott – "We did most of it at the Cape during our post-dinner geology sessions. Jack Schmitt was a major participant in one of them. Do you know why St. George Crater has it's name? 'Cause one night we were discussing what we would find at that crater – anorthosite or whatever. And Jack got in a heated discussion with somebody. [see explanation below] I forget who it was, Lee Silver or [Gordon] Swann or somebody. But, you know, when Jack takes a position, he takes a pretty hard position. And they got down to 'I'll bet you.' 'What do you want to bet?' 'I'll bet you a bottle of wine.' 'Okay, I'll bet you a bottle of wine, what kind of wine?' And somebody says, 'You know, one of Jules Verne's characters [Michel Ardan in Chapter 3 of *Around the Moon*] took a particular bottle of wine called Nuits-Saint-Georges to the Moon. Therefore, why don't we name the crater St. George Crater, after Jules Verne. And that's why it's St. George."

[Jones' later note – "Although I still do not know with whom Jack made the bet about anorthosite at St. George, Grant Heiken (co-author of this book and editor with David Vaniman and Bevan French of the marvelous *Lunar Sourcebook*) has provided a copy of an original, signed document that formalized a series of bets Schmitt made with Robin Brett (MSC) about the geologic material Apollo 17 astronauts would find."]

Scott – "That's what brought geology to life. That's the kind of thing that made it fun, it made it real, and it tied it in with other things. It got into the emotion of the time, which was good because it was great to hear somebody like Schmitt debate with somebody else [about] what was going to be there, because the rest of us learned a lot from that. You listen to a debate, you learn a lot, right? So those kinds of things were a very meaningful part of the methodology. How did we learn how to be quasi-geologists on the Moon? Because of things like that. How could I ever forget that Schmitt had this argument about anorthosite at St. George? I can't forget that. So I get up around St. George, and there is a [mental] hook and, hanging on that hook, is a bottle of wine. And, as a total sidelight, after the mission was over, I got a package from George Low [NASA Deputy Director during Apollo]. In the package was a cork, from a bottle of Nuits-Saint-Georges that he had on the night we landed on the Moon, and he'd signed it. Not bad, huh?"

Interviewer – "The publishers we've approached are all afraid of the size of this thing [the ALSJ]. Not a one of them doubts the proposition that, 300 years down the line, there are going to be some treasured copies at the community library at Hadley or where ever it is. And, there, people can read what you just said, and look out the window, at St. George, and know about Jack's argument and your mental hook. That's the kind of stuff that has to be in here."

Scott – "Absolutely. That's what makes it human. We can send robots, but that doesn't make it human. It doesn't give you the vigor."

In another anecdote, ALSJ contributor Harald Kucharek notes that the village of Nuits-Saint-Georges is justifiably proud of its double connection with the Moon. In 2001, Dave Scott wrote in an email that he and Jim and Al Worden made a visit there in conjunction with their attendance at the 1971 Paris Air Show. Dave writes that Nuits-Saint-Georges is "a marvelous little town, with marvelous wine, marvelous food, and marvelous people!!!" He says that they were able to "enjoy the 'fruits' of our labor . . . [and were] . . . honored by being made citizens of Nuits. Just wish we could go back for another round!!!"

Irwin: I can see the bottom of the rille. It's very smooth. I see two very large boulders that are right on the surface, there [that is, they are unburied] – on the top of the very smooth portion, of the bottom of the rille. And the one to southeast, I can see the track [on the rille wall] of where it's rolled downslope.

Allen: Roger, Jim. Copy. And is the bottom V-shaped or fairly flat?

Irwin: I'd say it's flat. Well, it's hard to estimate. I'd estimate maybe, oh, 200 meters wide of a flat area in the bottom.

Irwin – "[In 1971] most of the scientists were saying the rille's probably a collapsed lava tunnel. I don't know if that's still the theory or not."

Interviewer – "I think that's still the favorite one. There's some discussion that it might have been a fault zone; but, then, if you've got weak zone, it's also a place where you could have a channel for the lava to flow."

Figure 4.6. This site, near the margin of Hadley Rille and on the flank of Mt. Hadley Delta, is about 60 meters above the mare surface; the view is north and "downstream" in the direction lava once flowed along the rille. Here, the rille is 1.5 kilometers wide and 350 meters deep and has made a sharp bend from its west-to-east trend at the base of the Hadley Delta, beginning a south-to-north trend in the general direction of Hill 305 – a feature that dominates the far left horizon. The largest rocks visible at the bottom of the rille are about 15 meters in diameter. At this point, the crew (Dave is at the Rover) is nearly 4 kilometers from the LM. The Rover allows Dave and Jim to carry far more gear and many more samples than if they had been on foot, travel at roughly twice the speed (10 kilometers/hour by driving *vs* 5 kilometers/hour by running), and rest between geology stops. The boulder on the left may have been thrown to this location from a group of craters called the South Cluster at the base of Hadley Delta. This partial panorama was assembled by David Harland, author of many books about space exploration and contributor to the ALSJ).

APPROACHING ELBOW

Irwin: Oh, and I can see what we thought was Bridge Crater. And it definitely would not have been a place to cross Hadley Rille. It's just a depression in the west wall of the rille.

Interviewer – "If you had landed long – on the other side – would you have stayed on the other side?"

Scott – "I don't recall. Maybe that was the bridge to get back to St. George."

In 1989, Jim was also unsure if they had ever talked about the possibility of driving across the rille from a landing site on the west side. This suggests that it was never more than a point of casual discussion, if that.

Scott (from a 1996 draft review) – "We did discuss this possibility and that is the reason for the name. But, remember, the photos only had about 20 meter resolution – not really very good for serious planning."

Irwin: And I ... Boy, at this vantage point, there's sure a lot more blocks exposed on the ...

Scott: *(laughing)* Yeah!

Irwin: ... on the far side of the rille. I'm contrasting now the rille to the southeast ...

Scott: Oop, hang on, Jim.

Irwin: Okay. *(picking up his comment again)* ... and the side of the rille to the northwest.

Allen: Roger, Jim. Copy all that, loud and clear. And Dave, are the front wheels wandering off of straight ahead as you drive along there?

Scott: No, they're okay, Joe. It's just [that] there are a lot of craters and it's just sporty driving; I've just got to keep my eye on the road every second.

Allen: Roger. We understand that ...

Irwin: By golly, it's a real test for the Rover.

Allen: ... just want to get some engineering information here. Apparently, your front wheels are tracking straight ahead, is that correct?

Scott: That's correct. And, of course, when we turn, they dig in, and it makes the rear end break out [that is, start to swing around]. But it's okay; we can handle it.

Allen: Knew you could.

Scott: I might add to Jim's comment, that the near side of the rille wall is smooth without any outcrops, there by St. George, and the far side has got all sorts of debris. It almost looks like we could drive down in on this side, doesn't it?

Allen: Stand by on that, Dave.

Irwin: I'm sure we could drive down; I don't think we could drive back out *(they both laugh about Joe's warning)*.

Scott (from a 1996 draft review) – "This was a typical Joe Allen quick response – in a humorous vein – since we had discussed driving down into the rille many times before the flight, but not really seriously."

The wall beneath St. George is probably covered with St. George ejecta and/or other fine material that has tumbled down from Hadley Delta, piece by piece. Beyond the west wall, there is only the relatively flat mare; apparently, there has been insufficient erosion to completely bury outcrops in the steeper parts of the western rille wall.

Irwin: Oh, now I can turn around and look to the northwest – where the rille trends to the north. But I'll ... let me concentrate on Elbow for the moment.

Scott: Yeah, let's get to Elbow. *(pause)* *(Jim laughs)* Hang on.

Interviewer – "I had the impression you could turn your head in the helmet a fair bit ..."

Scott – "We did have the side panels. But he says he could turn around. He could probably turn around in the seat. Depends on the angle you're looking at and the angle you're driving."

Interviewer – "Sitting here in my chair without moving anything but my head, I can get around to 90 degrees and, with peripheral vision, get around quite a ways back."

Scott – "But he might have meant he could turn the suit."

Interviewer – "Even with the belt on?"

Scott – "Yeah. I think you could probably turn the suit a reasonable amount – if you're not driving. I mean, the belt keeps you here ..."

Interviewer – "You could really rotate the torso?"

Scott – "I don't know that I ever really tried. This is the first time I've ever discussed it. There was no joint, at that time [for that motion], but you could make it bend, twist if you will."

Interviewer – "Especially if you could grab something outboard with one hand and pull yourself around a little."

Interviewer Jones asked Jim Irwin about this issue as well.

Interviewer – "Jack gives me the impression that it's almost impossible to move the suit, itself and, consequently, to move the helmet. You can move your head inside the helmet, but then you're blocked from looking past about 90 degrees."

Irwin – "Well, I must have been able to look off to the right, there, to look northwestward."

Interviewer – "Would you agree with Jack's statement that you would not have been able to twist the suit?"

Irwin – "I don't think you actually twist the suit. *(pause)*"

Interviewer – "So it was a matter of the Rover being turned enough so that that direction came into your field-of-view."

Irwin – "Yeah. I'm not implying I was able to move my upper body that far, but I guess I could move it slightly and look in that direction."

Another possibility is that, by leaning his head forward and then turning it, he might have been able to look a bit more toward the rear.

Scott (from a 1996 draft review) – "You might have been able to 'bend' the suit – not 'twist' – from the shoulder to the waist if you could hold onto something. Perhaps video of the work at the ALSEP would show something."

Irwin: Okay, the range estimate must have been off for Elbow. Our map says 2.7 [kilometer range from the planned landing site to the east rim of Elbow] ; Joe said 3.2, I guess. You confirm that, Joe?

Allen: Our estimate, Jim, was 3.1 from your landing site [that is, from the presumed landing site near November Crater].

Irwin: I see, that's right.

Scott: That's the difference.

Irwin: Yeah.

Because they landed 600 meters north of the planned spot, they won't get to Elbow at a LM range of 2.7 kilometers, but rather at a true range of 3.3.

Scott: *(laughter)* Well, this is really a sporty driving course. Man, oh, man, what a Grand Prix this is! *(pause)* There's old Elbow. Isn't it? It's the real fresh one down here.

Irwin: No, Elbow's larger than that.

Scott: Yeah, but there's … Yeah, there's a nice fresh one there.

Irwin: Yeah, but you want to go a little farther east. See, that's Elbow out at 11 : 30.

Scott: Oh, yeah. Rog. Gosh, that's a long way away.

Irwin: Yeah.

Scott: Distances are very deceiving! [It feels] like we've been driving for an hour. *(pause)* Are you sure that's Elbow, Jim?

Irwin: Yeah. Yeah, you want to go farther east, Dave.

Scott: Okay. Down this little crater; back up. *(pause)*

Interviewer – "I take it you decided that you could drive down into the shallow ones?"

Scott – "Yeah, the vehicle's very mobile. It's just that with all the blocks all around, you had to watch it, every second."

Interviewer – "And make a judgment if they were shallow enough to go down in them or whether you wanted to detour?"

Scott – "I don't think there was much detour. I think the detour was around the blocks and the debris. I think we could motor right on through the craters. The fresh ones with all the debris we might have gone around, but the shallow ones, the Rover just moved right along, briskly. It would be interesting to get pictures of the tracks, to see how we traversed going down there."

Irwin: See, that's Elbow out at our 1 o'clock position.

Scott: Shoot, this is Elbow right here, I believe, my friend.

Irwin: Yeah, this is Elbow right here.

Scott: Yeah.

Irwin: Yeah, this large one.

Scott: The one we were just trending into.

Irwin: Yeah.

Scott: Yeah, that's some big fellow, isn't it?

Irwin: Yeah, it sure is. I don't know …

Scott: Take a look up here [for a parking spot], and we'll see how she looks.

Irwin: Maybe you can, you know, angle up hill here?

Scott: Yeah. *(pause)* How are we doing on time there, Houston?

Allen: Like gangbusters, Dave and Jim. Continue on, and we'll give you the exact number in a minute.

Interviewer – "You're at Elbow Crater and it's been 2 1/2 hours since you opened the hatch."

Scott – "A lot of overhead."

Interviewer – "But it's overhead that should be spread over three EVAs."

Scott: Okay. Do we want to stop at Elbow, or press on?

Allen: Stop! Follow the Checklist, just as planned.

Scott: Just as planned, okay. *(pause)* Okay, let's go right up on the ridge line there, I see some debris. Maybe we can get a fresh one in the rim [that is, find a small, fresh crater that has punched into the Elbow ejecta]. Be looking down-Sun [into Elbow with the TV]. *(pause)* Oh, look at this baby climb the hill.

Irwin: Yeah, climbing at about 8 clicks

Scott: Yeah, man. *(pause)*

Allen: Jim, can you get an Amp reading for us as you climb?

Irwin: Okay, Elbow's out there ... *(answering Joe)* Yeah, reading ... Oh, it's just about. It's 10 on Batt 1, Joe.

Allen: Roger.

Interviewer – "What they're doing, I presume, is trying to get a reading on Rover performance on this relatively modest slope in anticipation of the climb up to Station 2."

Scott – "I don't think they know we're going to climb to Station 2. See, they don't know the elevation [difference], really, because the photography was never that good. I think, when we got up the side of the mountain, we were all surprised at how far up we were. I don't believe anybody knew. I mean, we were going up to the Front but, with the photography we had, there was not a good relief map anywhere."

The EVA-1 contour map in the Apollo 15 Lunar Surface Procedures volume indicates only a modest climb of 10 meters from Elbow to the planned Station 2 location. Dave and Jim will actually make their Station 2 stop quite close to the planned location but will have a much steeper climb than anticipated.

Interviewer – "Was the lack of a good relief map due to the fact that the Lunar Orbiter high-resolution stuff was almost all in the equatorial band?"

Scott – "No. Hadley C had good, high-resolution stuff, but we had the 20-meter stuff. As we've discussed previously, one reason we didn't know where we were was [the fact that] the maps weren't good enough. They didn't have anything to work with; 20-meter resolution isn't very good. You can't identify many craters that way. So I doubt that there was much good understanding of the relief."

"I think it surprised everybody that the Rover could perform that well going up the hill [to St. George]. That's why we say we're going up and somebody in the backroom says, 'Boy, we better check and see if it's dragging the battery down.' Because 10 degrees is steep. But, in fact, the slope's probably not dragging the battery down and the Rover's doing a great job. The wheels have great traction."

"We ended up going up slopes that we don't realize were there. And I remember that clearly. Especially later on [during EVA-2]. All of sudden, you get off on this ... it feels like a cliff – which it isn't – it just feels that way. And we had no idea we were going up. 'Cause there's not much force. You're not leaning back or anything in 1/6 [g] ... You don't notice the lean. And you're focused on the direction and there's not a real horizon around. There are no vertical trees. As far as it feels, you're on a flat surface, going level, when, in fact, you're going up pretty steep.

If you go down into a crater, you can see you go down and up. You can feel it. [But not on a steady slope.]"

In the 1971 Technical Debriefing, Jim mentions that east rim of Elbow was "very subdued." Dave added that "there wasn't a raised rim at all." Dave and Jim have been driving along the "ridge" that runs parallel to the rille but were probably still below the level of the general mare surface and, here, are climbing the ejecta-covered slope on the northeast side of the crater up to the mare surface on the east side of the crater.

> **Irwin:** We got a good slope here about, I'd say 10 degrees; we're going up right now [to the Elbow rim].
>
> **Scott:** *(garbled, probably as they hit a small crater)*
>
> **Irwin:** I felt it.
>
> **Scott:** Did you feel that?
>
> **Irwin:** Okay, now we're up on the high part, and we're on the east rim – east rim of Elbow.
>
> **Allen:** Stupendous.
>
> **Scott:** Okay, this ought to give the folks back home something to look at. Right here. Okay, we're at our first stop. Okay. Power this beauty down. *(pause)*
>
> **Irwin:** And, Joe, here's some readings for you.
>
> **Allen:** Roger.
>
> **Irwin:** [Rover heading] 185, [bearing to the LM] 011, [distance traveled] 04.5 [kilometers], [range from the LM] 03.2, [amp-hours remaining] 105, 112, [battery temperatures] 085, 087, and … Gee, I'm reading the lower limit on the Motor Temps, both Forward and Rear. Don't look like that gauge is operating.
>
> **Allen:** Maybe they're still cool.
>
> **Scott:** Let's hope so. Okay, Joe, I'll give you FM/TV here.
>
> **Allen:** Roger. *(long pause with static)*
>
> **Scott:** *(static fades)* Okay, Joe. High gain's pointed. And we've got a fair amount of dust on the Rover. Very light, thin [cover of dust]. *(long pause)*

Houston starts receiving TV. Because of the complexities of the communications path that connected the Rover on the Moon to Houston, it takes a bit more than 10 seconds after Dave gets the high-gain antenna pointed at Earth before Houston actually sees a TV picture. The first images from Elbow show that the camera is pointed down and aft, giving us a view of the left-side battery cover. Details are relatively indistinct because of dust on the camera lens.

> **Allen:** And, Dave and Jim. We gained 20 minutes back. We're making money hand over fist on your driving.

Joe is mistaken, but later acknowledges his error. The drive took the expected 26 minutes. Because they landed farther north than planned, the average speed of the traverse, 3.2 kilometers in 26 minutes (7.4 kilometers/hour), was higher than the expected speed of 2.7 kilometers in 26 minutes (6.2 kilometers/hour) by 20 percent.

> **Interviewer** – "In some ways, [Apollo] 15 was the last of the development flights. Because you added the last major element to Apollo – the Rover."
>
> **Scott** – "The suits, too. There were two steps. The first step was doubling the amount of time you could spend outside in the suit. Big step. And the second step was putting the Rover on. And I would weigh them about the same, in terms of importance."
>
> **Interviewer** – "Now, the original PLSS had about a 6-hour capacity. That's a 4-hour EVA with plenty of margin. Whereas you guys had about a 9-hour capacity in the PLSS – a 50 percent increase – with the same 2-hour margins."
>
> **Scott** – "And there's really more than a 50 percent increase, because you'll have a fixed amount of overhead getting started, with either configuration. Which means that you have a much bigger increase in productive time. So, if you had a 6-hour PLSS with 3 hours of productive time, and you went to a 9-hour PLSS you'd have 6 hours of productive time, because the overhead would be no more. So the benefit of extending it was all productive time. And that was a major step, in my opinion, because it multiples the Rover performance. Without the step in the suit, you wouldn't have the benefit in the Rover that you had. So, if you try to say which was more important, I think I'd have them both about the same. It's a qualitative judgment."

During this first EVA, Dave and Jim demonstrated beyond doubt the remarkable increase in productivity that the Rover and the new backpacks made possible. It was a confidence builder and allowed the explorers additional time for observation and sampling. A bit over 4 hours into the EVA they were back at the LM, having driven a total of 10 kilometers and having spent well over an hour doing geology at Elbow Crater and on the nearby slopes of Mt. Hadley Delta below St. George. Their next job was to deploy the ALSEP array, and they had every expectation of getting that job done before the end of this first EVA. The ALSEP deployment proved troublesome – especially the lunar drill that was new to this mission – and, unfortunately, it cost them one-half an EVA's worth of Rover time.

Drilling into the lunar regolith at the boundary between the highlands and the mare surface, Apollo 15 was to collect a continuous sequence of layers that would help

Figure 4.7. This the final resting place for the Apollo 15 Rover, which was left parked on the mare surface between the heights of Mount Hadley and the Hadley Rille. Dave positioned the Rover about 100 meters east of the LM so that everyone following the mission could watch the LM liftoff, as captured by the Rover camera. The Rover proved itself worth every cent that had been spent to build it, and now will remain as it is for millions of years – unless it is hit by a stray interplanetary projectile.

provide scientists with a record of the Moon's history. However, this first effort in drilling proved difficult and frustrating for Dave Scott as he struggled with the newly designed equipment, confusing documentation, and the nature of the regolith itself.

The lunar module is alone and isolated against this mountainous backdrop that rims Mare Imbrium. This photo was taken from the end of the southernmost traverse, nearly 5 kilometers south of "home" – the LM. Nearly 35 years after the mission, this 500-mm photo is still Dave Scott's favorite.

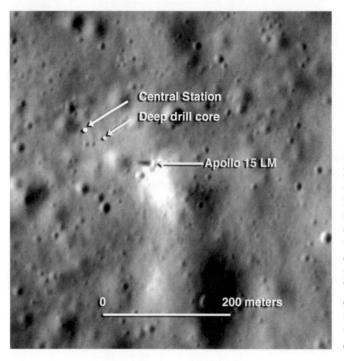

Central Station
Deep drill core

Apollo 15 LM

0   200 meters

In this close up orbital photograph of the Apollo 15 site, it is possible to see the shadow of the LM and an area of the regolith that was disturbed by the descent engine (the white patch south of the LM). The deep drill core was collected in a flat, regolith-covered area without any large, fresh craters.

5

Drilling Troubles

The driest, bleakest desert on Earth is a lush paradise when compared with the surface of the Moon. There is virtually no atmosphere, no water, and no sign of life.

Before there were footprints on the Moon, before there was a space race to emplace the first of those prints, and before there were rockets big enough to put even tiny objects into orbit around the Earth, visionaries and science fiction writers talked about trips to the Moon, and artist Chesley Bonestell (1886–1986), whose space art first appeared in *Life Magazine* in 1944, showed how it might look. Bonestell's Moon was angular and edgy. Although he knew better – knew from telescope observations that the mountains of the Moon are rounded, not jagged – Bonestell wanted dramatic backgrounds for his depictions of lunar exploration. It was the Bonestell image of the Moon that inspired many of those who made Apollo happen.

The real Moon's surface has a softened look, often likened to a landscape draped in snow. At the Hadley landing site, Jim Irwin said that he was reminded of Sun Valley Idaho (later, after he'd been invited to visit Sun Valley in thanks for that plug, he wished he'd mentioned some other world-class ski resorts). Although the Moon has no appreciable atmosphere and neither wind nor rain or even cycles of freeze and thaw, it has a weathered appearance, which is the result of eons of sandblasting. And the "snow" is a cover of mostly fine debris – a regolith that has been chipped and blasted out by countless impacts.

Early in their history, the Earth and Moon shared their part of this solar system with a large population of smaller objects – fragments, actually, left over from the formation of the planets. Impacts by the largest fragments have shaped major features such as the relatively dark mare basins that make up the appearance of a "man in the Moon," and many of the craters were later modified by volcanism. The mountains and ridges that surround these basins were thrown up in the same massive events.

Every impact with the Earth or Moon reduced the fragment population until, about 4 billion years ago, the Earth and Moon had swept up enough of the larger fragments to end the catastrophic activity that shaped the larger features of the

Moon's surface. Since that time, the rain of meteors has continued at a diminishing rate. The remaining fragments range in size from a few kilometers down to specks of dust finer than human hairs. They come at the Moon from all directions and almost always at high speeds.

Because the Moon has no atmosphere, even the smallest objects aren't slowed as they speed toward impact. When they hit, the rock and soil closest to the point of impact are melted and material farther out is shattered and pulverized. Large impactors are relatively rare and small impactors are far more numerous; therefore, most of the work is done by the smaller objects: combined, they produce craters that range from the size of a football stadium down to those that are literally microscopic.

Each of these impact craters – large and small – is surrounded by thrown-out debris. The overlapping ejecta from successive impacts have gradually covered the Moon with a layer of regolith, littered with boulders, that has an average thickness of a few meters.

At first glance, this slightly cohesive, dark- to light-gray "dirt" that covers the Moon seems not much more than a dusty nuisance, clinging to all surfaces, soiling space suits, and abrading moving parts. Although it is composed of fragmental material and the top few centimeters are soft and fluffy, deeper down the collection of particles has been shaken so often by impacts that the particles have settled into close contact with each other making the subsurface firm and quite compact.

From an engineering perspective, the compact nature of the regolith below a few centimeters depth may be its most important property; it certainly will be a great help in road-building for a future lunar base (as well as a source of construction material). However, it was a great hindrance to the Apollo crews when they tried to drill into it for core samples and placing scientific instruments below ground level.

From a scientific point of view, the overlapping layers within the regolith provide a rich history of impact activity, volcanic eruptions, and even the intensity of solar and cosmic particles. Unlike the geological evidence on Earth, the regolith's record is undisturbed by floods and windstorms and, therefore, offers a reliable means of evaluating the ebb and flow of particles and radiation in space through time (at least over the last 4 billion years). Layers in the regolith generally range from a few millimeters to tens of centimeters in thickness, and they are distinguished by changes in particle size and color.

Each layer contains clues to the origin of the lunar surface visible today; in fact, the regolith is a bit like the Rosetta Stone for lunar scientists: by understanding past events that shaped the Moon's surface, they can understand and prepare to deal with the hazards that must be considered in the design and operation of future lunar bases.

Apollo lunar surface crews probed the mysteries of the regolith in several ways. As the astronauts worked outside the LM, they disturbed the soil with their feet while setting up scientific equipment and collecting samples. Sometimes they uncovered material that was different in color than the surface material, and by using a long-handled scoop to dig trenches, they obtained soil samples from various depths. However, because the soil is so compact at depth and because it was so difficult to manipulate the scoop while wearing a pressure suit and gloves, it wasn't possible to dig a trench more than about 20 centimeters (8 inches) deep. To get deeper samples, the

Apollo 12 and 14 crews hammered thin-walled core tubes to depths of as much as 65 centimeters (just over 2 feet). After gently pulling these "drive core" tubes out of the ground, they capped the ends to preserve any existing layering for analysis back on Earth. To get deeper core samples, the Apollo 15, 16, and 17 crews used an electric drill and hollow drill stems to collect intact sections of the regolith to depths of nearly 3 meters (9.8 feet).

Dave Scott and Jim Irwin (Apollo 15) were the first crew to try using a drill on the Moon. In addition to drilling a deep hole and collecting a core, they also planned to drill two other deep holes and emplace instruments designed to measure the rate at which heat flows out of the Moon. The drilling tasks proved to be surprisingly difficult and frustrating but, in the end, the data gathered were worth the effort. Thanks to important lessons learned from Apollo 15, the drilling tasks went more smoothly on Apollos 16 and 17. Even 30 years after the last Apollo mission, scientists around the world are still extracting information from the three drill cores collected.

FALCON DRILLING COMPANY

The crew of Apollo 15 – CDR Dave Scott, CMP Al Worden, and LMP Jim Irwin – was the first all-Air-Force crew to fly to the Moon. In honor of the U.S. Air Force Academy's mascot, Scott and Irwin named their lunar module *Falcon*.

Dave Scott is tall, physically strong, outgoing, and a natural diplomat. He was a member of the third group of pilots chosen for Apollo and, as Neil Armstrong's co-pilot on Gemini 8, was the first of his group to fly in space. A stuck thruster cut that mission short but won both astronauts high marks for their professionalism in a very dangerous situation. He was one of the few pilot-astronauts who developed a real interest in geology and later, as Director of NASA Dryden Flight Research Center at Edwards Air Force Base in the California desert, he devoted some of his spare time to geology.

Al Worden, who stayed in lunar orbit while Dave and Jim were on the lunar surface, collaborated with geologist Farouk El Baz in naming the CM after the barque Endeavour, which Captain James Cook sailed on his first voyage into the Pacific. In 1974, Al published a volume of poetry titled *Hello Earth*, composed of his post-flight reflections.

Jim Irwin was selected as an astronaut in 1966; as Apollo 17 CDR Gene Cernan said, "Jim was one of those guys who, when you look around the room, you say 'He's a shoo-in, why am I here?'" Less gregarious than his fellow EVA crew member, Jim seemed content to defer to the mission commander in most matters and let Dave handle most of the discussions with Houston, but there was never any doubt as to his considerable capabilities.

THE HADLEY LANDING SITE

The site chosen for the Apollo first landing, by Neil Armstrong and Buzz Aldrin, was on as smooth a piece of ground as the Moon can provide. Although the site was littered with craters – and Armstrong had to take manual control of the lunar module to miss a crater the size of a football field for which they were headed – by lunar standards the area around Tranquility base was level and featureless. The Apollo 12 and 14 landing sites were hardly photogenic but demonstrated that, with targeting updates fed into the onboard computer during the descent, it was possible to land within 100 to 200 meters (330 feet) of a specific target. That proven capability allowed NASA to send the last three Apollo crews to locations that were more challenging to fly into but promised great scientific returns.

Dave and Jim landed at a stunningly beautiful site called Hadley. This spot is in a "bay" on the southeastern fringe of Mare Imbrium called *Palus Putredinis* (Swamp of Decay), hard up against the spectacular Apennine Mountains. The landing site itself is on a lava flood plain, which is bounded on the east by the 4500-meter (15,000-foot)-high Mt. Hadley and a series of lesser peaks named the Swann Range after Apollo geology team leader Gordon Swann. On the west, the site is bounded by Hadley Rille, a winding canyon 1500 meters (4900 feet) wide and 400 meters 1400 feet) deep. This canyon was probably carved by very fluid lava flows that erupted late during the filling in of the Imbrium Basin, about 3.3 billion years ago. On the south, the landing site is bordered by 3300-meter (11,000-foot)-high Mt. Hadley Delta, and on the north, by some hilly terrain and a collection of large craters, collectively called the North Complex.

On the earlier missions, the LM's final descent path was a shallow 15 degrees, which ensured a smooth, relatively safe landing and a minimum use of propellant. However, in order to clear the peaks east of the Hadley site comfortably and still land short of the rille, the Apollo 15 descent was a much steeper 25 degrees. The chosen target was about 1.5 kilometers (1 mile) east of the rille and about midway between the North Complex and Mt. Hadley Delta. Because the Rover allowed them to drive as far as 8 to10 kilometers from the LM, Dave and Jim had planned excursions to the Rille, to the lower slopes of Mt. Hadley Delta, and to the North Complex.

Dave was particularly eager to make the trip to the North Complex, where some geologists thought Schaber Hill (named after a member of the geology team) might provide proof of lunar volcanism.

THE STAND-UP EVA

The Apollo 11, 12, and 14 crews donned their backpacks for their first EVAs soon after landing. However, the timing of the Apollo 15 launch and Houston's desire not to disrupt their sleep schedule meant that, by the time they landed, Scott and Irwin had been awake for 11 hours. If they had tried to fit in a full 8-hour EVA, they would have put in a 26-hour day before they got into their hammocks. Consequently, they spent the next several hours working inside the LM and then rested for 8 hours before going out for the first time.

During those several hours, Scott and Irwin were busy giving the scientists back in Houston a thorough description of the surrounding countryside. Rather than restrict themselves to the views out the forward-facing windows, they donned their helmets and gloves for a SEVA, which Scott now wishes they had called a "Site Survey." Within 2 hours after the landing they were ready; they bled all the air out of the cabin and then Scott opened the overhead hatch. After moving the docking hardware out of the way (a daunting task in the tight confines of the LM), Scott stood on the ascent engine cover with his head and arms outside the spacecraft, bracing himself in the opening as he took photographs. By standing up in the hatch, Scott had a clear view all the way around the horizon.

At ground level, the rolling nature of the terrain was even more evident than it had been during the approach and – as might be expected from the lack of deep, fresh craters in the area – in the near field Scott could see no rocks bigger than a few inches across. "Trafficability," he said, looked "pretty good." They might be in for a bouncy Rover ride but, otherwise, it didn't look as if they would have any trouble.

In the far field, Scott had a clear view of the mountains and, as far as he could tell, the slopes were remarkably smooth. On Silver Spur, a feature that looked like a hogback ridge on the eastern flank of Hadley Delta, Scott saw lineations that he thought might indicate either structure or layering, but neither there nor elsewhere on the slopes could he see any large boulders. With no haze to obscure the view, his ability to pick out detail was limited only by his eyesight. Off to the north at a similar distance, he could clearly see rocks blasted out of the bedrock at 800-meter-diameter Pluton Crater, but on the mountains, the slopes were smooth.

After the SEVA, Dave and Jim drew shades over the LM windows to block out the sunlit landscape outside and slept soundly – but not as long as they would have liked. The evening's tasks had taken about an hour longer than planned and in the morning, Houston had to wake them up an hour early to check out a small oxygen leak. They soon traced the leak to an unclosed cap on the urine disposal line and fixed it quickly. As Dave told Houston, "The sleeping up here is really good; and if y'all ever see another little problem like that, why, we'd only be too happy to roll over and take care of it. I think, as a matter of fact, we'd even sleep better if we knew that you wouldn't mind waking us." Despite the late bedtime and the early wake-up, they got about 5 hours of sleep, a lot more than any of the earlier crews had managed.

DEPLOYING THE ALSEP – FIRST HEAT-FLOW HOLE

For this mission, rather than deploy the ALSEP experiments on the first EVA, Dave and Jim took immediate advantage of the Rover and made a 4-kilometer geology trip to the area where Hadley Rille curves to the west at the foot of Hadley Delta (a trek described in detail in Chapter 4).

Dave and Jim got back to the LM at about 4 hours and 20 minutes into EVA-1 and planned to spend the remaining 3 hours deploying the ALSEP equipment. However, for unknown reasons, Dave was using oxygen faster than had been expected and, once he was back at the LM, Houston suggested that he "do as little unnecessary moving around as possible." Dave was quick to point out that, for most of the remaining time, he was supposed to drill three deep holes into the lunar surface – two for a heat-flow experiment and one for a deep core – and, as everyone knew from training, the drilling promised to be hard work. Indeed, it was probably the most demanding physical work that either of them expected to do during their time on the lunar surface. Houston would keep a close watch on his oxygen supply.

In most respects, Scott and Irwin had relatively little trouble with the ALSEP deployment and, with the exception of drilling into the lunar regolith, they completed their work in about an hour and one-half. If things had gone according to plan, Scott would have spent about one-half hour doing the drilling, emplacing the heat-flow thermometers, and disassembling the six deep core sections; however, almost nothing about the drilling seemed to go right. Later, on Apollo 16, Charlie Duke drilled his first heat-flow hole to the full 2.5-meter depth in 1 minute flat, and, on Apollo 17, Gene Cernan drilled his in just a bit under 3 minutes. However, both Duke and Cernan had the advantage of equipment that had been extensively modified as a result of what turned out to be a very frustrating experience for Scott.

Transcripts from the drilling episode demonstrate both the intensity and frustration of exploration under tight time constraints. They also demonstrate that problems can be overcome with the flexibility and quick thinking possible only because this was manned exploration.

The voices here are astronaut Joe Allen (CapCom during the Apollo 15 EVAs), Dave Scott (CDR), and Jim Irwin (LMP). Author Eric Jones interviewed Irwin during a 3-day visit to Colorado Springs in December 1989; he interviewed Scott in Santa Fe, New Mexico, on several occasions during 1992–1993. Other quotations are taken from the Apollo 15 Technical Crew Debriefing done August 14, 1971 at MSC in Houston, Texas.

We join the crew just as Jim has deployed the SWS northwest of the Central Station. Dave is about to start work on the HFE (Heat-Flow Experiment). He plans to drill a hole 2.4 meters deep using a string of hollow bore stems with a closed bit on the bottom. The closed bit will prevent cuttings from entering the bore stems and, once the hole is finished, Dave will lower a string of very sensitive temperature-measuring devices into the bore stems to determine the rate at which heat is flowing out of the lunar interior. After Dave finishes placing the temperature probes in this first hole, he will repeat the process with a second hole about 10 meters away from the first.

Scott: Okay, Joe. Drill and the rack going to the first probe. It'll be the one on the right [northeast] today, because the rammer was packed in the one on the right, today. *(pause)*

Allen: Roger, Dave. *(pause)*

Scott: Check south, huh, Joe?

Allen: Check south.

The deployment sketch on the commander's checklist shows Dave drilling the western hole first.

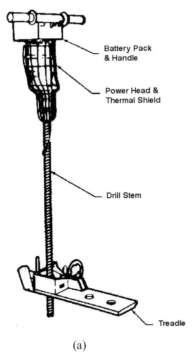

(a)

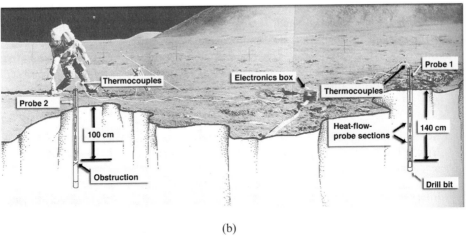

(b)

Figure 5.1. (a) Components for the Apollo Lunar Surface Drill, a rotary-percussive instrument used to drill two holes for heat-flow probes and one to collect a continuous soil column to depths of nearly 3 meters. Dave threaded the drill stems together sequentially as he drilled. When the drillstring reached maximum depth in the deep core hole, he planned to pull the string out of the regolith by hand. The labeled drill stems were then separated and capped for return to Earth. (*illustration is from Allton's Catalog of Apollo Lunar Surface Geological Sampling Tools*). (b) The Apollo 15 Preliminary Science Report provides a composite sketch and photograph showing the HFE emplacement.

Scott – "Taking a look at the two-probe parts of the box, I found that the rammer was in the right-probe box and, in training, it had always been in the left."

Interviewer – "Remembering how much of this equipment had never been on a mission before, it's probably not surprising that we had these little glitches."

Scott: Okay. I'm in a shallow depression here, Joe. And there's no way of getting around it. There's just nothing really flat. There's a little rim here, a little slight rise. Further to the north, maybe Mark [Langseth, principal investigator for the heat-flow experiment] would like it up there. *(long pause)*

Allen: Dave, drill it there.

Scott: Okay.

Allen: Right where you stand.

Scott: Thank you. Just checking. You know. Like, sometimes those guys have some neat, good ideas. *(long pause)*

Dave removes two drill-stem sections from the rack and threads them together; each is about 43 centimeters long, and the stem that will be deepest in the hole has a sealed bit at the end. Dave holds the combined sections by the bit end and threads the other end into the drill chuck. There is a single thread on each stem that goes around the shaft over a distance of about an inch; in shirtsleeves and bare hands during training sessions on Earth, threading is an easy operation, despite the fact that one's hand is about a meter above the drill chuck. Dave doesn't appear to have any trouble suited, either; but on Apollos 16 and 17, respectively, Charlie Duke and Gene Cernan will have noticeable difficulty. Dave sticks the bit into the soil at the desired spot and begins to drill.

Scott – (referring to the considerable difficulties he will have later with the drilling) "I blew it. I got all the smooth part done early, so I wouldn't have to worry about being smooth later on."

Interviewer – "It sure wasn't your fault. It was a poor design. And terribly frustrating, I imagine."

Scott – "I know it. And it cost us the North Complex." [when they ran out of time.]

Interviewer – "This and the vise. And the core."

Scott – "Whole experiment was a surprise."

Scott: Okay. First two [drill stem sections]. Down we go. Huh, it takes a little bit of force. *(pause)*

Figure 5.2. Training with the electric drill at the Kennedy Space Center, Dave Scott drills into a buried canister of simulated lunar soil. The drill stem rack to the right of the drill contains a cluster of four core stems. The wrench attached to the back of the rack becomes important in EVA-3. Despite all the tests and dry runs, there was nothing to compare with the arduous task of drilling into actual lunar regolith at 1/6 g.

Scott: As a matter of fact, it's getting a little stiffer. *(pause)* As a matter of fact, it's getting a lot stiffer. *(pause)* Wow. It's tough down there. *(long pause)*

Ed Fendell, who is operating the TV from Houston, pulls back on the zoom and, as he does, Dave comes into view just as he releases the drill handles. He now has both sections in the ground. It was only after the joint between the sections went into the ground that he noticed resistance. He turns the drill, trying to release it from the drill stem and then pulls the drill up. Neither maneuver works. Finally, he blocks the stem with his right foot, shakes the drill up and down rapidly a few times, and releases it.

Scott – "The drill was a chore, even on the best of days, because of all the mechanical interactions."

Allen: Jim, you'll get your feedwater tone shortly.

Irwin: Okay, I did. Time to go to Aux-Water, huh?

Allen: It's about that time.

Each PLSS has two tanks of feedwater for the cooling system: a main tank and an auxiliary tank. Jim's main tank is almost empty and he needs to throw a mechanical switch on the bottom of his PLSS to change to the Auxiliary tank. The switch is not easy to reach.

Irwin: Dave, can I disturb you to get my Aux-Water?

Scott: Be glad to, Jim. Here. *(long pause)*

Dave grabs the wire loop and puts the drill aside with the handles down on the ground and the chuck pointed up. Jim runs out to him, using a comfortable, foot-to-foot, loping stride. He has a good glide between steps. Dave puts his right leg out and bends his right knee inward so that he can get at the Aux water switch on the lower-right front corner of Jim's backpack. It takes him two tries to get into position.

Scott: Oh, wait a minute. *(pause)* Okay. Your Aux-Water is On.

Irwin: Thank you. *(long pause)*

Irwin: *(back at the Central Station)* Okay. I'm taking the LSM out, Joe.

Allen: Okay, Jim. And Min [feedwater] Diverter on your PLSS, please.

Irwin: Is that for startup?

Allen: That's affirm. *(pause)*

Irwin: Okay; it's Min. Although I understood it could start up on any position. *(pause)* Bet you're trying to make me sweat a little. *(long pause)*

As he returns to the drill exercise, Dave threads the new pair of drill stems onto the drill string in the ground without any particular difficulty, then picks up the drill and attaches it to the top of the stack. As Dave starts to drill again, he obviously has to put considerable force on the handles and, after a few seconds, is up on his toes, with nearly his full weight (only 60 pounds in 1/6 g) on the drill handles, which are at about shoulder height.

Interviewer – "You had your full weight on it, didn't you?"

Scott – "Which is a point that people made during and afterwards. I was pushing on it too hard, they said: 'that was the problem.' But you can't push on it very hard, because you don't weigh very much. You can't put a lot of force on it, even if you put your whole self on it, because your whole self isn't very much force."

Interviewer – "They were searching for explanations."

Scott: Joe, can you see the drill?

Allen: Yes, sir. We sure can. *(pause)*

Scott: Now I'll give you a better angle. *(pause as he goes around to the west side)*

At this point, Dave has drilled about another half of a section (25 centimeters) into the ground. The joint between the second and third section is probably now in the ground.

Scott: That's all I got.

Allen: We agree, Dave. *(pause)*

Scott: There are about three probes [means "stems"] in, Joe. And, my goodness, I think that looks like about the end of it. *(pause)*

Allen: Tough to argue with, Dave.

Scott: I guess the next question is do we dig a little trench and lay them [the heat-flow probes] in the trench, or do we just put [these] three in [vertically]? *(long pause)*

The cause of Dave's trouble was ultimately diagnosed as a faulty drill stem design. The drill stems for the heat-flow probes had external flutes to carry the cuttings to the surface but, because of the relatively low strength of the fiberglass/boron-filament laminate of which these stems were made – chosen for its thermal characteristics – the walls had to be made thicker at the joins by decreasing the depth of the flutes. Unfortunately, because the regolith is very compact at depths of greater than a few centimeters, the cuttings jammed in the shallow flute areas, causing the stems to bind in the hole. The re-designed stems eliminated the problem for Apollos 16 and 17.

Dave is standing still, with his arms slightly raised and his hands at about waist height as he rests.

Interviewer – "Were you waiting for an answer, here, or resting up?"

Scott – "I was waiting for an answer. My interpretation was that we were getting into some very heavy induration, somehow. 'Cause it certainly didn't appear to be regolith that I was drilling into – not knowing what the real problem was. Prior to the mission, we had discussed what you do if, for some reason, there was basalt or something right under this site and only a very thin layer of regolith? Well, you dig a trench and you put the heat-flow probes down parallel to the surface. So now, I'm saying I can't get all the probes in; so, what do we do? It's clear it isn't going to go in very much further. And it was surprising because, in training, it always went pretty well. So this was a very big surprise. And let me tell you, it felt to me like we'd hit concrete. Absolute concrete."

Dave planned to drill a 2.4-m- (7.9-foot-)-deep emplacement hole but has only reached a depth of 1.62 meters (5.3 feet). This is enough depth that he can place the entire probe beneath the surface, but not deep enough to bury the thermocouples on the cable that connects the probe to the HFE.

Allen: Dave, this is Houston. We agree that, if you've hit bottom there and think there's no way you'll go any deeper, just press on and put the probes in this hole.

Scott: Okay, Joe. I'll give it one more try here and see if I can get some more; but, ha, I tell you one thing, the Base at Hadley is firm.

Allen: Roger. So much for the "fairy castle" theory.

Scott: Yeah! *(chuckles)* Oh, I'm afraid that's it. I hate to say that because it [the drill]'s working good.

Allen: Roger, Dave. There's a lot of information right there. *(long pause)*

Dave tries the drill for a few seconds but gets very little additional depth. Joe is referring to a theory about the detailed structure of the lunar soil promoted by lunar scientists Bruce Hapke and Hugh Van Horn (Cornell University and University of Rochester, respectively), in which the soil is modeled as a very loose aggregation of particles containing mostly voids. In the minds of many Apollo participants, the "fairy castle" theory was associated with the late British radio astronomer, Tommy Gold, who was very vocal in advocating that the dust layer was deep enough and soft enough that it would not support the weight of a spacecraft.

Allen: [Jim], you had a good clean startup [of the Aux water tank]. *(pause)*

Irwin: Okay. I'm on Intermediate. *(pause)*

Allen: And, Jim ...

Scott: Joe, I can't get the [drill] chuck to reset.

Allen: ... could you verify all your [RCU warning] flags are off?

Irwin: Verified.

Allen: Thank you.

Scott: Joe, I can't seem to get the chuck to reset. It won't go counter-clockwise out of its seat. *(pause)*

Allen: Dave, try to rotate it 90 degrees both ways, and then push sharply down.

Scott: *(rotating the drill counter-clockwise)* I know that Joe. It won't rotate 90 degrees both ways. It'll only twist the stem [in the ground]. Maybe the drilling was so tight for it, it's locked up in there [that is, the stem is frozen in the chuck]. *(long* *pause)* It won't back off, Joe. It turns the drill ... *(correcting himself)* I mean, it'll turn the stem, but the chuck won't back off.

Allen: Roger. We copy, Dave. Maybe our sim[ulation]s haven't been so bad.

This is undoubtedly a reference to troubles they experienced during training. In fact, Dave's last ALSEP training consisted of two one-half-hour sessions with the drill – one on July 21st and the other on the 22nd, only days before launch.

Scott: Got any suggestions? *(pause)*

Allen: Dave, can you go either clockwise or counter-clockwise 90 degrees without the stem turning?

Scott: *(trying both motions)* Nope. *(pause)*

Irwin: Okay, Joe. The LSM [magnetometer experiment] is deployed. It's level and aligned, and the shadow is on the first degree on the plus side.

Allen: Okay, Jim. Thank you. Sounds good.

Scott: Okay, Joe. I'll stand by for your suggestions.

Allen: Roger, Dave. Take a breather. *(long pause)*

Scott – "When I got the first two stems in, why, it was apparent I was hitting something very hard which, subsequently, I really think was bedrock. But the first meter was quite easy to drill; and then it was very difficult to get the stem any farther. I go about two-and-a-half stems in and, in trying to remove the drill, the chuck has frozen; and I think that's because of the high amount of torque put on the stems themselves and the chuck just biting into the stems and locking up. We'd never seen this in training, nor had we ever seen any material that was compacted or as hard as that material I was trying to drill in at that time. The recommendation from the ground that came up subsequently was to drill slower – which was a good idea – and just let the drill do the work. We should have probably discussed that possibility before flight, because I really hadn't thought about it. That seemed to help get it in a little ways. It did on the second probe."

"In order to get the drill off the first probe, why, I had to get the vise – the little wrench – off the stand where the drill stems were, and get down on my hands and knees and force it off. I finally ended up physically breaking or bending the top half of that third stem to get the drill off."

Allen: Dave, this is Houston.

Scott: Go ahead.

Allen: Roger, Dave. We're requesting you spend a few more minutes on this experiment. We want you to take the wrench off the Rover. It's on the rack, as you know, and try to hold the stem with the wrench and turn the drill off that way.

Scott: Okay. *(pause)*

Scott: I guess you got to tell me where the wrench is on the Rover, Joe. I don't know of any wrench on the Rover.

The TV camera pans right, looking for Dave, but doesn't find him and reverses direction. What seems strange about this exchange is that in an earlier training photo we can see what looks like a wrench attached to the back of the drill-stem rack.

Allen: The wrench on the hand tool carrier, Dave. On the rack. I'm sorry. I'm giving you bad information here.

Scott: Yeah. There's no wrench. *(pause)*

Allen: Dave, that wrench is on the [drill stem] rack that's holding the [core] tubes, right along beside you. Sorry.

Scott: The what?

This is one of only a very few cases of serious communications problems between Joe Allen and the 15 crew. It seems unlikely that Joe is unfamiliar with the equipment on the back of the Rover; he had observed many of the training exercises. Here, he simply seems to be having an unusual amount of trouble finding the right words.

Irwin: Must mean the wrench I installed [on the back of the Rover, as per LMP-6] ... *(correcting himself)* Not the wrench, but the vise.

Scott: Oh, the vise! Yeah, why didn't he say the vise? Sure, sure, the vise.

Joe is referring to a wrench-like tool on the drill stem rack. It may be that Dave has never thought of the "wrench" in any context separate from the vise. For many years, author Jones was confused by Jim's and Dave's use of the word "vise" in this context. The vise, itself, is not what Dave will use to free the drill. He will use the separate part that has wrench-like characteristics. Evidently, Jim, too, thinks of it as part of the vise. However, a little later, Dave will start calling it a wrench and, for simplicity, we'll follow his lead.

Irwin: If that'll work. I don't know.

Scott: I was thinking of a pipe wrench, you know.

Irwin: Yeah.

Scott: That would probably be the thing to do. *(long pause)*

The TV camera has zoomed in on the magnetometer and is now examining the instrument and the cable connecting it to the Central Station.

> **Allen:** And, Dave, you're about a minute from a flag on your water.
>
> **Scott:** Okay. Should I go to Aux now?
>
> **Allen:** When you get the tone.
>
> **Scott:** Say again.
>
> **Allen:** Roger. Wait until you get the tone. Just wanted to warn you.
>
> **Scott:** Okay. You're clipping your transmissions a little bit in the beginning sometimes, Joe, and I miss it. *(pause)*

After pulling back on the zoom, the TV camera pans right and is watching Dave and Jim. Unfortunately, Jim is standing right in front of Dave and we can't see what he is doing to remove the drill from the stem. Because of a coating of dust on the TV lens, the picture is not particularly good. The picture is better when looking west but, here, with the camera pointed north, there is sunlight reflecting off the lens barrel and/or shade and onto the dust.

> **Scott:** It worked! It worked, Joe. Good thinking there in the backroom.

The deep-core stems were made of stronger materials than the heat-flow stems. The latter were not made to withstand the forces that Dave had to apply to free the drill and, consequently, Dave had damaged the top stem while removing the drill.

> **Scott** – "But it did come off, and that was a good call from the ground. I had never practiced that in training."

After removing the drill and, with some difficulty, the wrench, Dave gets the heat flow probe out of its protective bag. He drops the bag so he can insert the tip of the probe in the top drill stem, but then has to retrieve the bag to get the rammer.

Dave goes to his knee to get the bag. On his first try, he bends his knees and tries to reach down, but doesn't get close at all. On his second try, he sticks his left leg well out to the side, bends his right knee inward and drops almost to his knee, just missing the bag. On the third try, he actually touches his knee and finally grabs the bag. Once he is upright again, he removes the rammer and throws the bag away behind him. We see most of the trajectory and, in particular, the small puff of dust kicked up by the bag when it lands. The rammer is a thin telescoping rod – when fully extended, about 2 meters long – used to emplace thermal shields on the down-hole cable and to measure the depth of penetration.

> **Interviewer** – "It took quite a while to get the heat flow in the hole."

Figure 5.3. Dave Scott is leaning sideways to pick up the electric core drill motor. In front of him is a stand that held the core stems. The solar wind spectrometer, one of the instruments at the lunar surface science station, is in the foreground. Normally, leaning sideways to pick up a tool is simple; on the Moon in a stiff space suit, it required strength, good balance, and patience.

Scott – "I would say the people in the Control Center should be getting relatively impatient – especially those who have never watched the entire exercise – because it's looks like its taking an awfully long time, which it did. Other than the drilling part, this (ALSEP deployment) is going pretty good."

After emplacing the probe in the first hole, Dave begins drilling the second. Soon he is having the same problem with the stems binding in the hole.

Allen: And, Dave and Jim, to factor into your thinking, we'll be asking you to leave the ALSEP site in about 15 minutes.

Scott: Oh my. Okay, Joe. *(long pause)*

> **Scott:** It [meaning the subsurface material]'s even tougher here, Joe. *(long pause)*
>
> **Scott:** Whew! Boy, that's really tough rock. *(long pause)*
>
> **Scott:** Same problem. *(pause)* Okay. I got the same problem on the chuck, Houston. I think the rock is so tough that the chuck bites into the core stem and just won't release it without that vise.
>
> **Allen:** Roger, Dave. We copy.
>
> **Scott:** And *(laughs)* I'm afraid I'm going to have trouble getting the vise [wrench] off of this other piece here. *(long pause)*

Dave gets to his knees to put the wrench on the drill stem. In fact, because he has drilled so deeply, he has to rise up off his left knee in order to lean over to his right so that he can get his hand to the ground. Once the wrench is on, he stands up easily, using the drill as a support.

After some experimentation, Dave blocks the wrench with his left foot and puts a counter-clockwise torque (turning force) on the drill handles. As the drill breaks loose, the wrench pops off the drill stem. The motion of the drill and the wrench indicate that Dave is exerting considerable force to break the bond between the drill and the stem.

Dave has his left leg way out to the side and his right knee bent inward so he can get down to the drill stem and the wrench. In his effort to get the wrench off, he is not only turning it parallel to the ground but also flexing it up and down. In the process, he puts considerable stress on the drill stem.

> **Allen:** It looks like "vise" is a good word for it, Dave.
>
> **Scott:** Don't know what to tell you, Joe. *(pause)* Boy, Oh, boy. *(pause)*
>
> **Allen:** Dave, let us suggest to you that you go ahead and deploy the LR-cubed. And, Jim, another reminder on your [Central Station] antenna. *(pause)*
>
> **Irwin:** Okay. Well, the antenna is up now. *(pause)* I'll try and level it.
>
> **Scott:** Sure will, Joe. But let me put the second probe in. I got the vise [wrench] off, and all I got to do is put the probe in. Okay? *(long pause)*
>
> **Allen:** Dave, we'd like for you to stand by on that probe. We think we may be able to drill it deeper later on …
>
> **Scott:** Okay.
>
> **Allen:** … and let's ask you to go to the LR-cubed now.
>
> **Scott:** Okay; on the way! I'll stick the probe in the rack, if that's all right. *(pause while Joe gets an answer from the ALSEP Backroom?)*
>
> **Allen:** Sounds good, Dave. *(long pause)*
>
> **Scott:** Okay, you want the drill in the Sun, I believe. Don't you, Joe?

"The ground called a halt to the drill ... They said they would like to review it and see what would be the best thing to do. In my mind, I thought at that time we should go ahead and dig a trench and put the heat-flow probes in the trench, as we had discussed prior to the flight, if the drill didn't work. It seemed to me that the amount of time being invested in that particular experiment was already becoming excessive. Because of the ground calling, wanting a re-evaluation, I terminated the drilling at that time and proceeded to deploy the LR-cubed and take the ALSEP photos."

As a result of this Apollo 15 experience, new procedures and tools were developed for the drilling tasks that Charlie Duke and Gene Cernan were to perform on Apollos 16 and 17, respectively. A new wrench was designed specifically for the drill removal task; after getting the first stem in the ground, they would attach the wrench to the stem, block it with an ankle, rotate the drill off, thread on a new stem, attach the drill and, finally, hold the drill while removing the wrench. Obviously, after the last stem had been completed and the drill had been removed, the wrench had to be removed without benefit of the drill on the stems. However, because of a change in wrench design – and the fact that the drill stems were made of metal rather than the fiberglass laminate – removal became much easier.

A REST, A DRIVE, AND YET MORE DRILLING

The need to finish the second heat-flow hole hadn't magically disappeared, nor had the deep-core assignment; but, as had been discussed before Scott and Irwin bedded down, Houston was willing to defer those tasks to the end of the next EVA. As planned, they would spend the first part of the day driving south again in the Rover for about 4 hours of exploration and sampling. Scott believed that he'd gone to the Moon primarily to take advantage of the Rover, and the less time the Rover spent parked at the LM or the ALSEP site, the better he liked it.

After the rest period, while Dave and Jim were in the cabin getting ready for EVA-2, Joe Allen outlined the revised EVA plan for them. This kind of summary of activities was useful in keeping the EVAs structured and efficient because the astronauts and everyone at Houston knew the priorities and the general plan.

Allen: I'm going to start with our general rationale for the 6 1/2 hour EVA we're coming up on here, and then I'll get down to some details. I won't give you all the details of the traverse right now, but a lot of them I think we can pick up as we go along, depending really on what we see as we travel along. Basically, the EVA will last, as I said, 6 hours 30 minutes, and this is based on our experience (with oxygen usage) from yesterday. Consequently, the EVA-2 traverse distance has been shortened somewhat to provide good geological exploration with a minimum of travel time, primarily at the Front. We're going to strike out for the Front first, just as planned; however, we're going to skip Station 4 (at the South Cluster) for the time being, range along the Front, and we may very well pick up Station 4 and its corresponding activities on the way home.

Allen: [Along the Front] we're looking for craters like Spur Crater and Window Crater but I'm using these only as examples of craters that have plainly excavated Front material for us, and have provided a variety of fragments to sample. We want to return to the LM with about 1 hour and 30 minutes remaining. And Dave, we're going to ask you to invest some few more minutes on the drilling activity; we've got fairly detailed procedures for you to follow, and I'll go into those when it seems a reasonable time to do so. Jim, at the same time, we're going to ask you to carry out some miscellaneous tasks around the LM while Dave's out at the drill site. And finally, with about 45 minutes remaining – and this is a one-time-special, good deal for you, Jim – we're going to carry out Station 8 activities in the vicinity of the LM. In other words, we will not do our Station 8 activities [planned near Arbeit Crater] on our homeward-bound journey from the Front. Now I'll stop here and ask for questions.

Scott: No; no questions Joe. You're really talking our language today. Go.

Allen: Roger, Dave. On our way home, once again we'll skip Station 8. But don't get your hopes too high, Jim, because we're going to pick that up right before we ingress the LM, and we're just going to carry that out closer to the LM than we had previously planned. We're going to ask you to pick up the miscellaneous tasks around the LM, Jim, while Dave is out working at the ALSEP site. And finally, the two of you will start on Station 8 activities at the LM, together, after Dave finishes working around with the drill. And that, basically, is it. Let's see, let me go back through again, and comment on a few new activities we'll want you to carry out in addition to things on your checklist, listed under Station 6 and Station 7 And I'll have to un-key and shuffle papers here a minute, and I'll be right back with you.

By the time Scott and Irwin were out on the surface and were ready to drive off, it may have seemed as though their luck had turned a bit. They had been forced to spend a little extra time prior to depressurization cleaning up a few gallons of water that had spilled on the floor the previous night because of broken part on their water gun. They'd also had to use tape to re-attach Irwin's backpack antenna, which had broken off when he crawled back into the LM the previous night. But they'd been able to take care of both problems in short order and now, much to Scott's delight, they found that the formerly inoperative front steering was now working.

"You know what I bet you did last night, Joe?" said Scott to his CapCom. "You let some of those Marshall guys come up here and fix it, didn't you?" Perhaps, he suggested, Boeing had built a secret launcher so that engineers from Boeing and from Marshall Space Flight Center could come up to the Moon "to fix their Rover."

Three and one-half hours after wake-up and an hour after depressurization, they were rolling. It took Scott a little while to adapt to the new steering and indeed, after a few minutes, he decided to stop and switch to front-wheel steering alone. The vehicle, he said, was much too responsive, especially when he was going downhill. However, he

Figure 5.4. Regolith covers the Apennine Mountains much like a thick snowfall blankets the hills on Earth. The lunar module stands as a safe haven in this majestic but hostile environment. Because much of the Moon's history is preserved in the layers that make up the regolith, the later Apollo crews collected a deep-core sample at each of the landing sites.

soon discovered that, with front-wheel steering, the rear wheels seemed to be drifting rather than centering properly. Four-wheel steering was better than two, he decided, so he switched back and stayed with four-wheel steering for the rest of the mission.

To start this second trip, Scott drove almost due south toward one of a handful of medium-size, fresh craters on the lower slopes of Hadley Delta. With luck, they might find some of the unaltered samples of the ancient crust that had eluded them at St. George on EVA-1. The chances of finding fragments of bedrock depended on finding a crater that was both big enough that its impactor would have penetrated through the regolith and young enough that the fragments hadn't yet been re-buried by the same type of downhill movement of material that had smoothed the rille walls below St. George.

After an extensive tour along the base of Hadley Delta, Scott and Irwin were on their way back to the LM to resume work on drilling.

> **Scott:** Gee, it's nice to sit down, isn't it?
>
> **Irwin:** Oh, it is.
>
> **Scott:** *(laughing)* It's a good deal. You hop off and work like mad for 10 minutes and hop back on, sit down, and take a break. *(pause)*

This is one of the many advantages that the Rover provided. Not only do the crews have greater range and carrying capacity, but they also arrive at the geology stations rested. The difference is dramatic between this Apollo 15 experience and the experience of the Apollo 14 crew during their arduous climb to the summit of the Cone Crater ridge.

If the first part of the EVA made for the sort of geology-rich day that Dave Scott could really enjoy, the drilling promised to be his penance. And for Jim Irwin, the cost of the honor of having spotted the Genesis Rock – a piece of the ancient lunar crust they had hoped to find – was a set of soil mechanics experiments (the Station 8 activities Joe mentioned earlier) that he would perform while Scott did the drilling. Irwin almost wished that he could trade jobs; the instructions for the soil mechanics experiments filled five full pages on the cuff checklists (the tasks for a geology stop usually only filled two pages).

AT THE LM – SWITCHING FROM GEOLOGY TO DRILLING

Once back at the LM, Dave and Jim spent a little time doing some housekeeping, making the transition from geological exploration to the drilling and soil mechanics activities. They stowed the new samples in a rock box and then made sure they had all of the equipment they would need out at the ALSEP site. Because the drilling problems had forced a re-arrangement of the schedule, Joe talked them through the new plans. And there was a surprise in store for Dave.

> **Allen:** Dave, the only problem is, if we're able to get the deep [core] samples using the drill stems, we'd like them in the SRC [Sample Return Container]. I guess we'll leave it up to you, your choice. It may be better just to take bag 5 over there right now and forego that little nicety.
>
> **Scott:** Well, Joe, you didn't ... Just a minute, Jim. Just a minute. Now, Joe, you didn't say anything about getting deep cores. That's why ... *(to Jim)* Here, let's take 3 and put it over there. Keep it there. Let me take 2 back, because now that I know that they want to try and get the deep cores, we do need bag 2.
>
> **Irwin:** Yeah.

Scott: That's the first time anybody said anything about that.

Irwin: That's bag 5, Dave.

Scott: I mean 5 ...

Irwin: Well, Dave, why not leave 5 there!

Scott: Okay.

Irwin: If you get the cores, bring them back, and we'll put it in there. I just won't load it in the SRC.

Scott: We'll hold the SRC open.

Irwin: Yeah, that's right.

Scott: Yeah. Okay.

Irwin: I'll hold it open.

Scott: Okay.

Allen: Sounds good, Jimmy.

Obviously, Dave and Jim are well versed in the contents of the various bags.

Scott – "What they should probably say is, 'Here's the plan at the LM: want to get the deep core, want to move bags', and then let us go do it, 'cause we're probably more familiar with the bags and what's in 'em. For instance, one bag is almost full, and they don't know that. So, what they should have done was give us the objectives and then let us tell them what we did, 'cause then we can shuffle things around and, as long as they know what we did, they're in good shape."

Allen: Okay, Dave. We want you to park east of the ALSEP heading toward the west, and as far east as is comfortable for you. Once again with the dust problem in mind. And we want you to clean the TV camera and LCRU before you leave the Rover.

Scott: Okay. Park east heading west. And, I'll just get it fairly close to the Central Station and avoid the dust. How does that sound?

Allen: Roger. Just don't drive too far west. Keep it east, if you could, please.

Houston probably wants the TV pointing down-Sun to get good lighting but, as Allen indicates, they are primarily concerned about getting dust on the experiments.

Scott: Okay. Incidentally, at the Rover [meaning the LM], our bearing was 018 and range 0.2. That's pretty good for a trip like that.

Allen: Ain't it, though. *(pause)* And, Dave, as you climb off there and get the TV set and are ready, I'll talk to you about the next drilling job.

Scott: All righty. *(long pause)* Okay, I'm going to park right here. And if you get bored [watching the drilling], there's a big chunk of dark-gray breccia with white clasts right in front of the left wheel. Have fun looking at that, maybe.

Allen: Okay, Dave ...

Scott: Brake's on, power's off.

Allen: ... we know you're not too close to the heat-flow experiment holes now. We do want you to be particularly aware of the SIDE, which has had its dust cover pulled off of it, too. So it'll be particularly sensitive to dust. And, when you are ready, once again, I've got words on how this drill should work.

HEAT-FLOW REPRISE

NASA's experts had been thinking about the drilling problems ever since Scott had to abandon the effort the night before; Joe will pass along their suggestions.

Scott: Okay, Joe, I'm ready to go to work.

Allen: Okay, Dave, we want you to try to get the heat-flow drill in at least another section. We think that perhaps there might be an extra section added onto the unit you started yesterday.

Scott: All righty.

Allen: If you'll tell me what's there, I'll continue. If one or maybe two sections, however many you think you can put on and still drill. Once you get the sections on, we want you to use the drill again. And first, recycle the chuck several times, as you used to do in the simulations, and then start the drill, and put only a few pounds of force on top of the drill. And, while it's running, if you notice from the torque that it starts to seize up, we want you to try to pull it back out of the hole a bit to free it, as you drill. Over.

Scott: To free the probe, huh?

Allen: Roger, just try to free the torque that the drill is picking up from the soil. It seems to be binding around it. Soil and/or rock, that's binding around it. And we think that the secret to this ...

Scott: Okay.

Allen: ... may be not to put quite so much pressure on the top of the drill.

Scott: I see; I understand that.

Allen: And ...

Scott: We'll give it a go.

Allen: ... you might want to *(garbled)* before you get started.

Scott: Yes, sir. *(pause)*

Allen: Okay, Jimmy. Sounds good. We want you to deploy the [U.S.] flag after you finish the photography. And, we're wondering at the moment where the two empty core tubes are. If they are still in bag 5. we'll want you to carry them in your hand out toward the ALSEP station later on.

Dave gets another drill-stem section. He starts to thread it on to the stack but, for some reason, takes it off and puts it back in the rack.

> **Scott** (from a 1996 letter) – "It may not have started on smoothly."

Irwin: No, they are under my seat, Joe.

Allen: Okay, and unless I miss my guess, your seat's out near the ALSEP now. So that is beautiful.

Irwin: Yeah. *(pause)* Why do you want them out near the ALSEP, Joe?

Allen: Jim, I don't ...

Irwin: *(garbled)*

Allen: ... know how to break this news to you, but we're going to do Station 8 out at the ALSEP site, or nearby. Saving it especially for you.

Irwin: Ohh! *(hearty laugh)* Oh, thank you, Joe!

Allen: I knew you'd like it.

Scott: Hey, Joe, before we got out this morning, we figured you guys had a conspiracy against us, having Jim doing Station 8 and me drilling at the same time.

Allen: It may work out that way.

> **Scott** – "Jim didn't like Station 8, and I didn't like to drill. In relative terms *(dryly)* I mean, we liked everything we did, right? But they were both mechanical chores [and not nearly as much fun as collecting rocks]."

Scott: Okay. Okay, Joe, I've got the drill on one extra section now. Run through it again, please, just so I don't *(garbled)*.

Allen: Okay, Dave. We are interested in your starting to drill. We've got a lot of power left in the drill, just run it around several times; and don't bear down on it too much. Let's see how free – *(correcting himself)* or how freely – it moves in the surficial layer there, first of all.

Scott: Okay. *(pause)* Joe, I put very little force on it, and it binds up.

Allen: Roger. Any luck by trying to pull it back a bit out of the hole to free it?

Scott: *(chuckles)* No, it pulls me right on down with it.

Allen: Okay, Dave, stand by a second.

Scott: Okay.

Allen: Dave, is it possible at all to clear out the flutes on it by lifting up as you turn the power on?

Scott: I'll try. But it seems to want to pull me with it. *(grunts)* There, I got it up.

At this point Dave comes into view of the TV camera as he steps up to the drill and pulls it out of the hole perhaps 6 inches. The camera stops panning and zooms in on Dave as he lets go of the drill and steps back. During the following conversation, he stands relatively motionless next to the rack, leaning far enough forward that he can manipulate objects on the top of the rack.

Scott: Okay, Joe. Now I've got the drill partially out. Do you want me to try and take the drill off the probe [means the stem]?

Allen: Dave, we wonder if you can just hold it there. Begin it running, and ease it back down into the hole, but without a whole lot of force, down into the hole.

Scott: Okay!

Dave straightens up, hops to the drill, takes hold of the hand grips, and starts drilling.

Allen: And, just let it run for a while. There's a lot of power in that battery.

Scott: Okay, I'm not … *(hears Joe)* I'm not putting any force on it. I'm letting it do its own forcing.

Allen: Okay, let her run. We've got a lot of power to burn.

After about 6 inches of rapid penetration, progress slows dramatically. A good means of gauging progress is to compare the relative positions of the end of the grip in Dave's left hand and the tip of the rammer-jammer in the foreground.

Scott: *(chuckles)* It's a great massage.

Irwin: Hey! I want to come out and get some of that! *(long pause)*

> **Scott** – "I could just feel it vibrating. I don't remember the frequency but, yeah, you know it's turning. It feels good. And you don't have to push on it. Just put your hands on a nice vibrator."

> **Interviewer** – "Was there enough torque that you had to put some lateral pressure on it?"

> **Scott** – "I don't think so. I think it was torqueless. Was it?"

> **Interviewer** – "Gene's [Cernan's] experience was that he had to hold it pretty firmly because it felt like it was going to turn him. Now, I'll have to check whether that was the heat flow or the deep core."

> **Scott** – "You ought to check 'cause, early on, they tried to develop torqueless tools. I was supposed to do one on Gemini VIII. They have torqueless tools, now. I wonder if this was. I don't remember."

During Apollo 17, Gene was drilling a heat-flow hole when he commented on the significant amount of torque being exerted on his hands. He made no comments about torque while drilling the deep core.

> **Scott:** *(grunting as he pulls the drill a few inches upward)* It starts to bind up every once in a while.

During Joe's next transmission, Dave lets the drill run a few seconds and then pulls the drill up again; this time, he turns it counter-clockwise in the hole a fraction of a turn, turns it back to its original orientation, and lets it run again.

> **Allen:** Okay, Jim. When you get out to the ALSEP site, once again being very careful with your dust, and particularly the exposed SIDE experiment, we'd like for you to do a photo pan out there. And ... Stand by. Let's see; stand by. *(pause)*

Dave releases his grip and backs away from the drill to stand still and get a few seconds of rest. He stands leaning forward with his arms hanging down in front of him. He hadn't made any progress beyond the 6 inches of penetration he achieved on his initial effort.

> **Interviewer** – "It looks like you're resting your hands, which suggests you were having to grip a little bit."

> **Scott** – "Yeah. But, again, the hands are sore inside the suit. That could have been it, too."

Dave grabs the drill again, and pulls it up a few inches.

> **Dave:** *(subvocal)* Oh, man. *(long pause)*

Dave lets the drill run. We can see the grips jump under his hands, but still there is no progress. After a short while, he backs away from the drill, hops up and down a couple of times – perhaps to relieve some muscle tightness – and resumes his rest position.

> **Allen:** And, Dave, take a breather there.
>
> **Scott:** Yeah; *(chuckling)* it's tightening up again, Joe, and I'm not putting any force on it all. It pulls itself down in; and then it starts to bind up.
>
> **Allen:** Roger. Copy.
>
> **Allen:** ... as you can tell, that drill is going down. We're going to ask for about 2 more minutes and call it quits, probably. But just take a breather there.
>
> **Scott:** Oh, no. I just don't want to break it. *(long pause)*

Dave's tone of voice is quite normal and cheerful, an indication that frustration has not yet gotten the better of him. During this last transmission, he waves his arms back to loosen the muscles and then hops up to the drill.

Dave leans on the drill and, this time makes some noticeable progress. After a few inches of penetration, he moves the drill up and down, slightly, a couple of times. At first, progress remains slow but, then, just a few seconds before Dave's next transmission, the drill shakes in Dave's hands and the penetration rate increases dramatically. He gets the drill stems in another 6 inches or so in about 10 seconds.

> **Interviewer** – "If you watch the drill carefully here, it does look like it's jumping on you."
>
> **Scott** – "Yeah, it does. When it binds up on something, it's twisting in my hands."
>
> **Scott** – "You've got to remember this is 1/6 g; and I remember, after we got back, people saying 'Boy, you were really pushing hard.' Can't push hard. Haven't got much to push hard with, because of the light weight, not compared with what the visual image would convey to the viewer because of their association with 1 g."
>
> **Interviewer** – "The most you could put on it, by taking your weight off your feet, is about 60 pounds."
>
> **Scott** – "Yeah, and people get fooled by this visual image if they don't transport themselves to the $\frac{1}{6}$-g environment."

> **Scott:** Gee, Joe, I think I got through something. *(pause)* It's easier.
>
> **Allen:** Roger, Dave. And we're learning things.

Dave hops back to rest. During his next transmission, he waves his forearms from side to side.

> **Scott:** Whew! Yeah, I guess we are. Let me take a little break here. It just started easing up there and went down a little easier. Like *(laughing)* we might have got through a layer, huh?
>
> **Allen:** Roger. *(joking)* I hope we're not going to let the air out [of the Moon].
>
> **Scott:** *(laughing)* Yeah, and me.

In reality, on one of the occasions when Dave pulls up on the drill, he pulls one of the subsurface stem sections out of the one below it and, in this last stage of drilling, the upper section slides past the lower one. This means that the stems no longer form a continuous string; later, when Dave inserts the temperature probe, it will only go in until it gets to the bottom of the sections that form a continuous string to the surface rather than continuing to move deeper through the entire string.

> **Irwin:** Dave, I'm wondering, if we're really serious about Station 8, whether maybe we could get started on that, and I could be doing ...
>
> **Scott:** Yeah.
>
> **Irwin:** ... my task while you're working there.
>
> **Scott:** You ought to do that if you're really going to do Station 8.
>
> **Allen:** That sounds good to us, Jim. And Dave, I think maybe you could put another section on that.
>
> **Scott:** Yes, sir, I could. Let's see if I can get the drill off. *(long pause)*

Dave takes a step forward and then does a single long hop, which puts him next to the drill. He blocks the stems with his right foot and turns the drill 360 degrees, watching closely as the drill/stem assembly turns in unison.

> **Scott:** It's bound up again, Joe.
>
> **Allen:** Roger.
>
> **Scott:** Try the old vise [wrench]. *(pause)*

The wrench is on the ground next to the drill stems. Dave takes hold of the left drill handle with his left hand and drops to his knees. He reaches out with his right hand, grabs the wrench, and attaches it to the drill stems.

> **Allen:** ... [Jim,] you might go out and start to photograph the flow site, if that looks like a reasonable thing.

Irwin: Okay. Dave didn't get those pictures yesterday, huh?

Allen: Negative, we didn't . . .

Scott: I didn't get the heat flow, Jim.

Allen: . . . get those yet.

Irwin: Okay.

To get the vise on the drill stems, Dave has to reach well forward with his right hand. To do that, he extends his left leg behind him and supports himself with his right knee and the left hand. Apparently, he has some lateral force on the knee, because it starts to slide toward his left and, because of the weight of the backpack, Dave starts to rotate onto his back. To catch himself, Dave grabs the drill with his right hand and pushes up with his right foot. As he comes up, he gets his right leg under his center of mass and, with his left leg stuck out behind, spins through about 90 degrees. As he brings his left leg down, he makes a few short steps to the north and, finally, brings himself to a stop with a two-footed hop. The elapsed time from the start of this fall to the end of the final hop is almost 6 seconds; this episode is a dramatic (and instructive) illustration of the length of time one has to respond in 1/6 gravity.

Charlie Duke (Apollo 16) and Gene Cernan (Apollo 17) did the drilling on their missions and also had to contend with the problem of getting a wrench on the drill stems. Both of them used the drill for support but avoided the kind of tumble that Dave had here. In an interview, Dave said that there wasn't much interchange between crews on issues of this kind. However, Duke and Cernan undoubtedly learned a great deal by watching Dave as he worked.

Next, Dave tries blocking the vise with his calf. He attempts to turn the drill and, although there is very little handle motion visible in the TV, the fact that he is exerting considerable force is obvious because when the drill does finally break loose, the vise spins through nearly 360 degrees, hits the other side of Dave's leg, pops off the drill stem and lands about a meter away.

Interviewer – "There's a lot of energy in that thing."

Scott – "I finally got the drill off with a wrench [vise], again, and which required about as much force as I could give it. The wrench worked pretty good, though, at that point." [The wrench/vise gave Dave trouble again later when he started to dissemble the deep core.]

Dave puts the drill down and, for several seconds, manipulates the chuck.

Irwin: Okay, the pan at the ALSEP site's complete. I'll go out and photo the heat flow. *(pause)*

Allen: Roger.

Dave gets another drill stem from the rack and threads it easily into the ones in the ground. He then steps back, leans forward – perhaps to rest – and glances at either his checklist or his watch. He then grabs the drill and, in the next minute, gets no more than one-half meter of penetration. As he releases the drill, the handles jump noticeably. Dave hops back away from the drill for a rest.

While Dave drills, Jim walks slowly through the field-of-view past the heat-flow electronics package.

> **Interviewer** – "I'd never noticed that Jim was walking flat-footed, here. That's pretty rare. Is he doing it to avoid kicking dust?"
>
> **Scott** – "Yeah, probably."
>
> **Interviewer** – "So it's a conscious thing."
>
> **Scott** – "Yeah, he was trying to be careful and not kick dust."
>
> **Interviewer** – "I gather that it was actually harder work to walk flat-footed like that?"
>
> **Scott** – "Well, yes, based on the fact that we naturally would go to the easiest mode of travel. That's why we bounce around. So, if you don't bounce around, you're probably doing something that's harder than bouncing around. *(chuckling)* Basically, we're lazy, right? Take the path of least resistance."
>
> **Interviewer** – "That's how progress gets made. People get lazy."
>
> **Scott** – "This [drilling] is hard work. This is hard, physical work. Bending over and grabbing that stuff and holding on to it and putting the stems in and taking the drill off and all that. That's hard physical work."
>
> **Interviewer** – "I'll look at the heart rates in here."
>
> **Scott** – "I'll bet I'm running pretty high."

Because of the frequent breaks, Dave's heart rate is actually quite low. He also has very low rates while driving the Rover. Jim's normal heart rate is higher than Dave's.

> **Allen:** Okay, Dave, take heart. You've got just 1 minute of drilling left.
>
> **Scott:** *(chuckling)* Okay, Joe. The only things that give out are the hands. I'll tell you ... You know, all this working with the gloves on, after a while ... *(long pause)*

Dave turns to watch Jim taking pictures and after a few seconds, hops back to the drill.

> **Irwin:** Okay, Joe, this mag ran out. I'm going to go back and change.
>
> **Allen:** Okay, Jim. *(pause)*

Dave pulls the drill up a few inches but makes no further progress in a few seconds of drilling before Joe calls.

> **Allen:** And, Dave, we're satisfied with this drill hole. Suggest you stop, pull the drill off, and emplace the heat-flow probe.

Dave hops back from the drill.

> **Scott:** Okay. We made a little money, didn't we?
>
> **Allen:** Hand over fist. *(long pause)*

Dave hops to the drill and turns it briefly, just on the off chance that he won't have to use the wrench. The drill turns freely. To get down to the wrench, Dave rests his left hand on the top of the drill, which, this time, is about a meter above the surface. He spreads his legs, puts his feet behind him, and bends his knees. After his knees are on the ground, he leans to his left, grabs the wrench, and rises from his knees. Then, with his right leg still bent and supporting most of his weight and his left hand on the far drill handle for some added stability, he reaches down with his right hand and tries to put the wrench on the drill stem about 6 inches below the chuck and a foot off the ground. Either he is unable to get it in place or decides against using the wrench, but, in either case, he stands and goes to the rack, probably intending to stow the wrench. He decides against that idea and, instead, puts the wrench in a pocket on his right thigh and goes back to the drill stem.

> **Scott** – "Now that I've seen the video, I can remember doing this. I can remember this was hard work. Bending over and getting that stuff. But one of the reasons I was doing it was the time problem. I was trying to move along briskly, 'cause I knew they were way behind on time 'cause this thing was taking so much time. So, one of the reasons I'm pressing here – and I can remember, now, thinking at the time it was hard work – is that, in the back of my mind, is this clock problem. This thing is burning up all sorts of time we hadn't anticipated."
>
> **Interviewer** – "And what you really wanted to do was to go out and pick up rocks."
>
> **Scott** – "Yeah. But, in any case, when you get into a problem like this, with this kind of slow down, you have all these other things lined up behind it to do, procedurally, and you know you have a fixed amount of time. So you try to make up whatever you can by doing whatever you can. And that's probably why I'm bending over and going down on my knees and all that sort of stuff and working hard, 'cause I'm trying to make up. Not that it's physically exhausting work, it's just hard work. And it would be easier not to do that hard work. But, in order to make up some time – or to keep from losing any more time – I can remember working at it pretty hard to keep going. Keep going to get this done so we can get on with the other stuff."

Interviewer – "I don't remember that the prior crews went to their knees very often. Some, but not a lot."

Scott – "Well, it was a natural thing. It wasn't like 'I'm going to get down on my knees.' It was sort of like 'I've got to get the wrench; how do I do that?' So, without thinking about it, I bend down on my knees to get on with this thing. I've known since the early days in Gemini, 'Get ahead and stay ahead,' 'cause you know you're going to get behind on the timeline. So, in the back of your mind, you always have this nagging thing. 'Don't waste any time, especially if you know you're behind.'"

Interviewer – "Now, the impression I have is that there's not a lot of padding in this timeline. You guys were right up against it."

Scott – "Well, sure. And that was on purpose, 'cause we wanted to get as much done as we could. So fill everything that you can."

Interviewer – "And if you have to drop something, you drop something."

Scott – "Yeah. And you don't like to drop anything. So you try to find ways to make up when you're behind."

Interviewer – "I had forgotten that you did as much kneeling as you did, although you have the drill, here, which helps mechanically to get down."

Scott – "I'd forgotten that, too. *(chuckles)* Never having watched the video. But, after watching the video, I can remember doing it, and I can remember thinking, 'You know, I've got to get on with this.'"

During a conversation between Jim and Joe that is not included here, Dave examines the drill and stem and turns the drill a small amount. Next, he grabs the drill with both hands and shakes it vigorously up and down, and the drill comes off. He then puts the drill down and gets the heat-flow probe off the rack. He feeds the probe into the drill stem and gets the rammer; but, apparently, has trouble engaging the rammer on the wire.

Scott: Somehow, all this wire down close to the probes got wrapped with Teflon or something – makes it a lot bigger than we've been used to seeing [in training]. *(long pause)*

Dave finally gets the rammer on the wire, but is only able to get it about 1 foot into the hole. Dave steps away from the drill string and considers the situation. Just before his next transmission, he pulls the rammer out and takes it off the wire.

Scott: Joe, looks like … Hmm. We may have a problem. Let's see. Can always pull those out and put four more in. *(pause)* Joe, I don't think I got the probes all way down. I think that probably one of those cores may have been bent or something. What do you think about that?

Allen: Stand by. *(pause)*

Dave lays the rammer on the rack.

Post-mission analysis – adapted here from the Apollo 15 Mission Report – suggests that Dave's assessment was close to the mark. The individual drill stems were about 53 centimeters (21 inches) long, and there were three sections in the drill string in this hole. The entire length was approximately 159 centimeters (63 inches) long. However, when Dave inserted the probe, it only went in about 90 centimeters (35 inches), which means that the probe got to within about 15 centimeters (6 inches) of the bottom of the middle stem. Later examination of the TV and heat-flow-probe data indicates that, near the end of the bore-stem drilling operation, the first joint was inadvertently pulled apart when Dave moved the drill string up and down in an attempt to improve the drill penetration. Although Dave reported that the penetration was easier for approximately 6 inches – that was probably because the bottom of the second (or middle) section was performing more in a coring manner. Evidently, the bottom 15 centimeters (6 inches) or so of the middle stem filled with dirt. The change from boron/fiberglass to threaded titanium drill parts for Apollo 16 and 17 to take care of the flute/cuttings problem also helped prevent such separations.

At the time of the mission review, Dave was not aware of these conclusions or any of the engineers' discussions about the drilling problems.

Interviewer – "You never debriefed this?"

Scott – "We never debriefed this. Are you kidding? Again, this is the first time I've ever heard it. Well, we might have at some point after the flight. Nah, because all our debriefing occurred in the first week or so after we got back. So, if they [eventually] knew what happened, they probably didn't know it then."

Interviewer – "I'll find a place in the transcript to include this stuff but, the design of the drill stem was such that, in the join between various sections, they had a fatter envelope and the cuttings were jamming in there, which is why it was binding in the hole. Then, possibly when you managed to pull it back up, what you managed to do was to detach one of the sections from the one below. And then, when you started drilling again, it came down along the side of it and stopped."

Scott – "Oh really!?"

Interviewer – "Then, when you put the probe in, it didn't go very far. It went down a couple of sections."

Scott – "How'd they figure that out? Or did they?"

Interviewer – "Analysis, obviously; not direct knowledge."

Scott – "I mean, it's a diagram there."

Interviewer – "After the flight, they used what data they had. But you're right."

Scott – "I mean, people should know somebody guessed so that, when they look it up, they don't say, 'Oh, those guys know exactly what's up there and that's what happened."

Interviewer – "Good point. It's conjecture."

Scott – "Right. The diagram you showed me makes it look like conclusive evidence. I haven't read the narrative, but it makes sense that that's what happened. But nobody ever talked to me about it. Nobody asked me, 'What did it feel like, or could this have happened? Did you disconnect it?' Or whatever."

Interviewer – "Let's see, you didn't spend any time in quarantine after the flight. That's right, isn't it?"

Scott – "Yeah, unfortunately. Which we have talked about. When you get back, you need to focus on the job and not go out and go to parties and satisfy the whole world signing pictures. What you should do is sit down behind closed doors, with the scientists and engineers who put it all together, and scrub down as much as you can [that is, do a thorough debrief], while it's fresh in your mind and get some rest. Rather than going out and trying to do everything for everybody, with no time and already being tired."

Even though it was improbable, some early concern about contamination of Earth by lunar life forms had resulted in NASA's LRL being set up as a quarantine facility. Here, the unadulterated samples could be protected and examined and the crew were isolated so they could be tested and observed. Although the three-week quarantine was onerous for all concerned, it actually provided a valuable time period during which the astronauts could be debriefed by scientists and engineers, just as Dave commented, while the experience was still fresh in their minds.

DAVE DRILLS THE DEEP CORE AND JIM DIGS THE STATION 8 TRENCH

Meanwhile, back at the ALSEP, Jim gets some good news from his friends in Houston ...

Allen: ... we've decided it's about time you start on your Station 8 trench, if you would, please.

Irwin: *(deadpan)* Thanks a lot.

Scott: *(resting while he talks)* Hey, Joe, it won't go in any further than that without really trying to force it. I could try and push it with the rammer if you want, but I suspect the first thing *(chuckles)* would be the rammer would collapse.

Allen: Stand by.

Scott: I could try. *(pause)*

Allen: Okay, Dave, our reading is: using your calibrated arm, put about 15 to 20 pounds of force on it, and we'll be satisfied with whatever we get.

Scott: Okay. *(pause)* No way.

Allen: Okay. That's good.

Scott: *(resting as he talks)* It's stuck, Joe, and I think where it's stuck is where the third probe joins the second probe. And, you know, you can never get those things apart. But I got four other probes in here; if you want me to pull out the four we got and stick the other four in, in hopes that it gets down some distance, we could try that if you like.

Allen: Stand by. *(pause)* Dave, we'd like the rammer-jammer reading and have it pulled out, then we'll take what we got. It's a good job.

Scott: Okay. Sorry I couldn't get it all the way in, because that sure isn't very far. *(reading a mark on the rammer where it goes into the top drill stem)* Bravo 9.

Allen: Roger. *(long pause)*

Dave carefully removes the rammer so that he doesn't pull the probe up.

Allen: And, Dave, could you give …

After Dave disengages the rammer from the probe wire, he puts it along side the drill string and positions it so that the "Bravo 9" mark is at the top of the string. The location of the rammer's tip corresponds to the top of the probe in the drill string and, from the TV camera view point, it appears to be about 6 inches above the surface.

Scott: About like that.

Allen: … us an outside reading.

Scott: An outside reading? What do you mean?

Allen: Never mind, we got it already off the TV.

Somebody in the Backroom realizes that Dave has just given them the desired information, albeit nonverbally.

Allen: Of just how high the pipe comes up above the surface. We'd like for you to make sure the dirt is solid against the outside of the pipe, and then ask you to police the area as best you can of foreign objects.

Scott: Okay.

Dave had started to measure the length of pipe sticking out of the ground but, when he realizes that Houston has the information they need, he stops and begins to tamp the dirt around the drill string by planting his right foot and pressing firmly on the surface.

Allen: And as you ...

Scott: The other thing I'm going to do, Joe, is re-align the ... *(stops to listen)*

Allen: ... leave, we want you to make sure the cables are lying as flatly as possible on the surface.

Scott: Okay. And I'll also make sure that the box [meaning the HFE] is aligned.

Allen: Good idea. *(long pause)*

Scott – "After the initial alignment and all the shuffling around there with the probes and all, and one point I tripped over one of the wires to the probe and I moved the electrical box from its alignment position. I think the ground called up and asked some question relative to the position of wires or Boyd bolts around the electrical box. Maybe they were trying to get data, and the thing wasn't properly aligned. I did re-align it after we went out the second time."

Dave picks up the rack and carries it off-camera (toward the right) to the place where he plans to drill the deep core.

Scott: And I'll leave this here; maybe the next guy can fix it.

Allen: Roger. *(long pause)*

Dave puts the rack down, removes a UHT and heads back to the HFE.

Allen: And, Dave, you'll be interested to know that probe went in as far as it will go. That's as deep as the hole was.

Scott: Really?! You'd have fooled me.

Allen: Roger.

Scott – "I wonder if this is a correct statement. 'Cause it didn't look right."

Interviewer – "It doesn't seem right to me, either. After all, there's another drill stem down there. Maybe what they mean is that the probe went to the same depth as the rammer jammer [about 100 centimeters below the surface.]."

After Dave finishes tidying up around the heat-flow holes and the electronics package – all in the interest of removing anything that might get hot sitting out in the Sun and influence the measurements – he helps Jim get started with his Station 8 activities and, with time running out on EVA-2, drills the deep core. That goes smoothly and, about 15 minutes after he started, he's nearly done.

Allen: Dave, are you working on the last stem there?

Scott: Yeah.

Allen: You are one fast worker. *(pause)*

Dave's breathing can be heard again. After about 57 seconds of drilling, he relaxes his grip and hops back to rest. When Fendell pulls back on the zoom, we see Dave leaning forward with his arms hanging down.

Allen: Okay, Dave. Take a breather. And I've got one last instruction for you here. Using the drill, we want you to break it loose [meaning, free the drill string in the hole] and then let the drill and the stem sit there in the surface and we'll pull it out later.

Scott: Okay. Let me finish it off. *(long pause)*

Dave starts drilling again and completes the remaining 20 to 30 centimeters. When he steps back, the drill chuck is only a few centimeters off the ground.

Allen: And just leave the drill on [the] stem handle [pointed] away from the Sun as long as [the threads] pull free.

Scott: Yeah. Well, we'll try that now.

Dave leans forward and grabs the drill handles. He pulls up sharply, raising the drill perhaps 20 centimeters.

Scott: *(grunts)* Ahh, we can get it! Okay, Joe; we're in good shape.

Allen: Okay, Dave. We want the handle away from the Sun. And we're ready for you to get back on the Rover.

Because of the drilling problems and the number of lunar-surface tasks being performed for the first time, by the end of the second EVA Scott and Irwin were running 1 hour and 40 minutes behind the mission timeline. When they climbed back into the LM at the end of the EVA, they had only 22 hours left before they were scheduled to go back to orbit. In that short time, they still had to complete their post-EVA-2 tasks, get some sleep, conduct EVA-3, and then, 4 hours after climbing back in the LM for the last time, launch and rendezvous with Al Worden in the CM.

DAY 3 ON THE PLAINS AT HADLEY

Allen: And, Dave and Jim, basically the EVA is going to last somewhere between 4 and 5 hours [depending on how quickly they get outside], so it will be a short EVA. I'm told that we checked off the 100-percent science-completion square sometime during EVA-1 or maybe even shortly into EVA-2. From here on out, it's gravy all the way; and we're just going to play it cool, take it easy, and see some interesting geology. It should be a most enjoyable day. Over.

Scott: Okay, Joe. Thank you. We're looking forward to it.

Allen: Okay. Roger, Hadley Base. Taking it from the top, we're going to ask you to stop first at the ALSEP site and spend a few minutes recovering the successfully drilled core tube and, then follow that with the Grand Prix photography. [This is a planned "slalom" in front of the hand-held 16-millimeter movie camera; the film will be used later to evaluate Rover performance.] From there, press on towards Station 9, as planned. We're going to skip the Delta stop in between.

Allen: Station 9 is exactly as we planned it. From Station 9 up to Station 10, exactly as we planned it; and at Station 10, we're going to hit a branch point. We can update you there when you arrive at Station 10. The two options are basically, to head north for the Complex, although we think it's more probable we'll just want to loop back towards the north across Alligator Chain doing good mare sampling, and wind up at Quark West Crater, that's the western crater of the Quark triplet, and use that as a Station 14 stop. Over.

Just 3 hours after wake-up, Scott and Irwin were back out on the surface. They made good time loading their backpacks and the Rover with tools and sample bags, and just 45 minutes after cabin depressurization, they were ready to leave the LM. However, before they could drive to the rille, there was the not-so-minor matter of retrieving the deep core; they could only hope that the bad luck they'd been having with the drilling had ended.

EXTRACTING THE DEEP CORE

Scott: Okay. Now to the drill. *(pause)* We last left our friend ...

Irwin: Now it's "our" friend, huh?

Scott: Yes, it is. And if you could ... [Let me] see, if I can ...

Irwin: Okay. Check me out on it.

Scott: Well ...

Irwin: What should I do there? *(pause)*

Scott: Ahh! The object is to pull it out of the ground. But I'm not sure we can do that without driving the drill.

Irwin: I don't think so either. Why not just drive it a little bit to break it loose.

Scott: It's broke loost [sic].

Irwin: Huh.

Allen: Dave and Jim, this is Houston ...

Scott: *(lost under Joe)* pulled it up yesterday.

Irwin: Well, one of us get on one side ...

Allen: ... we're standing by for Rover powerdown and TV remote.

Scott: Okay, Joe. I didn't know you wanted the TV out here, too. *(long pause)* Okay. TV. *(brief static)* Okay, down-Sun position is a bad one for the old Earth, but I'll do it on the AGC [Automatic Gain Control on the LCRU].

Scott: Oh; you ought to have it, Joe.

Allen: Okay. Thank you, Dave. Appreciate that. And by watching, maybe we can give you a few words of advice on this drill.

Irwin: Dave, I'm thinking, maybe if you get on one side, I'll get on the other. And maybe the two of us, by hooking an arm under that, can lift it out.

Scott: Okay. Let's try it.

Scott – "Each of us had a handle of the drill under the crook out our elbow, and we got it up to the point where we could put our shoulders under it."

Irwin: Okay. You say when. 1, 2 ...

Scott: When.

Irwin: ... 3. 1, 2, 3. *(grunting on "3")* A little bit.

Irwin: Let me get down a little bit. I'd like to get down and get a ...

Scott: But it's got a long way to go.

Irwin: Yeah. I know it. But if we break it loose ... 1, 2, 3. Okay. 1, 2, 3.

Scott: Okay. That's enough. Hold it. I suspected as much. Joe, do you think ... Looks to me like the only answer is going to be to back it off with the drill.

> **Allen:** Roger. Let's do that.
>
> **Scott:** Okay. *(pause)* Except that treadle is going to go.

The treadle is a flat plate with a hole in the center. When Dave started drilling the core, he put the first core stem through the hole. If he had run into rock, he could have steadied the treadle – and, therefore, the core string – with his foot. For Apollos 16 and 17, a jack was added to the treadle to make core removal easier. Although Dave thought he was drilling into rock, he was actually drilling into very dense, compact soil.

> **Irwin:** Yeah, watch the treadle.
>
> **Scott:** I know it.
>
> **Irwin:** Don't want me to push it [meaning the treadle]? Can I push it down and stand on it? *(pause)*
>
> **Scott:** No, no, no, no. The other problem is you got to break ... I guess we could pull the whole ...
>
> **Irwin:** Pull the whole thing out, huh?
>
> **Scott:** ... thing out and put it up on the vise and take it apart.
>
> **Irwin:** Yeah.
>
> **Allen:** Sounds good.
>
> **Scott:** That's about the only way we're going to do it. Now we got to get it [perhaps the treadle] back down. *(pause)* That'll do. Let me turn it around and stand ...
>
> **Irwin:** Stand clear of the treadle.
>
> **Scott:** Can't stand clear of it, because ... Should have left the treadle down there. *(long pause)*
>
> **Allen:** Dave, this is Houston. Is the treadle on the ground now?
>
> **Scott:** Hey, Joe; stand by, will you?
>
> **Allen:** Okay. We're standing by.
>
> **Scott:** *(garbled)* this thing. And if we get hung up, we'll let you know, and ... We'll give you a progress. If you keep asking questions, we got to stop what we're doing and talk to you. Let us work the problem.
>
> **Allen:** Roger. We understand. No problem.
>
> **Scott:** I know you're anxious, but I guess we've had as many hacks at this drill [as anybody] ... We'll ask you if we get hung up. *(pause)* There, now. Little ... *(laughing)* Sucked me right back down. That was a good idea, but that didn't

work either. *(laughing)* What happens, Joe, is that, when I turn the drill on, the drill drills, like all drills should.

Irwin: There you go, Dave. Put a little angle on it. I was just going to suggest that.

Scott: Whew! Oh, brother.

Irwin: I think you broke it loose.

Scott: Let's see, I've got a flag. I mean a tone. No flag. Ninety percent on the oxygen. *(pause)*

Irwin: Everything else okay?

Allen: Looks good to us, Dave. You might check your cuff gauge.

Scott: See if we can get it to … *(responding to Joe)* Yeah, I did; it's okay.

Irwin: Let me get an elbow under it.

Scott: I've got another tone. I don't think it's worth doing, Jim. We're not going to get it out.

Irwin: I could put a lot of pressure on it this way.

Scott: Yeah. Let me try, too.

Irwin: Okay. 1, 2, 3. *(grunting)* Here it comes. 1, 2, 3. How many more [sections] do we have?

Scott: Just one more [try]. 1, 2, 3. Okay. Now we're making a little progress. Tell you what we're going to have to do is, I guess, break the drill, take the drill off, and then break the stems off one by one here and put the drill back on and pull it up again. Are you guys that interested in this thing? In Houston? *(no answer)*

Irwin: I'll get the wrench.

Scott: Yeah, get the wrench.

Allen: Dave, how many inches has it moved upwards?

Scott: Well, we've got it up about 3 feet. And I think we can do it piecemeal, if you're really that interested. *(chuckles)*

Allen: Roger. We copy.

Interviewer – "This isn't the only time you ask them if the core is worth all this effort."

Scott – "Right. 'Do you really want to do this? Let us get out of here.' And I remember thinking 'Look at what we've already put into this thing! How long are we going to go before we call it off?' At some point you got to stop screwing with it because you know it's compromising everything else. That's why, when we had dinner with Grant [Heiken in 1992], I asked him, 'Was it worth it?' This is just eating into oxygen, energy, time, frustration. 'Do you really want to do this, guys?'"

Grant Heiken is one of the editors of the *Lunar Sourcebook* and, at the time of the mission, was a NASA geologist in Houston. During one of Dave's visits to Santa Fe, he and author Eric Jones had dinner with Grant and David Vaniman, a co-editor of the *Sourcebook*. Their opinion is that the Apollo 15 deep core, along with its Apollo 16 and 17 counterparts, is one of the most valuable samples returned from the Moon.

Irwin: Here's the wrench. Bringing it.

Scott: I'll tell you, you sure invested an awful lot in this thing. *(pause)* I got a little bit. *(pause)*

Irwin: You don't think there is any chance of us pulling it all the way out, Dave?

Scott: Well, let's try again.

Irwin: Yeah. If we could just get our shoulder under that.

Scott: Okay.

Irwin: Let me get down here, and get a shoulder under it.

Scott: Okay, me too. Hold it? Ready? Oop. Wait a minute. 1, 2, 3. Oop, slipped off. Wait, maybe you can get an arm under it, now. There you go. Do this.

Irwin: Oh, this is like isometrics.

Scott: Yeah. Okay.

Irwin: 1, 2, 3.

> **Scott** (from the 1971 Technical Debriefing) – "Then, with each of us with one handle of the drill on top of our shoulders, we pushed as hard as we could – it must have been at least 400 pounds [of force] and finally got it to move and got it out."

Scott: Okay. One more.

Irwin: That's too high for me. I think I can grab it here. Okay. You tell me when.

Jim may be taking hold of the drill string with both hands.

Scott: Okay, 1, 2, 3. 1, 2, 3. Let me help you down there.

Irwin: Oh, we can ...

Scott: Easy, easy, easy. Don't auger it; just hold it now. Don't bend it. I'll never get it apart. 1, 2, 3. Okay, 1, 2, 3.

Irwin: It's kind of stuck there.

Scott: Let's take a break.

Irwin: Yeah. *(both breathing heavily)* *(pause)* Why don't you go to Max cooling?

Scott: Yeah. Yeah, I just thought of that. *(pause)*

Irwin: Joe, you having trouble with your TV?

Allen: Oh, you better believe.

Scott: Yeah, they're hung down.

Allen: And, Jim, why don't you take a breather and tip it up for us, please. Thank you.

Irwin: Dave's tipping it for you. *(pause)*

Dave raises the TV camera and goes back to work. We now have a view of the left flank of Mt. Hadley.

Scott: Nothing like a little PT [physical training] to start the day out. Try it again, here. Okay.

Irwin: I'm ready.

Scott: 1, 2, 3. Okay, it's coming. It's coming. Okay. Let me get underneath it here. Okay, 1, 2, 3. *(chuckling)* Okay.

Irwin: One more.

Scott: 1, 2, 3, okay. I've got it. Okay. Let me have it now.

Irwin: Okay. You've got it. Good.

Dave is holding the drill stem at about shoulder height. The treadle is at about eye level and, off-camera, the drill chuck is another meter higher. They started pulling the drill stem out.

Scott – "Now that I hear it, I can hardly believe it: that we put that much effort into it. I knew at the time it was [a lot of effort]. But, now, getting to hear all the huffing and the puffing ..."

Interviewer (reviewing the charts) – "I'm looking at the heart rates here and I don't believe this for a second! They've got you at 100 beats per minute and Jim at 125, and I don't believe for a second."

Scott – "I don't believe it, either! I mean, we were working max. This is max output. I mean, you can just tell from the huffing and the puffing and the breathing. This is absolutely max output."

> **Interviewer** – "150, 160, somewhere in there."
>
> **Scott** – "This has to be our max heart rate. Well, I know it isn't, cause I know on the treadmill they take you off at 180. So I know I've been to 180."
>
> **Interviewer** – "That's fiction, there."
>
> **Scott** – "That is fiction. And it makes you wonder, is any of this data good?"

Dave thought about the heart rates for a moment.

> **Scott** – "You know, it may be right. Because we're not in a cycling mode, like you would be on a treadmill. You're pushing and then backing off. You're pushing and then backing off. So it's not like running the heart up at a high rate over a long period of time."
>
> **Interviewer** – "But I hear you breathing hard."
>
> **Scott** – "I know. You're breathing hard, but that doesn't mean that you've got a high heart rate. You may have some peaks, but you're breathing hard 'cause you're straining. We're doing isometrics and we're not exercising the heart very much. We're not pumping a lot of blood through the system, because we're not doing dynamic kind of exercises. Pretty static exercise, really. Just pushing; but it's hard. Pushing with everything you've got. I'm just amazed they kept letting us go."
>
> **Interviewer** – "If you believe it, I will, too."
>
> **Scott** – "There are different kinds of heart rates, and high rates are over a longer period of time."

During the time Dave and Jim spent on the Moon, they both experienced occasional heartbeat irregularities called *arrhythmias*. After the mission, the flight surgeons discovered that both Dave and Jim had become potassium deficient, in part because of the hours they'd spent training in the suit in the hot Florida sun with only air cooling and – this was before restorative sports drinks became available – only water in their in-suit drink bags. In addition, Jim's drink bag wouldn't work during his time on the Moon, and he became at least partly dehydrated during all the EVAs. Jim's arrhythmias were more frequent and more severe than Dave's; consequently, during the rest periods between EVAs, the flight surgeons choose to monitor Jim's heart rather than Dave's.

> **Scott** – "They may have had concern about Jim, 'cause they had him put on his new sensors. And then we're out here doing this stuff and the surgeon must have been going bananas, too."
>
> **Interviewer** – "But it was Houston who suggested that Jim pull it [the core] out."
>
> **Scott** – "Yeah. I wonder why they did that? Probably knew I was getting real

frustrated and might just throw the sucker away. They probably didn't want the thing broken off at the bit."

Although Dave does not mention the fact to Houston at the time or discuss it during the Technical Debriefing later, he has injured his right shoulder as a result of "muscular/ligament strain." During the rest of the mission, Dave will take a total of 14 aspirins to help with the pain. According to the *Mission Report*, after the flight his shoulder "responded rapidly to therapy."

> **Scott** (from a 1996 letter) – "They used ultrasonic therapy on it, as I recall – which was new stuff then."

> **Interviewer** – "You put an amazing amount of physical effort into the core extraction."

> **Scott** – "Yeah. It was just a brute force exercise."

> **Interviewer** – "And you strained your shoulder doing this."

> **Scott** – "Somewhere. I didn't know it at the time. We were too involved in this thing. 'Let's turn this thing off and get on with it!' You can hear me. Picture it. The rille's sitting out there; the North Complex is sitting out there. We already know we have a big time problem. And how much time did we put into the drill already? And it's tough for us to make the trade-off in terms of science value . . ."

> **Interviewer** – "Are you constrained in the suit? Is the [pressure] bladder away from your body?"

> **Scott** – "Pretty much"

> **Interviewer** – "So you've got motion in there? Or is your shoulder reasonably constrained?"

> **Scott** – "Oh. It's close to the suit. There is some clearance, or you couldn't get the suit on. It's not the suit's fault. The suits do well, especially taking this beating. I mean, 'What are these guys doing to these suits?' The suits were not designed to do this with the drill. It shows how tough and durable the things were. I'm really surprised that somebody in the back row in Houston didn't get real squeamish about all of this. I'm surprised some boss didn't just say, 'Hey, just knock that off.'"

> **Interviewer** – "Which they might have done had the TV camera been working."

> **Scott** – "Yeah. Could have, 'cause they could hear us grunting and groaning – two guys on the Moon in pressure suits doing this kind of stuff. In retrospect, not smart, from a safety point of view. Not smart. But, nevertheless, there we are."

> **Scott** – "We don't pay any attention to this sort of stuff – safety hazards and all that – when we're doing this. You've already erased all of that from your mind and, only if a tone comes on do you do something. As long as there are no tones, you work as you would work on the Earth and you never really think about what I've been

talking about. Houston, that's their job. To sort of pace us and guide us and all that sort of stuff, 'cause, once we're out in the suits, boy, it's very comfortable."

Scott – "People have said, 'Boy, didn't you worry about having a leak?' Nope. Never worried one bit. Boy, you're out there and the stuff works and press on. If something's going to be wrong with it, the ground'll tell you or you'll get a horn. And then you stop and take care of whatever you have to take care. And, in the mean time, you just go do whatever you got to do."

Through long-term familiarity, Dave and Jim were aware of the suit's limitations and, when caution was advisable, they were cautious. The work they did at the Station 6a boulder is a case in point. The only striking example of someone taking undue risk during an Apollo EVA is that of Charlie Duke falling backwards onto his PLSS at the end of Apollo 16's EVA-3. He was trying to see if he could jump high enough to execute a full rotation around the long axis of his body before coming back down – rather like a figure skater's axel jump, but from a standing start.

Scott: Okay. Now if you'll close the gate.

Irwin: Okay. *(pause)* Gate's closed.

Scott: Okay.

Irwin: Want me to put up the vise here?

Scott: No. Just leave it alone.

Irwin: Okay.

Scott: It's up.

Irwin: Yeah, you don't ...

Dave pulls up on the drill stem with his right hand but is unable to move it.

Scott: Come and help me pull again, Jim. I thought I had it.

Irwin: I thought you did, too.

Jim grabs the drill stem at about waist height with his left hand and bends his legs.

Scott: It's a two-man job. Okay. 1, 2, 3. Oop. There we go!

Irwin: We almost flew with it.

When they pulled, the drill stem broke free; it yanked Dave up onto his toes and Jim off the ground by an inch or two.

Scott: I've got it.

Irwin: Okay. *(pause)*

Scott: Man, oh man!

Dave rotates the tip up to a horizontal position and takes the core to the Rover.

> **Scott** (from the 1971 Technical Debriefing) – "Because of the significance of drilling in the bedrock, [spending time extracting the core] was probably the way to go. We could only accept the ground's evaluation but, at the time, it seemed like we were investing an awful lot of energy and time in recovering one small experiment, however important it may have been. But, at that stage, I guess we had so much invested in it that we couldn't afford to leave it. It sure was expensive."

THE VISE THAT WOULDN'T GRIP

Irwin: Okay, let me see. I'll get the caps for that.

Scott: It's in bag 2, I think.

Irwin: Yeah.

Jim goes to get SCB-2 (Sample Collection Bag), which is under his seat. Dave lays the core across the top of the geopallet at the back of the Rover.

Scott: Put all the stuff back here, if you can. I can just work the problem right here.

Irwin: Okay.

Allen: Jim, all the gear's in bag 2. And as you pull it apart, we want you to put the filled stems back into bag 2, please.

Irwin: Yep. We understand. *(pause)*

Dave is standing at the right side of the geopallet, behind the LMP seat, and tries to move the treadle up the drill string toward the chuck.

Scott: Okay. I'm going to need you to help me get this treadle up to the front, Jim.

Irwin: Okay.

Jim is taking pieces of equipment out of SCB-2 at his seat.

> **Scott:** Bring the whole bag back here, so I can work it.
>
> **Irwin:** Oh, okay.
>
> **Scott:** Just bring the bag back here, and I'll just work it like I usually do. You can be doing something useful, instead of just standing.
>
> **Allen:** Jim, we need pictures of your beautiful trench there and the collapsed wall. And we'd like, I guess, a photo pan around this remarkable core hole.

Jim takes the bag to the back of the Rover and takes up a position behind Dave's right shoulder.

> **Scott:** *(to Jim)* Here. Come on; bring it on over here, Jim.

Dave motions toward the other side the geopallet, behind the CDR seat.

> **Scott:** Jim, bring it on over here.
>
> **Irwin:** Yeah.
>
> **Scott:** Joe, just stand by until we get this settled down, and then we'll come at you for what is the next task.
>
> **Allen:** Okay.
>
> **Scott:** *(to Joe)* You're going to have to just hold off on jumping ahead of us, because we always have to come back and ask you what you said anyway. *(to Jim)* Okay ...
>
> **Allen:** Read you loud and clear.
>
> **Scott:** ... Jim, if I could get you to help me take the treadle off.
>
> **Irwin:** Okay.

This public show of irritation is most unusual for Dave and is indicative of his level of frustration. Throughout the extraction, Joe makes only a few short transmissions, undoubtedly to avoid disturbing Dave and Jim's concentration. His quick response to Dave's suggestion that Jim find "something useful" to do indicates that he has a list of proposed tasks waiting. However, when he realizes Dave's level of frustration and that Dave still needs Jim's help, he backs off with an immediate "Read you loud and clear."

After putting SCB-2 on the back of the Rover to Dave's left, Jim comes around to Dave's left.

Scott: So, if you'll get on the other side, we can jiggle it and move it up towards the drill.

Irwin: Okay. *(pause)*

Jim comes around to drill and stands next to his seatback.

Irwin: Okay.

Dave and Jim both take hold of the treadle and rotate it back and forth around the drill-stem axis, trying to work it up toward the chuck.

Scott: You can't put any up or down on it. It's got to be sideways. *(not making any progress)* Nah, that's ...

Irwin: Let me try it. *(pause)* It's locked on there, isn't it?

Scott: No, it shouldn't be. *(pause)* *(frustrated)* Boy.

Irwin: Uh ...

Scott: Boy. *(pause)* I tell you, we'll do it section by ... If you could hold it up there, Jim. Hold it on the handtool carrier.

Irwin: Yeah. Let me get on the other end.

Scott: Let me hold it. You just twist the drill off.

Irwin: Okay. *(pause)*

Dave moves to his right and holds the drill-string with both hands. Jim steps around to grab the drill handles in both hands.

Scott: Okay, now. Just rotate the drill left. That a boy. Just rotate it left. Easy does it. Easy does it. *(pause)* Keep it straight if you can. There you go. One more. Just a little ... Keep it straight. Ought to break in a little bit, here. *(pause)* Hold it up. There. Now. *(pause)* Ease it off. Easy does it. *(pause)*

The drill comes loose and Jim takes it to his footpan.

Scott: I got to take my camera off. Okay. Just hold *(garbled)* there.

Dave takes his camera off to get it out of the way and takes it around to his seat.

Irwin: You want me to stand by here, Dave?

Scott: Yeah. Just hold on while I ... *(long pause)*

Dave returns to the back of the Rover.

Scott: [Get the] handy dandy caps out of here. Well, listen, I can get it from there, I think, Jim. Go ahead *(garbled)*

Irwin: Okay; well I'm going to take these pictures that Joe requested. And if you need any help, just holler, and I'll be right back.

Scott: Okay.

Irwin: Because I'm right here. *(pause)* Here's my trench, yeah.

Scott: Okay. *(long pause)*

Jim goes to the trench and starts the trench-sample "after" photography with a down-Sun stereopair. He then moves to the north end of the trench to take a cross-Sun stereopair.

Scott: Okay, Joe. On the drill top end goes [core cap] Alpha.

Allen: Copy Alpha.

Scott: On the bit [end] goes Beta.

Allen: Roger. *(long pause)*

Jim lopes around to the west side of the trench and positions himself directly in front of the TV camera as he takes a cross-Sun stereo-pair from south of the trench.

Irwin: Okay. I have the photos of the trench. Did you say you wanted a pan from this location, Joe?

Allen: Roger.

Irwin: Okay.

Now, Jim moves off-camera to the right and, once again takes up a position north of the trench and, as is usual, starts the pan with a down-Sun. He takes three pictures and then, after he makes another slight turn, slightly bends his right knee, bends his left knee even more and gets up on the toes of his left foot in order to raise his aim. He holds that position for several seconds before he resets his feet and turns around to get some sunlight on the top of his camera, which has failed again.

> **Scott:** Golly, there's some stuff in there! *(pause)*

Dave has broken off the top drill stem and, looking at the bottom end of the section before he caps it, sees that there is soil in it.

> **Irwin:** How you coming on that?
>
> **Scott:** Coming. Okay, Joe. On the [bottom of the] top section goes [cap] Charlie.
>
> **Allen:** Roger.
>
> **Scott:** And I ... Let's see. The other bag ... which are going to have to go in here.
>
> **Irwin:** Going to grab your camera, Dave.
>
> **Scott:** Yeah.
>
> **Irwin:** Mag's jammed. *(garbled under Dave)*
>
> **Scott:** Is it?
>
> **Irwin:** That's the one that jammed yesterday, isn't it? Yeah.
>
> **Scott:** No, It worked ... Is that right?
>
> **Irwin:** It was working there for a while, and then it jammed again.

Jim puts the magazine back in the camera and then takes the camera off.

> **Scott:** Okay. Hey, Joe, what bag do you want these core stems to go in?
>
> **Allen:** Bag number 2, Dave.

Jim puts the LMP camera on his seat and goes around the back of the Rover to get Dave's camera off the CDR seat.

> **Scott:** Bag number 2 doesn't have any pockets.
>
> **Allen:** No problem.
>
> **Scott:** *(sounding doubtful)* Okay. *(pause)* There will be a problem when we start working in the bag.
>
> **Allen:** Negative, Dave. That's an extra bag now, and we'll keep that in mind.
>
> **Scott:** Okay. *(pause)*

Dave pulls on the drill string to get the treadle away from the Rover and starts to work it off the stem.

Scott: Now let's see. Get the treadle off. *(pause)* [Try] this *(pause)*

Dave hasn't had any success in getting the treadle off and gets the wrench to try to secure the drill string. Before using the wrench, he puts a core cap on the open end of the drill string.

Scott: *(to Houston)* Okay. Delta is the cap on top of the next section. *(long pause)*

Scott: Let's see. *(long pause)* Jim, did you get the vise on right?

Irwin: Sure did.

Scott: Oh, it's backwards.

Irwin: Can only go on one way, Dave.

Scott: Really!? *(pause)*

Scott: It's not working. There it goes. *(grunting)* Ohhh!

Irwin: Okay. The pan's complete here, Joe.

Allen: Super.

Irwin: And I think I'll take advantage of the time and put a black and white [magazine] on my camera.

Allen: Sounds good.

Jim comes around the back of the Rover to his seat. Dave is still turning the drill string with the wrench, trying to figure out why the vise won't hold the section seated in it.

Scott: I hate to tell you, Jim, but that ... Oh boy! This vise is on ... I swear it's on backwards. *(pause)*

Irwin: The holes on the handtool carrier only line up one way. *(pause)*

Scott: *(discouraged)* Doesn't work then.

Dave slides the drill string to his left.

Scott: *(to Houston)* How many hours you want to spend on this drill, Joe? *(laughs)* Like the vise doesn't bite strong enough to get a grip so I can break the sections.

Irwin: Dave, if you want, I can get on the other end and hold it steady.

Scott: Well, you can try ...

Irwin: Yeah.

Jim puts his camera down on the LMP seat and goes around Dave to try to hold the stems in the vise.

Scott: ... [and] see if it does any good. But the training ones hold good. There's never any problem with them.

Scott – "The vise on the pallet just didn't work. At first, I thought it was on backwards. I knew darn well we'd discussed it before [the flight] and that it could only go on one way, but I just couldn't believe it was that bad. It just didn't grip at all. The hand wrench worked fine. It would grip the stems and hold them very well, but the one mounted on the pallet provided no torque at all."

As it turns out, the vise mounted on the back of the Rover *was* on backwards, but not because of any crew error. The following explanation is taken from the *Apollo 15 Mission Report*:

"The sections of the core-stem string could not be separated using the vise and wrench because the vise had been mounted on the pallet backward. The configuration of the core stem vise is the same as that of the core-stem wrench head. The vise is mounted on a bracket on the lunar roving vehicle aft-chassis pallet, located on the right-hand side on the vehicle (behind the LMP seat). The core-stem wrench head is similar to a conventional pipe wrench head, with one fixed jaw and one pivoted jaw. The throat width is not adjustable and is designed to fit the outside diameter of the core stem."

"As mounted, the vise would hold the core stem so that the joint could be tightened by rotating the wrench (up and forward) on the adjoining section. However, the vise would not hold in the opposite direction (down and aft) so that the joint could be loosened and separated. Working on the inboard side of the vise (that is, by switching the core string end for end so that the top sections were behind the CDR seat and the bit end was behind the LMP seat), the core stem could have been held properly for loosening; however, there is insufficient clearance on the inboard side of the vise for wrench rotation and (as well) the distance to the other side of the lunar roving vehicle is greater than the length of a core-stem section."

"The installation drawing of the vise was in error and was corrected to assure correct orientation of the vise for Apollo 16. The training vise was installed backward from the erroneous drawing, but correct for loosening the stems."

An obvious question is: did the person who installed the training vise notice the drawing was backwards and, if so, why didn't they call attention to the fact?

Irwin: Okay. Let me get on the other end. Some caps behind your left boot.

Scott: Oh, shoot! I knocked them off again.

Irwin: Here I'll get them.

Scott: *(dismissively)* No, go get the vise! I mean the drill [stem]. I think we're about through with this.

Irwin: Okay. I've got it.

Scott: Okay. *(pause)*

The fact that Dave snaps at Jim is the clearest possible indication that he is very frustrated. His painful shoulder is probably not helping his mood either, especially because he hurt himself getting the deep core out of the soil. Characteristically for both of them, Dave recovers his composure before his next transmission and Jim graciously ignores the outburst.

The stems turn freely as Dave uses the wrench. The treadle is still attached to the top section and its motions give us a good idea of how the stems are turning.

Scott: Won't bite. Don't hold it that way. Hold it straight in the vise, if you can, Jim. See how it sits?

Irwin: Yeah.

Jim may have been holding the stems at an angle to the vise in order to improve their chances of getting it to grip.

Scott: Okay. See, it doesn't grab. There, that's got it there. Okay. Hold it like that. *(pause)*

This time, the treadle moves in jerks as the stems bite in the vise and Dave starts to get the top section loose. The treadle was pointed down when the vise began to grip.

Scott: Man, oh man! Okay. Hold it right there.

Allen: Dave, the treadle may be jammed against the fender. There it moves away. *(pause)*

Scott: *(to Joe)* Yeah, I know. No problem. *(pause)* *(delighted as the stem breaks free even more noticeably)* There!! *(pause)*

The treadle has gone through a complete revolution and, once again, is close to the right-rear fender.

> **Scott:** Trouble is ... Okay. Let go a minute. [I'll] get it out here a little ways. *(pause)*

Dave lifts the stems and pulls them out 6 inches or so to get the treadle clear of the fender.

> **Scott:** Hold it again, Jimmy.
> **Irwin:** Okay. *(pause)* Okay, I got it. *(pause)*
> **Scott:** Okay. Oh boy!

Dave completes another half turn with the wrench, puts it down, goes to his left, and stands facing the Rover so that he can grab the treadle and turn it to finish removing the second stem from the string.

> **Scott:** Joe, I haven't heard you say yet you really want this that bad. *(pause)* Tell me you really want it this bad.
> **Allen:** *(with a touch of wistful sympathy in his voice)* It's hard for me to say, Dave.

After a three-quarter turn of the treadle, the second section comes free.

> **Allen:** Beautiful.
> **Irwin:** I'll get those caps for you, Dave.
> **Scott:** Okay.
> **Irwin:** You need those now?
> **Scott:** Yeah.
> **Irwin:** Okay. *(long pause)*

Dave watches while, off-camera to the right, Jim gets the core caps off the ground. Dave reaches forward and down to take them from Jim.

> **Scott:** Okay. Thank you. *(pause)* Okay. Cap number Echo.

Dave has put the cap on the bottom of section 2.

> **Scott:** Is in the next section. *(perhaps putting the caps in the SCB)* Okay. *(to Jim)* Now, old buddy, if you think you can have some luck taking that [treadle] off. I'll tell you what, we got to break it again.

Off-camera, Jim is helping Dave get the treadle off section 2. Although we can see Dave's PLSS and upper torso and head, his hands are off-camera and it is impossible to know exactly what he and Jim are doing.

> **Allen:** Dave, how many more sections to come apart?
>
> **Irwin:** If you can get the wrench, I'll hold the treadle.
>
> **Scott:** Okay. *(to Joe, with clear annoyance)* Oh, stand by, Joe. *(counting the remaining sections)* We've got 1, 2, 3. 4.
>
> **Allen:** Thank you.

Dave is probably annoyed because Houston should have been able to keep count.

> **Irwin:** Okay, I've got it there if you can ...
>
> **Scott:** Yeah. *(long pause)* *(grunting)* Ahhh! *(pause)* Nope, wrong way. (long pause) That doesn't look right either. Let's go the other way. Man, how did that treadle get like that? *(long pause)*

During the struggle to remove the treadle, Dave edges off-camera to the right.

> **Irwin:** It's moving.
>
> **Scott:** Yeah.
>
> **Irwin:** It's broken.
>
> **Scott:** That's got it. *(pause)* It's really jammed.
>
> **Irwin:** See if I can get it out.
>
> **Scott:** Work it out towards you, because of the cap. See what I mean?
>
> **Irwin:** Yeah.
>
> **Scott:** Take that ... The end of your right hand should come through, while I work on the rest of them, here. *(pause)* Okay. Foxtrot on the next section.
>
> **Allen:** Roger. *(long pause)*

Dave is back in view and has put a cap on the top of the third section. He pulls the string out until the fourth section is seated in the vise. Dave starts turning the wrench but, without the treadle on the string, it is difficult to tell if he is getting any grip from the vise.

> **Irwin:** If I had known [about] this, I would have left my cover gloves on, Dave.
>
> **Scott:** Well, don't mess with it then. Don't mess your gloves up. I'll do it.

Evidently, Dave is still wearing his cover gloves.

> **Irwin:** Oh, they're okay. I'll take it.
>
> **Scott:** This vise just won't hold. There's something wrong with it. *(long pause)* The vise doesn't work, at all! *(to Jim)* I'll have to have you hold it, the ... *(pause as Dave glances at the watch on his left wrist)* How about that: 1 hour and 15 minutes into it [meaning the EVA] already. We're still fiddling with this thing.

> **Scott** (from a 1996 letter) – Ah ha! I did use the watch!! And I recall this.

They have been working on the deep core for about 30 minutes.

> **Irwin:** Okay. The treadle's off.
>
> **Scott:** Beautiful! [I'll] stick this [stem-section 2] in here [SCB-2]. *(pause)* Now, hold that section for me ...
>
> **Irwin:** Okay.
>
> **Scott:** ... [Like you] did before. *(pause)* It's just not gripping. The ...
>
> **Allen:** Dave and Jim, this is Houston.
>
> **Scott:** Go ahead.
>
> **Allen:** Roger, troops. What's your best guess?
>
> **Irwin:** *(under Joe)* Try to bring that back that way, Dave?
>
> **Allen:** Do you think you can turn off [remove] the bottom-most drill section?

Dave takes the drill string off the back of the Rover and holds it up for Houston to see.

> **Scott:** Joe, you can see on the TV. That's what we got. Now, the vise ... *(correcting himself)* My hand wrench works okay. The one on the back of the handtool carrier doesn't seem to want to work for some reason. It may just be because of the threads

on the stems. I just can't get them broken apart! And that's the main problem. The wrenches don't work. *(pause)*

Dave puts the string back in the vise and resumes his efforts with the wrench.

Allen: Dave and Jim, put that section on the ground, if you would, please.

Dave takes the drill string out of the vise without hesitation once he hears Houston's decision.

Allen: We'll pick it up on the way back. And we want you to continue on with the Grand Prix.

Scott: Good enough. Do that.

Irwin: Stick it in there, Dave. We might be able to return it [to Earth] just like that.

Scott: Yeah, I think probably so. I don't know where we're going to put it in the Command Module. *(Jim chuckles)* I'll think of something. Let me see. *(looking around)* Let me put it someplace where we don't ding it. There's no place to put it. I'll lay it right here on the treadle. *(pause)*

Dave goes off-camera to the right. He is looking for a way to prop the string up so that he can see it easily when he arrives back at the site at the end of the traverse.

Scott: I guess we ought to take it back [to Earth]. There's more time invested in that than anything we've done. Okay. *(to Jim)* Get your [16-millimeter-movie] camera.

Jim comes into view at the back of the Rover and goes to the LMP seat. Dave is at the back of the Rover.

Irwin: Okay. Let's see. We don't want the drill on here. What do you want to do with the drill?

Scott: Oh. Just leave it right here.

Irwin: Put it on the surface here?

Scott: Yeah. *(long pause)*

Jim lifts the drill and merely turns so that he can put it down far enough out that Dave won't hit it when he starts driving. Jim then removes the 16-millimeter camera from the post next to his inboard handhold so he can use it to record the Grand Prix.

During this event, Jim stands facing northwest and takes movies with the data acquisition camera while Dave drives through a pre-determined set of maneuvers. They had originally planned to do the Grand Prix at the end of EVA-2 but ran out of time due to drill-coring problems. Unfortunately, the movie camera jams and the engineers back home will have to wait until Apollo 16 to see the Rover in action on the Moon.

ONE LAST DRIVE ACROSS THE LUNAR SURFACE

And here, we pick up the conversation again as Dave and Jim are about to depart on their final geology traverse.

Scott: Hey, Joe, you never did tell me that drill was that important. Just tell me that it's that important, and then I'll feel a lot better. *(pause)*

Allen: It's that important, Dave.

Scott: Okay. Good. Because then I don't feel like I wasted so much time.

Allen: No. Quite seriously, Dave and Jim, that's undoubtedly the deepest sample we'll have out of the Moon for perhaps as long as the Moon itself has been there.

Scott: Well, that sounds good. Okay, Joe. I'm torqued to 292. And, Jim, I'll take a left turn out of here. *(pause)* Making sure I don't get any dust on your experiments. *(pause)* Okay. You're in a good spot.

An hour and one-half into the EVA, they are finally on their way in the Rover. Now that they know almost exactly where they landed, they know that the drive to Hadley rille will be a quick one: a trip of about 2 kilometers that would take 10 to 15 minutes, depending on the sort of terrain they encounter. During the SEVA, Scott had gotten a pretty good look at the countryside in all directions except the west. For the same reason that the Apollo 12 crew had been slow to recognize that Head Crater sat just down-Sun of them, Scott hadn't been able to see much detail in the direction of the rille and, now that he is driving toward the West, he and Irwin can only assume that they will find the same sort of terrain that they found south of the LM; consequently, their encounters with a series of three large depressions came as something of a surprise.

This visit to the rim of Hadley Rille will provide not only glimpses into the structure of the bedrock underlying the landing site but also a new perspective on the structure of the regolith. Although the thickness of the soil layer undoubtedly varies from place to place around the landing site, evidence from the seismic signals generated as the astronauts walked around the LM/ALSEP area as well as evidence from the depth of craters that brought up bedrock fragments suggests that the regolith is typically about 5 meters thick. However, as Scott and Irwin approached the rille edge, they see clear evidence of a thinning of the soil layer.

Figure 5.5. The Swann Range of the Apennine Mountains. On the moon, a lack of atmosphere makes shadows very intense and, with a low sun angle, the relief is spectacular. Because there is nothing for scale, however, these mountains (reaching 2 to 5 kilometers above the mare plains) don't appear to be all that high.

Because the Rover had only 14 inches of clearance, Scott and Irwin park well back from the edge of the rille. After they finish describing and photographing the far wall of this lunar canyon, they move down toward the boulders along the rim. Scott is quite confident that these are bedrock; Houston is so interested in them that the planners readily agree to drop a more conventional mare sampling site scheduled for the trip back to the LM in favor of a time extension here.

Time is running out. Only 55 minutes after they stopped, Scott and Irwin are again headed north, going to a spot another 200 meters along the rim so that Scott can take a second series of photographs for a stereo view of the far wall. It will be a very short stop – for photography only – because Houston wants them back at the LM in no more than 45 minutes. Lift-off is barely 5 and one-half hours away and there is a good deal still to be done. Scott understands the situation perfectly, but is still a bit disappointed. "Okay," he says. "Shoot! No time to go to the North Complex, huh?"

Allen recognizes the question as rhetorical and maintains an appropriate silence while Scott and Irwin go about the business at hand; 14 minutes after they stopped, they are moving again. They are still faced with the prospect of separating the bottom sections of core, but are determined that, somehow, they will get the entire core home. (In the end, Houston decided that there would be room for the three connected sections of core in the CM, and that is the way it was done.)

Scott: Hey, Joe, are you planning a mare stop on the way back?

Allen: Dave, we're standing by for a mark when you're rolling. And we'd like for you to press on back towards the drill site. We've got a procedure for you to separate two sections of the deep stem from the other two sections, and we're going to carry the two halves into the LM that way.

Scott: Okay.

Scott: Boy, it's just over hill and dale, isn't it?

Irwin: Yeah. *(to Houston)* Go ahead, Joe.

Allen: Roger. If you can take your ...

Irwin: Go ahead.

Allen: ... eyes off the scenery and the road for a moment, I could explain the core-stem separation procedure, if you'd like it now.

Irwin: Fire away.

Allen: Rog. Your choice. I can give you the procedure now or we can wait until we get past ... Oh, okay. Sorry, didn't understand. We think you can put the four sections of the core stem in the vise. Jim, you move it horizontally, put horizontal torque on the stem and tighten it up in the vise, while, Dave, you use the wrench and try to separate it.

By "horizontal," Joe means that Jim should push one end of the string either fore or aft to force the vise to grip.

Scott: Well, we tried that, Joe, but we'll try it again. It's worth another go. That was just about what we tried the last time, by trying to put horizontal and vertical forces and everything else on the vise, and the vise just doesn't seem to grip like it should, not nearly as well as the wrench does. And I don't really know why. It might be a little loose.

Allen: Okay, Dave. Copy that. It may also be that it's quite dusty, and the brush might help you out there.

Scott: Okay. We'll try that. Didn't try that.

Allen: And to set your minds at ease, we think that even if you can't separate it into two pieces, we can get the whole thing into the LM. So we'll bring it home anyway. We'd like it in two separate pieces, though.

Scott: Well, I guarantee you we're not going to leave it here now *(laughing)* after we got that much invested into it. Oop! *(pause)*

We rejoin the crew as the Rover approaches the LM.

Scott: Jim, I'll go around the north here and avoid the dust on our trusty ALSEP. *(pause)*

Scott: Hey, that's ... That's a pretty nice picture right there, Jim. Let me point you ...

Irwin: Swing around there, I'll take some.

Scott: That's really pretty! What kind of mag ... Oh, you got black and white. Better change that mag, buddy. *(pause)* Okay; there you go, right there. *(pause)* Got it?

Irwin: Yeah, I'll change it out when we stop. *(pause)*

Scott: I'll drive slow in here. There's our trusty drill.

Irwin: I thought you left it in the erect position.

Scott: No. No. *(long pause)*

Irwin: Friendly plains of Hadley.

Allen: Dave and Jim, this is Houston.

Scott: We've stopped, Houston! We're at ALSEP. *(pause)*

Allen: Roger, troops. We're thinking that when you ...

Scott: Okay. We're reading.

Allen: ... arrive back at the drill site: Jim, why don't you hop off and pick up the three important items there: the drill stem, the treadle, and the wrench. And, Dave, you can drive on back and park by the LM. Over.

Scott: Okay, give me your camera, Jim.

Irwin: Got it, Joe.

Scott: In work.

Irwin: Okay, I'll just skip the navs [that is, the readouts].

Scott: Yeah. *(pause)* Here. *(pause)* Okay.

Irwin: Let's see you have all the other stems in the bag.

Scott: Yeah.

Irwin: You have the wrench in there, too, don't you?

Scott: Yeah. I think so. If it's not there, I've got it in the bag.

Irwin: It's not here by the treadle. So all I'll do is pick up the treadle? *(pause)*

Scott: I don't know why we need the treadle.

Irwin: I don't either ...

Allen: Just take it on back. We might use it as a wrench.

Scott: *(responding to Joe)* Yeah, I guess you could. Allright.

Allen: And that's Walter Mitty plan number 2. *(long pause)*

Irwin: Okay, I have the treadle [and] stems, and I'm heading back. *(long pause)*

Scott: Okay, I'm off the Rover [at the LM], Joe. *(pause)*

Irwin: When I get up there, Joe, I'll give you the readings.

Allen: Okay, Jimmy. We're standing by. *(long pause)* Dave, are you arriving at the LM, now?

Scott: If you have that stem there. Jim? Okay, keep going the way you're going. Let me have the stem. Don't bother with the treadle yet. Let's ... *(pause as Dave looks in the SCB on the back of the Rover)* Oh, shoot! The tool isn't in there. Must be in your seat. You didn't see the tool out there?

Irwin: No, all I saw was a *(pause)* UHT.

Scott: Must be in your seat, then.

Irwin: Is it?

Scott: Let's see. *(pause)* Ah, shoot; I don't see it. *(pause)*

Irwin: I'll whip out there [to the drill site] again. But the only thing I saw was the UHT lying on the surface out there. Did you have a UHT in that area?

Scott: Yeah!

Irwin: Yeah. That's all that was out there.

Scott: But, you know, I thought I put the tool in the bag with the stems.

Irwin: Maybe it just ...

Scott: How could it get out and the stems not get out?

Irwin: How many stems did you have?

Scott: Two. *(pause)*

Irwin: I don't know, Dave.

Scott: I don't know either. I guess we'll have to use the treadle. *(looking under the CDR seat)* It's not in my side, either. Darn. Whew! Man, that's hot! I tell you what. The treadle.

The treadle and drill stems have been out in the Sun at the drill site since the start of the EVA and could have become quite hot; however, Dave doesn't remember to what he was referring.

> **Scott** – "Something I touched was hot – through the suit – so it must have been pretty hot."

Irwin: Want to hold it, and I'll get on the end there with the treadle?

Scott: Yeah, okay. *(pause)*

Irwin: How am I going to get the treadle over the cap, though? I'll have to come in from the other end.

Scott: No, the other end's got the same kind of cap, Jim.

Irwin: Don't guess it will hold with the cap on.

Scott: I'll take the cap off and go gently. I'll put the cap back on. Let's see. Which way should the treadle ... That's right. I guess it ... Easy does it. *(pause)* No, we'll never ... No, wait. Don't do that. Don't do that. We'll never get the treadle off. No, don't put the treadle on it. We'd never get it off. We've got nothing to get it back off with. Back off. Pull the treadle off. Bet thing ... Joe, will this stem fit in the LM someplace. I think it will.

Allen: We think so, Dave.

Scott: I think that's what we're going to do. We're going to take the stem with us, just like it is. I think if we try and fiddle with it, we're going to mess it up. *(pause)*

Allen: We hear you, Dave. And our camera's turtled up again.

Scott: Okay. Ahhhh.! Now! Okay.

Irwin: Did you get it?!

Scott: I got one off.

> **Scott** – "We fiddled around with the treadle some. That was somewhat of a chore, also, but I think that is inherent in the design of the equipment. If the drill works as

advertised, it really isn't bad. But, in summary, the ground being very hard tightened up the drills stems much harder than we'd seen before, and the vise not working on the back of the Rover complicated the separation of the stems. Finally, we had number 4 stem off about half way; and I finally, just in gripping the things, unscrewed it by hand."

Irwin – "I'd taken my protective covers off my gloves before I even went out on EVA-1; so, of course, they were off for this operation. I was kind of reluctant to grasp that drill [stem] very hard, afraid I might rip the gloves."

Scott – "That's a good point. I had to leave mine on the whole time because of the drill. The protective covers can restrain you hand movements even more than the gloves. [Consequently,] I had sort of degraded mobility because of those protective covers, all the way. I finally took them off after we got through with the drill."

Irwin: Okay, I'm going to work with the 16[-millimeter] camera here and see what I can do.

Scott: Good.

Allen: What did you do, Dave?

Scott: Boy, I tell you, my hands *(garbled)* done. Well, Joe, I just decided it was time to take that drill [stem] apart, and I took it apart.

Allen: That sounds easy enough.

Scott: So, now we have a three-stem section and three one-stem sections.

Allen: That ain't half bad.

Irwin: *(laughs; pause)*

Scott – "I just twisted them off."

Interviewer – "Put a hand on either side of a joint ..."

Scott – "Just twisted it off. I remember doing that. Quit screwing with it."

Interviewer – "You must have a hell of a grip!

Scott – "Comes a time when you've got to get it done."

Allen: Dave, we're standing by for a cap number on that ...

Scott: And Golf on the single stem *(garbled)*.

Allen: ... and, Jim, we need some help with the [TV] camera.

Scott: Here's the cap. *(garbled)* is here. Hotel is the upper part of the three-stem section.

Allen: Copy. *(long pause)*

Scott: Those caps aren't on there very good, Jim, so gotta be awful careful of it [while taking the core stems into the LM cabin]. Awful careful.

Irwin: Put it in the bag.

Scott: Can't.

Irwin: Oh, I see what you mean, yeah.

Scott: That one.

There is a break here as the crew loads equipment and samples into the LM and then climbs in to prepare for their launch and return to Earth.

Scott: *(garbled)* Brush you off. *(pause)* Man, that stuff really accumulates, doesn't it?. *(pause)* Getting it off your ... *(pause)* Okay; when you're ready, you can dust me off. *(pause)*

Allen: Jim, while you're dusting there, how many suitcases have you carried up?

Irwin: *(garbled)* dirty *(to Joe)* Oh, I only have two up there. There are two more down here plus the ETB ...

Allen: Okay, and ...

Irwin: ... and the core stem up there on the porch ...

Allen: Okay; good.

Irwin: *(garbled)* porch.

Allen: Good. And that core stem will go inside and lie on the floor against the midstep.

Irwin: Okay; we understand.

Scott: Man, I'm glad we got that core!

Irwin: *(laughing)* Davy's core!

Scott: I'll tell you.

Irwin: Okay, Dave, I think that's as good as we're going to get you.

Allen: How you doing, Dave?

Scott: Getting the last one right now, Joe.

Allen: Oh, boy. *(long pause)*

Scott: Okay. *(long pause)*

Irwin: Dave, just a reminder on the stems there.

Scott: Yeah. Let me get you this bag.

Irwin: Yeah. *(pause)*

Scott: There. Got it?

Irwin: Got it.

Scott: Good. Those are good little holders, there.

Irwin: Yep.

Scott: Last thing will be the stems. *(garbled) (pause)* Go, partner! Get them?

Irwin: Yeah, I got them.

Scott: Okay. Watch it. The caps are on not very tight; so be careful.

Irwin: Lie it on the floor, here.

Scott: Why don't you put it back up in the ... That's all right.

Irwin: Well, I'm afraid it'll get bumped there, Dave.

Scott: Yeah. I guess we'll just have to be careful and not step on the thing. *(pause)* Okay.

AND WAS IT WORTH IT?

The following is from an exchange of e-mails between author Jones and Dave Scott on November 25, 2003.

Interviewer – "Grant Heiken and I have been discussing your drilling activities at Hadley and we wonder if you have any thoughts about how hazardous an activity it was. If, for example, a drill stem snapped and you fell on the exposed end, would the fall in 1/6 g have been enough to compromise the suit? I know you strained your shoulder lifting the core out. Was that stressing the suit?"

Scott – "Well, falling on the exposed end would not be good, but I believe the suit would have taken it, especially since it would have been a glancing blow. Also, I doubt that the small force I was able to put on the drill would have broken it – remember, at 1/6 g. and in a static mode, I could not put much force on the drill; I did not weigh much [regardless of how it looked on TV!!]. And the shoulder [injury] was probably due to the upward force on the drill handle when we were pushing upward with our legs, it was not due to the suit *per se*."

Interviewer – "Neil [Armstrong] was once asked to rate the risk of the landing on a scale of 1 to 10. 'Thirteen,' he said. When asked about the Apollo 11 EVA, he said, 'One,' without hesitation. How would you rate the drilling exercise?"

Scott – "Based on that scaling (13 high for landing and 1 low for EVA), about a 3, risk-wise, primarily due to the difficulty in pulling the stem out and then breaking its sections, which I had to do by hand, exposing my gloves to the flutes in a compressed twisting mode. If everything were nominal, I would say a 1 or 2."

The troublesome drill core stems, some still coupled, were returned to Earth safely. From the stem weights, it was clear that the Apollo 15 crew had been successful in sampling the regolith to a depth of about 2.4 meters (7.9 feet). When the samples reached the LRL in Houston, the ticklish question for astronauts and scientists alike concerned the integrity of the sample: was it a good sample of the layers that made up the regolith or a tube full of jumbled dirt and rock?

Author Heiken vividly remembers what happened when the cores were brought to the LRL. Immediately after the Apollo 15 mission there was pressure on scientists in the LRL to produce justification for the intense effort and time spent by the crew to collect the drill core sample. 'During earlier missions, we had developed a procedure for x-raying the cores within their metal tubes, before any dissection or sampling. The procedure gave us a feel for what we would be analyzing and sampling and also provided information on myriad details of the regolith layers. Early in the preliminary examination of the Apollo 15 samples, we carried the unopened cores (in sterile, sealed bags) over to the JSC medical facility for x-ray imaging. By that time, the technicians there had grown accustomed to weird requests from the scientists in the LRL, and concerns among the principal scientific investigators about sample damage by the x-rays had also been answered; so on we marched, along with several security guards. After several hours in the clinic, we carried the samples and the x-radiographs back to the lab. We quickly analyzed the x-radiographs and found that the regolith layers were well preserved. Eventually, we identified 42 textural units (layers) that made up the 2.42 meters (7.94 feet) of sampled regolith. This was a truly significant geological sample.

We rapidly traced what we could see in the x-radiographs of each core tube and assembled a full-scale ink drawing of what we saw. Several poster boards were taped together, and the drawing was mounted on them. We raced the drawing to the Space Center auditorium where the post-mission Apollo 15 press conference was in progress. Just as a reporter was asking Dave Scott about the value of the time-consuming drilling effort, we ran down the aisle with the drawing. After looking at it for a few seconds, Dave grinned and said something to the effect that it was indeed worth the sacrifice.'

The core sections were finally opened extremely carefully in the sterile environment of the LRL's stainless steel glove boxes to protect the samples from organic contamination. Each metal core tube was cut along the long axis by a special milling machine. The grooves left on each side of a tube after the milling were only a few thousandths of an inch thick, leaving a delicate foil that protected the sample for the next step, which would allow the scientists to remove the top half of the tube. The trickiest part came next: using a scalpel to cut the thin remaining metal film on the sides of each tube. Those working with the cores were very nervous about this aspect of the operation – one slip would have been catastrophic because it could have destroyed the delicate layering and thus compromised the integrity of the sample.

Figure 5.6. This photograph, taken at the LRL in Houston, shows regolith layers (from a depth of 2.3 meters) near the bottom of the drill-core sample collected by Dave Scott. For scale, the core is about 2 centimeters in diameter. Dust-covered rock fragments are embedded in mostly fine-sand-size material – composed of rocks, minerals, and glass formed as products of impacts into the lunar surface by meteors traveling at high speeds.

After the top one-half of the cut tube had been lifted and placed to one side, the exposed regolith core was described and photographed.

What followed was a period of many months of description and dissection (years, in fact, for the entire 2.4-meter-long core from Apollo 15). Looking at an exposed dirty grey cylinder of soil was not immediately inspiring. However, by systematically dissecting the exposed half of the core (along the long axis), grain by grain, scientists eventually had a view of a remarkably varied sequence of layers. Samples from each layer have held data for every aspect of lunar research – from geotechnical engineering to planetary chemistry. After dissection and minimal sampling, the lower half of the split core tube was sealed to provide pristine samples for future generations of scientists, who will undoubtedly possess new analytical tools.

The dedication of the Apollo 15 crew of explorers to the Moon's surface resulted in a complex sample of the lunar regolith – one that has been and will continue to be the source of scientific surprises for generations. Collecting sequences of the lunar regolith was a part of all of the J missions (15, 16 and 17).

During the Apollo 16 mission, Charlie Duke and John Young would collect a drill sample with improved drilling equipment, which made the job considerably easier. In addition to continued regolith sampling, their goal was to explore the Descartes Highlands, which many scientists had interpreted as a volcanic field.

Having reached the rim of North Ray Crater, Charlie Duke is getting his camera out from under the Rover seat and is anxious to document this momentous occasion. He and John Young are about 4.5 kilometers north of and 200 meters above the LM. Later, Al Bean used this photograph as the basis of one of his paintings, adding John at the right side.

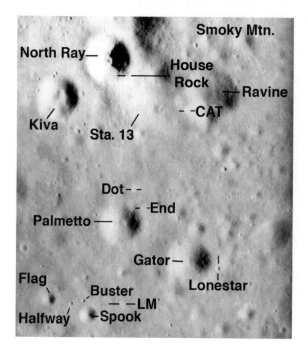

For the last EVA of the Apollo 16 mission, the crew drove the Rover north from the LM, past Palmetto Crater, and up on to the rim of North Ray Crater – a distance of about 4.5 kilometers. Although North Ray is a young crater and threw out some very big blocks that are easily visible in photos like this one (taken from the CM), the traverse route is surprisingly free of small rocks and craters.

6

The Descartes Highlands – High Land But No Volcanoes

During the decade before the Apollo expeditions, there had been an increased interest in Astrogeology (the study of the surfaces of other planets). Technically, *geology* is the study of the Earth, but the addition of *astro* was accepted, mostly because the entire Astrogeology Branch of the USGS was created to map the surface of the Moon. After the Apollo missions, the term *planetology* became the most common usage for the study of all planets.

NASA realized that photographic reconnaissance of the lunar surface was required, not only for evaluating the surfaces at proposed landing sites, but also to provide a scientific framework for further exploration of the Moon. Preliminary data came both from telescopic observations and from early orbital missions to the Moon; the most complete imagery was collected by the five Lunar Orbiters, which were flown over a 12-month period from August 1966 to August 1967.

The 1960s were good years for those who chose to study the Moon. As well as support for the USGS, there was research money available for geologists and geophysicists at universities and NASA centers. It became obvious that geologists were also needed to help train the astronauts themselves. In the early years, that training consisted mostly of classroom instruction in geology – which many of the astronauts disliked – and occasional excursions by large groups of astronauts to view geologic showpieces like Iceland and the Grand Canyon. However, as the first landings drew closer, the effort shifted from a general introduction to geology to the specific skills and knowledge the astronauts would need to perform well as field geologists on the Moon. Qualified instructors were needed and decisions had to be made about what the astronauts needed to know. Perhaps not surprisingly, differences of opinion, priorities, and organizational responsibilities between various groups of geologists had to be resolved.

Before and during the first several Apollo missions, a tension had developed between several geological groups. The USGS declared that they had the mandate for *all* astrogeology – a territorialism that did not create many allies within the academic and NASA geology teams. This tempest within what should have been a tightly knit research team was counterproductive. Fortunately, much of this competition was between the managers of the various groups; most of the staff geologists were able to work effectively with their counterparts in the other organizations. To establish a formal advisory group and to create referees in this academic melee, NASA hired a group of geologists who would work for Bellcom (a subsidiary of Bell Laboratories) at NASA head-quarters in Washington, DC. Noel Hinners, Jim Head, and Farouk El-Baz, of the Bellcom group, were superb not only as scientists but as advisors to the space agency.

APOLLO 16 FACTLIST

Crew	Commander, John Young; CMP, Ken (aka TK) Mattingly; LMP, Charlie Duke
EVA CapCom	Tony England
Command Module	Casper
Lunar Module	Orion
Launch from Earth	17:54 GMT 16 April 1971
Lunar landing	02:24 GMT 21 April 1971
Popular site designation	Descartes Highlands
Lunar stay	71 hours 2 minutes
Splashdown	19:45 GMT 27 April 1971
Command Module lunar orbits	64
Time outside the LM	3 EVAs, 20 hours 12 minutes
Maximum distance from the LM	4.6 kilometers
Distance walked or driven	27.0 kilometers
Lunar samples	96 kilograms
Lunar surface photographs	1787
Surface equipment	563 kilograms
Highlights	First landing in the lunar highlands

The astrogeology community resolved many of their differences during the earlier missions; during Apollo 14 and 15, the Apollo geology team and those examining the lunar samples in the LRL worked well together. By the time of the Apollo 16 mission, scientists had studied the additional high-resolution orbital photography from earlier Apollo missions and a great deal had been learned from the returned samples as well. It was becoming evident that high-velocity impacts and not volcanism had shaped most of the lunar surface.

The USGS astrogeologists were predicting that the Apollo 16 landing site was of volcanic origin. The geologists who constructed the pre-mission maps inter-preted the Cayley Formation (plains units) as "volcanic materials of high fluidity, either ash flows or lava flows of low viscosity." Maps and reports also described the Descartes Mountains as volcanic cones or shields. Newell Trask and Jack McCauley, two members of the USGS group, weren't so sure. Although they had published a paper in a geologic journal on the importance of the site, should the highland be volcanic, they wisely hedged their bets, saying that "Photogeologic interpretation alone cannot rule out the possibility that all the hilly and gently rolling terrain belongs to one or more of the hummocky ejecta blankets surrounding the large circular basins."

APOLLO 16

As Don Wilhelms mentions in his book *To a Rocky Moon* in 1993, the Apollo 16 crew had one field exercise a month between September 1970 and February1972. The geology principal investigator, Bill Muehlberger (University of Texas) and the NASA mission science trainer Fred Hörz (NASA MSC) weren't totally sure about a volcanic interpretation for the landing site and made certain that the crew saw some impact craters and their deposits as well as volcanic features during those field sessions.

Before Apollo 16, one interpretation of the site geology concluded that the rounded hills at the site were lava domes. On Earth, such domes are distinct features formed during the eruption of extremely viscous lavas with high silica contents. They are often shaped like a well-developed soufflé, but some have surface depressions (like a failed soufflé). The eruptions that create such domes are slow, tedious extrusions, and these features may take many decades to form (as is the case with ongoing eruptions on the Caribbean island of Montserrat). If such domes were found on the Moon, there would have been clear implications about the history of the lunar interior and the generation of the lavas that had erupted on the lunar surface.

Based on a volcanic interpretation for the Descartes region, it was logical to plan Apollo 16 astronaut training exercises at an accessible line of rhyolitic (silica-rich lava) domes. One such site was Mono Craters – a line of spectacular domes that extend south from Mono Lake, California, toward the town of Mammoth Lakes, a popular ski destination. From the air, the domes look like enormous muffins. Training traverses involved sampling, photographing, and describing various geologic features and materials in a way that would best characterize the site. Although the crew had some geological training, it was much more important that they be objective observers, mastering the ability to describe and sample a site in the shortest possible time. The training stressed that they not spend too much time interpreting what they saw, yet it was important that they have some sense of why they were exploring a particular landing site. Aerial photos of volcanic domes on Earth, taken by Gordon Fullerton (Apollo 16 support crew member) from his NASA jet, were used in training the CMP as well as the astronauts who would walk on the surface.

The major objective of the Apollo 16 mission was to determine the origins of the hilly (purportedly) "volcanic" features as well as the basin-filling deposits of the Descartes Highlands. The landing site provided access to both types of landforms. The crew landed less than 100 meters (330 feet) from the planned landing point, and their traverses covered an area of about 18 square kilometers (7 square miles).

Because of a problem detected in the Command Module shortly after the LM crew undocked in preparation for the descent to the surface, the Apollo 16 landing was delayed by 6 hours. John Young and Charlie Duke had planned to conduct three 7-hour EVAs during the mission and, thanks to Houston's careful revision of the remaining mission timeline, the first two EVAs lasted for more than 7 hours each and the final EVA was cut by only 1 hour. The primary goal of EVA-3 was to collect samples at the rim of North Ray Crater and – because all other stops were eliminated from the plan except one on the way back to the LM – John and Charlie would be able to spend nearly 2 hours at North Ray. During the 5-kilometer drive north from the

Figure 6.1. In the early years of Apollo, one hypothesis for the nature of the lunar highlands at the Apollo 16 sites was that they included large lava domes similar to those on Earth, which are composed of silica-rich, very viscous lavas. In June 1971, the Apollo 16 crew were given a chance to examine such features closely during a training exercise at Mono Craters domes in eastern California. These large, biscuit-shaped volcanoes are quarried for the "pumice rock" boulders commonly used in landscape gardening. In this photo, taken near the base of one of the Mono domes, John Young examines a sample while on the move. It is unlikely that John took his pipe to the Moon! (*photo taken during the June 10–11, 1971 field trip, probably by NASA photographer Pat Patnesky.*)

LM, they would cross the rim of Palmetto Crater, named for John's native state of Florida, spot Cat Crater, the planned location of Station 14 (named for Charlie's sons, Charles and Thomas), and see Dot Crater, the planned location of Station 16 (named for Charlie's wife, Dotty).

North Ray is a 1000-meter-diameter impact crater that was blasted into a ridge adjacent to Smoky Mountain. The final portion of the drive involved a climb of about 75 meters in a distance of about 1 kilometer. The climb was similar to the one Shepard and Mitchell made on foot to the rim of Cone Crater during Apollo 14; but the Rover made it a lot easier. Although there had been no high-resolution site photographs available before the mission, there were obviously hints of big boulders on the rim of

North Ray. Actually, during the landing, Duke had snuck a peek up north and had assured himself that there were, indeed, some very promising rocks at the rim of the crater. The geologists wanted samples from a single, huge boulder, if one could be found. What they hoped for was a rock big enough to show multiple igneous (volcanic) units. Specifically, they wanted – in the words of the procedures volume – a "crystalline rock, larger than 5 meters."

TRAVERSE TO NORTH RAY CRATER

The following transcripts and interviews quote John Young (Apollo 16 CDR); Charlie Duke (Apollo 16 LMP); Tony England (the EVA CapCom), a scientist astronaut with degrees in geophysics who now holds dual professorships in the Departments of Electrical Engineering and Atmospheric, Oceanic, and Space Sciences at the University of Michigan; and author Eric Jones.

Duke: Okay. We're all set to go. And off we go.

England: Good show.

Duke: Okay. First heading out of here, John, is 030. 030 supposedly. Okay. The DAC [data acquisition camera] is On.

Young: Shoot, that thing [possibly the DAC] drifts.

England: Charlie, can you turn that DAC off for another 19 minutes?

Evidently Houston wants to make sure that the DAC doesn't run out of film during the approach to North Ray Crater.

Duke: Okay, Tony. I'm going to be spending ... *(stops to listen)*.

Charlie may be saying that he will have to hold the DAC to keep it from moving.

England: I'll give you a mark [when to turn the DAC on].

Duke: Okay. *(listens to Tony)* Okay, it's Off.

Young: I'll tell you, this ridge up here would be a good place to park the Rover, Houston, if you want [to watch the launch from the north] ... We're reading ... I think it's [meaning the Rover Nav system] working. It's reading 162 and 0.1 now. And that's how far away we are.

England: Good show.

> **Young:** This ridge would be a good place to park the Rover. Up north of it. I don't know if you can stand it thermally.
>
> **Duke:** I don't think they can pan the camera like that, when we lift off. *(pause)*

At the end of the EVA, John will park the Rover about 100 meters east of the LM so Houston can watch the lift-off. Because the Rover has its own batteries, independent of the LM, a TV camera can be mounted on the Rover and, thanks to the two-way communications link provided by the Rover's high-gain antenna, can be remotely controlled from Earth. With the Rover parked east of the LM, Ed Fendell, the Houston operator of the Rover's TV camera, will only have to pan the camera up to follow the LM. Charlie's thought is that, although the ridge might be a good vantage point, Fendell would have trouble panning left to right to follow the LM once it pitches over at an altitude of 800 feet during the westbound ascent. John's concern is the higher thermal load on the front of the Rover if it is parked facing south rather than west.

The following is taken from a conversation author Jones had with Ed Fendell at his Houston office on August 13, 1996.

Interviewer – "My reason for being here, Ed, is to ask you to talk about how you got prepared for doing the J-mission TV and anything you think we need to know about how the Apollo TV system evolved."

Fendell – "Well, let me give it to you this way. I'll tell you what I can remember, because I've got a good friend of mine who was one of the senior flight directors at that time – Pete Frank. Pete and I are pretty good friends and we worked on Habitat for Humanity together and different things. And he just retired from Loral [Space Systems]. And we were talking about remembering things and he says, 'It's not only the point that you forget things, it's the point that you remember certain things and then somebody else says something and you find out that you remembered it wrong.' And this conversation went on yesterday; it had nothing to do with you. A good friend of ours passed away and we had a funeral we had to go to."

Fendell – "Anyways, let me see if I can kinda go back and start with the right place and see what I can help you with."

Fendell – "Prior to Apollo 10, all the communications equipment was spread out in the Control Center in different areas. For example, the Lunar Module communications equipment belonged to the Lunar Module guys. You know, the TELMU [Telemetry, Electrical, and EVA Mobility Unit] guys. And the Command Module equipment belonged to the Command Module guys. And the erectable antenna we had to go on the Moon and the backpack stuff and all that belonged to the experiments people. On 10, we tried to separate [meaning undock] the two vehicles but we couldn't get a Go on the comm between the Lunar Module and the CSM. We couldn't make it work. And we fooled around for 2 or 3 hours – I can't remember how long – and, finally, we got it to work and found out that we were just configured wrong. The Lunar Module guys [at Mission Control] were not talking correctly to the Command Module guys and vice versa."

Interviewer – "This was pre-flight?"

Fendell – "This was during the mission. We were out of configuration. The switches were not matched up between the two vehicles. So we got that straightened out and we separated and went on to do the mission. When the mission was over, Kranz came up to me – and I go way back; I'm an ex-Capsule Communicator back at the Cape and I'm an ex-air traffic controller. Anyway, Kranz came up to me said 'We're going to form a communications section, and you're going to run it.' See, I was an assistant Flight Director in those days. And I said, 'Yeah'. And he said, 'We're going to take all this gear out of all the different areas and we're going to put it all into one section and you're going to run it.' And I said, 'Who am I going to run it with?' And he said, 'The following people.' and there was only three guys who knew communications out of the twelve or so people. Anyways, we took it over and we started doing it on [Apollo] 11 and then it progressed down the line and we built this thing up and we started coming along."

Fendell – "We flew 11, we flew 12, and we got to 13 … And, in the meantime, a guy by the name of Bill Perry, an engineer over in the telecommunications division, came up with a design for actually controlling a [TV] camera – from the ground or remotely. It [meaning Perry's concept] went through a PRCB (Program Review Control Board) to decide whether or not the program wanted to do this. And each directorate had to come in and make their pitch."

Interviewer – "When was this?"

Fendell – "I can't remember exactly when this was; but it's kind of an interesting story. Anyways, I, being a communications guy, went up to make the pitch for Flight Operations Directorate [FOD]. At that time, [Chris] Kraft had just moved from the head of Flight Operations to the head of the Center and he was there at the Board. And I had a 'one chart'. My pitch was one slide. 'FOD has no requirement for remote TV.' That is what it said. That was it. Well, we got going on this thing and just ahead of me was a fellow by the name of Tony Calio, who later on ended up becoming a director up at NASA Headquarters. Well, Tony headed up Science and Applications Division, which were the guys who controlled the geologists and everything else. And he got up there and started making the pitch for them; and he was saying how they had no requirement for this thing. Well, Kraft got a hold of him and ate him a new butt hole. I mean, he was chewing him … 'What do you mean? This is stupid. This is your chance … This is no foresight …' And he was eating him alive; and I'm sliding down in this chair, trying to figure out how I'm going to get out of this."

Fendell – "There was a guy by the name of Mel Brooks, who's over in Europe now, and he was the Assistant Director of Flight Operations and he comes running back to me and he says, 'Get rid of your slide! Get rid of your slide!' And I said, 'I can't; it's already up with the projectionist.'"

Fendell – "So, anyways, he ate on Tony for about 45 minutes. I don't know how well you know Chris, but Chris can just eat you alive. He's a good friend of mine; but he's the most outstanding person I've ever met in my life."

Fendell – "Well, he gets all done with him [meaning Calio] and I walk up there and the guy throws up my slide. And the place just came apart. Everybody just started hollering and I thought Kraft was going to die."

Fendell – "Well, that's how it all started. The end result is we were directed to go off and do this. They put together what they called a Tiger Team and it works out that the guy who did the actual procurement job – he was the procurement officer – is running this building [that houses Ed's 1996 office]. And there was this guy Bill Perry, and another guy by the name of Olin Graham out of telecommunications, who was his boss. And there was – oh – about eight of us. And it was sole-sourced to RCA-Camden [New Jersey] to build the LCRU. It [meaning the development of the TV and the LCRU] was a two-pronged thing: they [RCA-Camden] were going to build the LCRU and then, [for construction of the TV camera], a procurement went out and an RFP [Request for Proposal]. We had two bidders [on the camera] – RCA-Hightstown [New Jersey] and Westinghouse – and, after doing the Source Board, RCA-Hightstown was picked over Westinghouse. Westinghouse had the Command Module camera."

Fendell – "We started traveling back and forth and working it. And we worked it a little different from normally; they basically took Bill Perry's design on the camera controls [TCU, Television Control Unit] and implemented the whole thing together. Both companies bid, basically, on Bill Perry's control design [for] the camera and the motors and the gears."

Interviewer – "I know the setup from a crew operations perspective where they mount the LCRU on the front of the Rover and then they mount the TCU and then the camera."

Fendell – "So, that was all put together and we started training. And the real question came about: 'Well, how are we going to do this?' One, we had to put in commands in the Control Center to operate it. And what we did in the Control Center was we built two types of commands. Pan Right was a command; Pan Left was a command; Pan Stop was a command. Tilt Up; Tilt Down; Tilt Stop. Zoom In; Zoom Out; Zoom Stop. The other thing we did was we built what we called 3-degree commands. 'Cause the way we did that was, underneath each one of the depressions of the PBI [Push Button Indicator] was a Pan Right Increment command. So, when you send Pan Right Increment, the control went up to turn the camera to the right [and then] it waited the appropriate amount of time, and then automatically sent Pan Stop."

Interviewer – "And it's called a '3-degree command' because that's how far it goes, about half the width of the 6-degree field-of-view [at minimum zoom]."

Fendell – "Yeah. So we had increments for left and right [and] zoom in and out. So that you could tweak [the pointing and zoom]. See, you've got to understand you had that delay time on the camera [response]."

Interviewer – "Absolutely."

Fendell – "So we built the commands to do that. [For example], on the thing you saw when the Lunar Module lifted off, those commands were not sent by watching the Lunar Module. They were sent on time [that is, at pre-planned times relative to the instant of lift-off], because we were moving the camera to where the Lunar Module was supposed to be. [There was] a fellow named Harley Weyer [pronounced 'wire'], who worked for me and is now retired from NASA, and Harley had figured all that out. And that's the way that worked. We watched it going up [in elevation], then opened up [on the zoom], and then [watched it] going down [as the apparent elevation decreased with increasing LM range from the TV]. And we only got to do it right once [on Apollo 17], because the [TCU TV] motors froze the first time [on Apollo 15], etc."

Duke: Okay, Tony. We're topping out of the little ridge. We can now see Dome and Smoky. On top of the ridge, there are boulders much like we seen yesterday. [The boulders are] half a meter or so. Cobbles [cover] about 5 percent of the surface. Looks like a lot of secondaries though. The boulder population is really concentrated around the secondaries; and we'll get some pictures of that. The regolith up here is identical [to that around the LM]. You can see these little lineations, which is, I think, a function of Sun angle.

Young: I think the boulder population is starting to thin, Charlie.

Duke: I do, too. They're getting smaller, and the cobbles are getting smaller. Looks like we could be just out of this ray. We don't see any – maybe one or two – of the half-a-meter-size boulders now, Tony.

England: Okay, we copy that. There are a couple of those mounds mapped about 200 meters off to your left. We were wondering if you could see those.

Tony may be referring to features similar to the meter-size conical mounds of soil that Pete Conrad and Al Bean found near the Apollo 12 ALSEP deployment site. However, to be even marginally visible in the photos available at this time to the geologists in the Backroom, mounds would have to be at least 2 to 3 meters tall.

Duke: Nope. I got 179 – 180 now – at 0.3; and really on top of a ridge here. And . . .

Young: Better go over this way more, Charlie.

Duke: Yeah. There's North Ray right up there. Look at the big rocks [on the rim of North Ray], John.

Young: Yeah.

Duke: Okay, Tony. You got a good view of North Ray here, and as I look at it, there's a northeast-southwest line of boulders that come out from the southwest rim and go up – *(correcting himself)* the northeast rim – Smoky Mountain.

Young: See, there's Palmetto, too.

Duke: Yeah, we see Palmetto. *(pause)*

Palmetto Crater has a diameter of about 1000 meters and the southern rim is about 800 meters north of their present position.

Duke: Coming down the ridge now. We look like we're going into a big sag-type area. It's at 12 o'clock and 300 or 400 meters, and we're now at 188 at 0.4.

The "big sag-type" area may be a shallow feature about 400 meters in diameter. At about this time, John angles his track to the east and goes around it, thinking they have reached Palmetto Crater.

Young: We may be into Palmetto right now, Charlie.

Duke: Huh?

Young: [This] may be Palmetto.

Duke: No. That's over there on the rim, isn't it?

Young: Huh?

Duke: That big thing right there? No, Palmetto's at ...

England: You should be about halfway to Palmetto.

Duke: ... 2.1.

England: You're looking [as if you are] right below [meaning, a bit south of] Turtle Mountain, we bet.

Young: Yeah. That's where we are.

Duke: Okay. Up at Station 11 and 12 ...

Young: Any resemblance between this and the topo is going to the devil.

John is probably referring to the large site model they used in training. Before each mission, the USGS used the best available photographic data to build a model that covered much of the site that would be visible at pitchover during the descent. The model was about 4 meters north-south by about 7 meters east-west and was ceiling mounted in an area adjacent to the LM simulator. A mobile TV camera would move under the model in response to the simulated motions of the LM and the resulting view was presented to the crew on monitors mounted in the simulator windows. Alternatively, this Landing and Ascent (L&A) Facility could be hooked up to a stationary Rover simulator to give the crew a feel for the major features they would be seeing during the traverses. Because of limitations on how close the camera could get to the model, the view they had in the monitor was equivalent to a height above the ground of about 100 to 200 feet. (Photos of the Apollo 15 model being delivered to the Cape and the Apollo 15 crew doing a simulated traverse are available on the ALSJ site.) After the end of Apollo, most of the site models were discarded; fortunately, the Apollo 15 site model was salvaged and is on display at Brevard (County) Community College in Florida.

Duke: I'll tell you. It sure is. Okay; we're going down slope now, Tony. [It's] about a 5-degree slope, and we're going to go down perhaps 50 or 60 meters before we start climbing back out again towards Palmetto. *(pause)* And up around North Ray, we see two tremendous blocks at about Station 11 and 12 that appear to be black in color. Black with white spots. And we're just about out of the ray material now. We only see a few cobbles left.

England: Okay. It might be a good idea to try to get there [near the large boulder] for our Station 11.

Duke: That's what I thought we were going to pick: those two big – big rocks.

England: Good show.

Duke: It's right up on that ridge. *(pause)* That might be ... That's Palmetto right there, I guess, off to the left there, isn't it, John? Course, we've only been 0.6 though. Tony, we're at 195 at 0.6, and there's a big depression off to our 2 o'clock position on a heading of 030, with some white boulders on the inner rim. It's a very subdued feature, but it does have stuff, at least around the rim ... The south rim ... Wow! Great!

John may have just bounced the Rover by hitting a small crater.

Young: [We're] all right, Charlie.

Duke: The east side is a very shallow slope into this pit.

Young: How far are we supposed to go this way before we turn back?

Duke: Just keep going. Straight out. *(pause)*

England: You should be heading of about 356.

Duke: *(lost under Tony)* turn ... *(stops to listen)* Yeah. Now ... At 0.8, you can turn to about 356.

Young: Okay.

Duke: Okay. Tony, this big depression off to the left that I was describing . . . on the east side it's a very shallow slope into it, about 4 or 5 degrees. But, on the far end, the west side and the southwest side, it has very steep walls, 20 degrees or so.

Young: *(garbled)*

England: Right. Understand. You're looking right at the base of Turtle Mountain.

Duke: *(lost under Tony)* piece of cake, John. *(hearing Tony)* Okay.

Young: Listen, Houston. *(laughs)* I hate to tell you this, but these mountains don't look the same [as they did in the training model].

Duke: Which mountain? Those straight . . .

Young: Where's Turtle Mountain. Right here?

Duke: It's off to the left, way off to the left. We just passed it. We could do a 360 [-degree turn] and get a pan of it.

Young: How about that rock there, Charlie?

Duke: And it's got some lineations in it, huh?

Young: Yeah. Look at the size of it.

Duke: Oh, this big one coming up, you mean?

Young: Yeah.

Duke: Yeah. *(pause)*

Duke: Hey, Tony, it seems to me this is a more subdued [meaning, smoother] surface over here than going towards South Ray. Not as many craters. It's almost . . . Except for 3- or 4-meter-size craters, it's all subdued and just hummocky and rolling.

Young: Yeah, that's true. It's much better driving. We're doing 10 clicks.

England: Outstanding.

Interviewer – "If they targeted you up north of where they did, would there still have been a problem finding a level place for the LM?"

Duke – "No; I think there would have been some nice landing spots up there. It was like a big depression. And I don't remember any major craters out to the northeast of Palmetto."

Interviewer – "No hummocks or anything."

Duke – "No. Well, it was just gently rolling and I think it would . . . We make some further comments about it, later on; but it was real nice. Not any blocks."

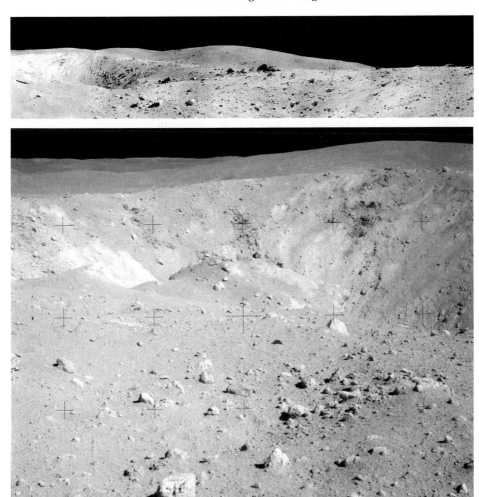

Figure 6.2. (*top*) John took this panorama shortly after he and Charlie arrived at North Ray Crater. John is standing below the Rover near a dramatic change in slope, downwards into the crater. At this point they are 150 meters above the lunar module on the plain below, enjoy a great view into the relatively fresh North Ray Crater, and have an overview of the geology of the Descartes Highlands – exciting exploration! (*bottom*) This is a closer view of the 1-kilometer-diameter, 250-meter-deep North Ray Crater. Because the inner walls become quite steep not far inside the rim, John and Charlie decided that they wouldn't go far enough in from the rim to see the bottom of the crater. An understanding of how impact craters form indicates that the samples collected from this site come from deep below the original surface and the samples collected along the North Ray rim are mostly impact melts and fragmental rocks, not silica-rich lavas. Later studies confirmed that both the hilly Descartes material (excavated by the impact that formed North Ray crater) and the plains below were formed by earlier impacts and not by volcanic activity.

England: Could we have an Amp reading? *(pause)*

Duke: Twenty [amps]. Hey, there's about a 4-meter boulder off to our ... With a good fillet ...

Young: Oh, that's nice. That's been there for a while.

Duke: ... we're just passing at 195 at 0.9. It's rounded.

England: Okay. How are we doing now on the rounded versus the angular boulders?

Duke: And we can see into Ravine ... *(stops to listen)* Okay; most of them over here are, I'd say, probably a good half and half. Rounded to angular.

England: Okay.

That is, about one-half of the rocks are angular and one-half are rounded.

Duke: There are some small, indurated [means, dug into older craters with high-standing, consolidated rims], secondary craters. And as we approach Palmetto, the boulder population is beginning to pick back up.

Young: This Palmetto right here, Charlie?

Duke: Yes, it is. We've got to ... I think it is, John. Yeah, right up there to the left. We've been ... I think that's 1.1 [kilometers], and we're not quite there yet. We've got to go on this heading for ... *(pause)* Okay, Tony, as you look to the northeast, you get a spectacular terrain view of rolling hills occasionally pock-marked with large boulders. The craters are very subdued, and the hills almost appear smooth, off to the northeast. Occasional craters ...

England: Okay. We copy that, and ...

Duke: ... That might be a function of the Sun angle, though.

England: ... according to our track, you're a little bit east of – correction, west – of course, and probably a 005 heading would take you right along the rim of Palmetto.

Young: Yeah. That's what I figure.

Duke: I think we're coming up on the rim now, John.

Young: You're right, Charlie.

Duke: There it is.

Young: There it is. Beautiful.

Duke: Okay. Tony, we topped out on the rim of Palmetto, and hit it right on the nose at 1.2 at 189.

Duke: And it's a tremendous crater. The walls to the south . . . south . . . northeast – correction, northwest – south . . . Wow!

Young: Sorry, Charlie.

Duke: Beautiful. That's great.

Young: I got to keep my eye on the driving.

Duke: That's great. And to the southeast here are steeper than the walls to the northeast. Apparently, it looks like it's almost breached to the northeast.

England: Okay. *(pause)*

Duke: John's cutting away from the rim now, because it's a little bit easier going.

Duke: There's a good ejecta blanket of one-half-meter-size boulders around the rim of Palmetto into some of these secondary craters here.

England: Okay. Do you have an estimate of the coverage [meaning, the percentage of the surface covered by rocks]?

Young: *(to Houston)* Palmetto is as big as Meteor [Crater in Arizona], isn't it? It's . . . *(stops to listen)*

Palmetto is about 1000 meters in diameter; Meteor Crater is about 1200 meters across.

Duke: Okay; of cobble size – in my usual size of being cobbles – I'd say 30 [to] 40 percent of the surface [is covered]. Let's make it 30 percent; and the one-half-meter size, [there is] maybe one for every 10 square meters.

England: Okay.

Young: Okay. We're just traveling right around the . . . We're traveling about 100 meters inside the rim, and we're at 195, 1.4 [kilometers], now.

England: Okay.

Interviewer – "The pictures you took of Palmetto don't show the kind of detail that we have of some of the other big craters. Did John turn the Rover so you could get pictures? Or did you basically keep going."

Duke – "No; we just kept going."

Young: 1.7 is the distance.

John is saying that they have driven 1.7 kilometers (a little over a mile) since leaving the LM. There is a difference between the 1.4-kilometer range and the 1.7-kilometer distance driven because he has not driven in a straight line from the LM.

Duke: Okay; and to the northeast, Tony – northwest, correction – you can see large blocks on the rim on the . . .

Young: Hey, Charlie, there's Dot.

Dot Crater is a fresh, sharp-rimmed crater at CN.2/81.3, just north and slightly east of Palmetto. It is surrounded by a light-colored ejecta blanket.

Duke: Yeah, I see Dot. Great. Hanging right in there, right on the rim. You won't be able to see the . . . You wouldn't be . . . Yeah, you'll be able to see into Palmetto from there. Okay; the large boulders over there seem to be 3 or 4 meters [in size and they are] to the northwest [means northeast] on the flank of Palmetto, but I think they came from North Ray. Over.

Charlie may be underestimating Palmetto's size because, in the blow-ups of high-resolution photographs, it appears to be about the same size as the LM.

England: Okay. *(pause)* Understand they're angular?

Duke: Angularity is sort of rounded.

England: Okay.

Charlie is using *angularity* in the sense of geologic characterization, which includes a spectrum of angularities from "angular" to "rounded."

Duke: Apparently, the only thing preserved here is . . . There's the large blocks out of North Ray, but [we] don't see very many small ones. I think trafficability is going to be excellent, though it looks like a steep slope climbing [to] that [North Ray] rim, doesn't it?

Young: No, not too bad. It's not near as bad as Stone Mountain.

Duke: Okay. The boulder field out of North Ray does not reach Ravine, Tony. It stops on the outer flank of Ravine about a tenth of a [Ravine] crater diameter away.

England: Okay. *(pause)*

Ravine is a large, shallow crater on the southern flank of Smoky Mountain. In the high-resolution orbital photos, a number of LM-size boulders are visible just west of the Ravine rim. There are also a few on the floor of Ravine and on the inner, eastern wall, but these may be hidden from Charlie. There is only one visible large boulder in the area south of Ravine, and that LM-size boulder is about a kilometer south of the southern rim.

Duke: Okay. Most of the rocks here [at Palmetto] are rounded. Have some real good secondaries. The types are very difficult to identify as we go by. We're now at 193 at 1.7. The Nav system seems to be working super.

England: Outstanding.

They are about 150 meters south/southeast of End Crater, which is a small feature on the northeast rim of Palmetto.

Duke: Palmetto has a very definite raised rim to it, and we're going to be going off the rim down probably a 5- to 10-degree slope into a valley before we start climbing up to North Ray.

Young: That's a real valley, too, Charlie.

Duke: Yeah. I'll say. And these valleys in here, Tony, tend to trend towards Big Sag.

England: And Charlie, could you go ahead and ...

Young: *(to Charlie)* Hey, see Cat [Crater]?

Cat Crater is on the southwestern rim of Ravine and was the planned location of Station 14.

England: ... put that DAC on that?

Duke: Yeah, there's [Cat] ...

Young: Yeah, put it on now, Charlie.

Duke: *(straining to reach the camera)* Okay. Wait a minute. *(pause)* Okay; it's running.

England: Good show.

Young: It's not pointing up ...

Duke: And I got it pointed off to the right ... *(correcting himself)* or left. Excuse me.

England: *(joking)* Your other right.

Interviewer – "When you were out on field exercises driving the 1-g trainer, did you try to do this kind of terrain description?"

Duke – "That's what we did; that's what we practiced. The navigation is a lot easier than it was on EVA Number 2; and so I had a chance to talk more."

Interviewer – "Whereas, on the first two [EVAs], you'd spent a fair amount of time looking at the map."

Duke – "Uh-huh."

Duke: Okay. We're now in an area, Tony, that's at 195 at 1.9, that is [covered with] about a one-half-meter-size boulder every 5-meter square.

They are passing significant numbers of one-half-meter boulders. As John mentions in a moment, they are traveling east to find some smoother ground.

Duke: Some of these blocks are angular; they're fractured. They appear to be grayish in color, dust covered, and most all of them have fillets. Man, look at that slope! That's End Crater right there, John, just over that rim, there, just to your left. And, Tony, End Crater is on about a 10- to 12-degree slope, pointed toward North Ray.

England: Okay.

They had planned to do Station 16 at End Crater; fittingly, as the last stop, it was also the "end" of the traverse geology.

Young: And we're traveling due east here for a while to pick up a little smooth ground.

Duke: What do you say? We're going down a 5-degree slope here or 10?

Young: Yeah, Charlie; 5 to 10. More like 10.

Duke: Yeah. Okay, Tony. We're about maybe a half-a-crater diameter to the northeast of Palmetto, about a 10-degree slope, and the boulder population is about 5 degrees [means 5 percent] here. And the small cobbles have just about disappeared. Very smooth regolith, except for these 20- to 30-centimeter boulders, which are not very numerous. We're really moving out, downslope, at about 12, 15 kilometers an hour.

England: Okay.

Duke: It's remarkable how subdued all these craters are. It's almost a smooth plain except for a few of the 5-meter craters or so. The 1-meter size, and all, and smaller, are just about gone. Apparently. Very subdued. *(pause)* Okay. John, we're at 2.2 at 195. We'll swing the [16-millimeter] camera around toward the Sun. *(pause)* It's looking off to the right now.

Young: Let's get a better heading here from 2.2 to [means "and"] 195, Houston.

England: Okay.

John is asking Houston for advice about the heading he should take from their current position.

> **Duke:** Okay. End Crater was 2.1. You should ... What they want is about ... Just directly north, John.

Charlie may be using the contour map. As shown on that map, they had planned to drive 2.5 kilometers (1.5 miles) north on a heading of 002 to a point where the bearing and range to the planned landing site is 185/4.4. Because they landed only 200 meters (656 feet) north and 80 meters (262 feet) west of the planned spot, the actual bearing and range will be about 185/4.1. The mobility that the Rover provides is a key factor in making such variances unimportant.

> **Young:** Forward?
>
> **Duke:** Yeah. It looks great to me, that heading.

Although North Ray is now hidden by the slope they are about to climb, John and Charlie have been keeping track of horizon features and the large boulder on the rim of North Ray and have a good idea of the heading they need to follow. The distance itself is not a problem if they head in the right direction. They may reach North Ray before they expect to, or after, but they will find it if they are on the right heading.

> **Young:** Yeah. It looks like ... Well, we're down to about where the rock population is almost nonexistent. I hope it stays that way for a while.
>
> **Duke:** It is.
>
> **England:** Y'all are making some outstanding time there.
>
> **Duke:** It's really easygoing, Tony. *(hears Tony)* Well, he's got it full blower at 11 clicks, and we're just going over an undulating terrain. The ridge lines here predominantly trend east-west, and they are about 5 meters in [vertical] relief. And really, the only significant craters that you have out here are the ones that are 5 meters and larger, and they only maybe cover 30 percent of the surface. Look at that view!
>
> **Young:** Look at those boulders [on North Ray]!
>
> **Duke:** Look at those rocks! Tony, there are some tremendous boulders on North Ray, and they get bigger as we go nearer them.
>
> **England:** Okay.
>
> **Young:** Tell you one reason why North Ray looks like – in the photos – it had such

steep walls on one side, is because the rim is raised on one side higher than the other. Don't you get that impression. Charlie?

Duke: Yeah. Sure do.

England: Do you think you'll be able to recognize the edge of the continuous ejecta blanket?

Duke: Hey, John, it looks like ... *(stops to listen)* Well, we'll give you a try at that. Right now, I can't ...

Young: I think we're starting to get into it right now, Charlie.

Duke: Well, the cobbles and boulders are picking up. We're at 2.6, Tony, at 192, and beginning to pick up a high frequency, maybe 10 percent now, of cobbles and boulders.

Duke: John, I think it looks like those ... See that white boulder dead ahead? Looks like the greatest variety of boulders is going to be over there. But that's farther east and our Station 11, or farther north than Station 11 is called for. It's almost at the foot of Smoky.

Young: *(garbled)* Let's go up on the rim and see what we've got up here?

Duke: Okay. I'd love to.

England: Okay; and you may get a caution flag on Battery 2 temperature ...

Duke: Okay, Tony ... *(stops to listen)*

England: ... just reset it, and press on.

Young: Understand. Reset and press.

Duke: Okay. Tony, in this area now for 192 at 2.7, we're getting in a greater frequency of 1-meter-size craters, and it's making it a little bit bumpier ride.

England: Okay. You might watch for a change in soil ...

Young: Okay. The Battery 2 temperature is reading what? You can read that, can't you, Charlie?

England: ... color, or albedo as you go along there [as an indication that they are on the North Ray ejecta blanket].

Young: That's what we're watching for, and the real change comes up ... Oh, man.

Duke: Uh-oh. Oh, man.

They have probably just encountered a fresh, sharp-rimmed crater and must either go through it or slow down and then turn quickly to avoid it. In the Apollo 15 Mission Report, Dave Scott observed that, even on level terrain, if he tried to make a quick turn at a speed greater than about 5 kilometers/hour, the Rover's back end would break out and start to swing around.

Young: [We got] a little closer to the ... That was ... *(laughs)*

Duke: That's one of those sharp craters ... They map sharp out here in the plain. [Charlie may be saying that with few craters visible on the map, these that do are sharp/fresh.] John, I don't think we're going to be ... Go straight between those two big rocks. I think we're going to have to ... Looks like to me that's a pretty steep slope. If we swing a little bit east here, and then go up just on the edge of that boulder ray right there, we'll make it.

Young: *(garbled)*

Duke: Okay. *(pause)*

Duke: Okay. Tony, now that we get over here and can see down off the flank of North Ray, we can see good boulder rays out of North Ray that go for perhaps ... I'm going to say a half-a-crater diameter. Boulders greater than 1-meter size.

England: Okay. Could you take a look up at Smoky area there, and see what kind of structure and texture you can see on the face?

Duke: Can't ... [I've] been looking at that. Can't see anything except for a couple of rays of boulders out of North Ray that trend ... One goes almost into Ravine that I described, and one goes on up to the top [of Smoky Mountain]. In the northeast wall of Ravine, you can see the lineation. To the northeast, they're horizontal. To the north, they are dipping east about 30 degrees.

England: Okay. Can you push your camera up that far to get a picture of that?

Duke: I don't want to break my RCU bracket; I don't think I can ... Wait a minute, I'll take the camera off and do it.

Young: Charlie, don't do that.

Duke: No sweat. *(long pause)*

John is probably concerned that Charlie will drop his Hasselblad.

Young: *(chuckling)* Take a picture of that crater the road had us going through.

Duke: Oh, yeah. I did.

Young: That's a nice one.

Duke: Okay. Tony, there's a ... That, to me, looks like just a big sink feature, John. Tony, the road had us ... the map had us going ...

Young: Okay. We're definitely in the regolith [meaning the North Ray ejecta] right now, Houston, because, see how these blocks are all laid in there?

Duke: Yeah, I do.

Young: Remember how it was up at that crater? At Schooner [a nuclear explosion crater at the Nevada Test Site].

Duke: Yeah.

Young: Those rocks are laid into the ejecta blanket.

Duke: Yeah.

Young: That's where they came from.

Duke: Okay, Tony. At 191, at 3.1, we're coming into some good-size whitish-looking rocks that are 3 and 4 meters across. They're fractured. And there's probably a permanent shadowed sample ... No, that wouldn't be ...

Young: If you didn't know better, you'd say that they were bedrock outcrop, but they're just laid in there, I'm sure, from North Ray.

Duke: And as we go to the southeast side of North Ray, there is a big sink feature, a big pit, that's elongate east-west, and we could drive in it from the east. But once you get to the south of South Ray [means North Ray], it is really a deep pit, Tony. And that ridgeline that we saw from the LM is on the west side of that deep pit. It's probably 100 meters below the rim of North Ray. Over.

Charlie is referring to a ridge line that extends south from North Ray Crater.

England: Okay. We copy that. And on the boulders you are looking at now that you think might be thrown in, you might talk about the fillet size away and towards the crater and see if that corresponds with the secondary.

Duke: Okay. Well, we ... Okay. We're not close to any of them right now. We're in a very smooth area. At 3.4, at 190, we're down in this area where I've just described it, that goes into that big pit off to our west.

England: Understand.

Duke: About a crater diameter from North Ray off to the east, I see some 3-meter boulders that are all rounded and sitting in the ejecta with – or, in the regolith – with good fillets. Okay. Now here's one, Tony, off to the right ...

Young: My bag fell off again, Charlie.

Duke: ... at 34. It did?

Young: Yeah.

John's SCB (Sample Collection Bag) may have come off and he may have seen it fall in his shadow. It's empty and they have spares.

Duke: That's not supposed to happen. Okay. That's a 2-meter-size boulder with a fillet that looks like sort of equi-dimensional around the boulder ...

England: Okay. Understand.

Duke: ... Equal height [on all sides], I should say.

England: Do you see any clasts in these boulders?

Duke: We're just passed another 1-meter ... *(stops to listen)*

Young: *(garbled)*

Duke: *(to Tony)* They just look whitish to me.

England: Okay.

Duke: But that was another one we just passed, 1-meter size, that had the biggest fillet upslope [meaning toward North Ray].

Young: Here's a new crater right there, Charlie.

Duke: Right to the right?

Young: Yeah.

Duke: Yeah, I know it. Tony, these craters that we call secondary, that are indurated, I frankly think are very, very fresh craters, because it looks very cloddy around them, and the other ones that are secondaries do not appear that way. Over.

England: Okay. Understand.

Young: Yeah, I'm not sure that this isn't such an old crater, that the secondaries aren't eroded down. We've really got good going right here.

Duke: Yeah.

Young: Now, before we get too far along, let's study this thing and see if we can figure out a way to get up that rim without going through all the boulders in the world ...

Duke: Okay, John. See that big one off to the right over there?

Young: Yeah. I see that.

Duke: Okay. I think up that slope looks to me to be the best. Of course, it might be straight ahead might be best.

Young: Well, I don't see any rocks straight ahead.

Duke: All right. Let's go ...

Young: Let's try straight ahead.

Duke: Straight ahead.

Duke: Okay. Tony, we're heading about 300 and [the bearing to the LM is] 187. The large boulders, Tony, will be off to our right. There's a black to brownish looking one, and then there is a solid white one that's right at the base of Smoky Mountain and North Ray. That might be worth a little jog over there if it's not too far. It's the most unique white boulder we've seen.

England: Okay.

They are near Shadow Rock, where they will collect samples (at Station 13) on the return trip.

Young: That is a beauty, isn't it?

Duke: Yeah. See that white one over there, John?

England: ... We'll keep that in mind on the way back. Give us bearing and range again.

Duke: Okay, we're at 3.7 and 186. And we just passed some very frothy ... two frothy-looking boulders. The biggest one is perhaps 5 meters across, and they have vertical joining or fracturing to them, and they have a frothy appearance to it. And I'm about 20 meters from it now.

There is a cluster of boulders immediately east of this location; from the length of its shadow, the largest appears to be LM-size. There is an even larger rock about 350 meters east-southeast of them.

England: This sounds really great.

Duke: *(laughing)* Man, that is a big rock!

Young: Yeah.

Duke: Okay, Tony. They're are not any house-size rocks, but the biggest ones are maybe 5 meters. *(pause)* And it's really smooth except for these big rocks out here. It's smooth going.

Young: Why don't you leave the ...

Duke: There's a real fresh little crater right there. See the rays ...

Young: Yeah.

Duke: ... off to the left?

Young: Yep.

Duke: It's about a meter size.

England: Hey, could you use a couple more words ...

Duke: And 3.9, we're at 183.

England: ... to describe that frothy rock?

In the detailed orbital photos, there are at least two boulders visible about 100 meters east of that position. At this range from the LM, the uncertainty in the position given by the Nav system is at least 100 meters. Unfortunately, the two boulders are near the limit of resolution in the orbital photos, so it is not really possible to say which one Charlie is describing.

Duke: It's got a hackly surface to it ...

Young: It's black colored. Right, Charlie?

Duke: Yeah.

Young: Okay; we're going up a pretty steep slope right now, Houston. We're doing ... I think we're almost at the rim, Charlie.

Duke: Yeah, we are. Looks like we're just about 20 meters from the rim.

Young: I'm going to slow down here.

Duke: Yeah. *(laughing)* You think you'll be alright. How about hooking a right, over here, John.

North Ray is about 1000 meters (3280 feet) across and 200 meters (656 feet) deep and John wants to approach the rim slowly enough that he can stop quickly if he finds himself at the edge of a steep drop-off into the crater.

England: We got you at about 400 or 500 meters from it yet.

By this time, Houston has range and bearing readings at enough well-defined locations that they know almost exactly where the LM is; they also have an excellent fix for End Crater – all of which shows that the Nav system is working well. John and Charlie are, indeed, about 500 meters southeast of the rim.

Duke: Tony, coming to the rim ... *(stops to listen)* I don't believe it, but ...

England: Okay; we'd like you ...

Duke: Well, you might be right.

England: ... to go to 12 frames per second [on the DAC].

Duke: Okay. You got it.

England: Okay. This is going to make some great pictures.

Young: Okay. *(pause)* We're on a relatively flat surface now.

Figure 6.3. The crew is heading back toward the lunar module, following the tracks made when they ascended North Ray Crater. The boulders above the camera to the left were almost certainly ejected from North Ray when it formed. The somewhat unusual, shallow crater immediately left of the camera's sunshade may be filled with pieces of regolith breccia – lumps of lunar soil compressed in the crater-forming impact. Pieces of regolith breccia examined by the Apollo 15 and 17 crews were fragile enough to break up when squeezed between the fingers. Charlie is in the right-hand seat with the TV camera immediately in front of him. The camera and its box-shaped control unit below are covered in gold-colored, reflective mylar to prevent overheating; the top surface is mirrored, to promote cooling. At each stop on the traverse, they wielded the large dust brush (lower left) to keep the electronic equipment on the front of the Rover clean. Before they climbed back into the LM cabin, they used the brush again to clean their suits and stamped their feet on the ladder rungs to eliminate more dust.

They have just driven onto a sloping shelf on the flank of North Ray but are still below the rim. The slope they are on is about one-half as steep as the first part of the climb: 25 meters up over about one-half a kilometer distance.

Duke: Okay. The rocks here, Tony, are all rounded. Well, most of them. Seventy percent of them are rounded, and the other ones are subangular [meaning that the sharp edges and corners have been partly eroded], mostly dust covered, grayish in color. The big rocks are not on the rim, Tony! The big rocks are farther away from the rim. At least, we can't see any big rocks as we approach the rim, but we're still climbing upslope.

England: Okay. *(pause)*

Duke: Man. Look, there's a tremendous one. There's a 10-meter boulder off to the right over there, John.

Duke: There's a fresh crater, really fresh one, that has a white interior that's punched in about 2 meters deep, and that was at 181 at 4.0. *(pause)*

Duke: Okay; it looks like to me we're ... The rim ... Hey, there's some beautiful white ones over there ...

Young: There we go, Charlie.

Duke: ... John, at 2 o'clock. Think this is the rim, right here?

The black 10-meter-diameter boulder on the right is House Rock, which they will visit just before they leave North Ray. They will park the Rover near the white boulders straight ahead; they are looking for crystalline bedrock, and the white color of the rocks ahead suggests that they are composed of anorthosite.

England: We still think you're about 500 meters from the rim.

Duke: We'll be able to sample these white ones. Here's some ... *(stops to listen)* We are. There's the rim up there.

Young: Sure is.

Duke: Sure is, Tony. You were right. We just climbed ... What we thought was the rim [was] one of these little hummocks.

England: Right ...

Duke: Little hummocks! It was a pretty steep hummock.

England: ... Just like mountain climbing, there's always another ridge.

Young: *(laughing)* I'll be darned!

Duke: Okay. I'm going to pan the DAC around to get to that boulder field that goes up to Smoky Mountain. It's really tremendous. The boulders are very angular over there. They're dark gray in color, and some of them are almost solid white. The most unique ones appear to be solid white. Up on the rim here, they appear to be almost white; none of the dark ones. *(pause)* And we're at 180 at 4.1. Smooth regolith.

Duke: John, the rim's left.

Young: The rim is right there!

Figure 6.4. John and Charlie are sampling a light-colored, almost white boulder near their parking place on the rim of North Ray Crater. The tongs are stuck in the ground (lower left) as a substitute for the broken gnomon, and John has the rake. As part of the set of photos taken to document a particular sample collected from the white boulder, Charlie turns to get this "locator" photograph. Such shots are designed to show a recognizable object in the background – usually the LM or the Rover – so that the analysis team can later determine a map location for the sample site. In this case, it is easiest for Charlie to take a photo showing House Rock, the large boulder beyond John's right shoulder. Because the Moon has no atmosphere and because there are no familiar objects like trees or houses to give scale, it is often difficult to judge sizes and distances. For example, how far is House Rock from this sampling site and how big is it? Actually, it's about 150 meters from John and is 16–20 meters across and 12 meters tall – equivalent to a three-story house!

Duke: No, sir. I bet it's over there to the left where those rocks are. But you might be right. *(pause)* That's too far away. You're right. That's probably too far west.

England: We think the most direct route from where you are ...

Young: Okay; we're at 180 at 4.0 ... *(stops to listen)*

England: ... to the rim would be about a heading of 350.

Duke: Okay! We're headed that way. And there's some beautiful ... Those white rocks are ...

Young: Right on the rim, Charlie!

Duke: ... right on the rim.

England: Outstanding. Can you see on around to see if there are any black rocks around at 3 o'clock in the crater?

Duke: Well, we can't see in the crater. But around at the 3 o'clock position, yeah, there's a biggie. The biggest one [which is House Rock], Tony, is this 10- to 15-meter boulder that is on the rim; and it's blackish.

Some geologists' pre-flight concept of the site involved a variety of volcanic features overlying an anorthositic (plagioclase-rich) bedrock. If the North Ray impact were big enough to have penetrated to bedrock, then John and Charlie might have found both black volcanic rocks and white anorthositic rocks. As it turns out, although House Rock is very dark, it is an impact breccia and did not have a volcanic origin.

England: Okay. Is there any chance of working around towards that contact? And if we could get both the white and black in one stop, that would be really fine.

Young: It's ... No way!

England: Okay.

Duke: Right. Pretty far. I think we can do it with a short stop over there. And when we get up there, Tony, we might be able to find a black rock.

England: Okay. Fine

Duke: Okay. We're going through a ... We're definitely on the ejecta blanket here. And, oh, within 100 meters or so, I think, of the rim.

England: Right. We have you about 100 meters from it ...

Duke: These rocks are just white! Crystalline, white looking. *(hearing Tony)* Man, you guys are right on. If we copied, I think y'all are right on.

Young: Yeah. We're 179 at 4.4 right now.

Duke: Okay. That's great, John. *(pause)* [As per LMP-30] he wants us to park 360 [that is, with the Rover heading north]. Go right the rim! Okay; that's a breccia. That white one's a breccia.

Young: There's the rim.

Duke: Yeah. There it is. Okay. I think we can get over there and maybe get them a picture. We're headed about 360, aren't we?

Young: Yep.

Duke: That is ... That big ... I can't believe the size of that big black rock over here. And I don't think that's a breccia, John, because ... Although it might be. I see some large white clasts. *(pause)*

Duke: Oh, spectacular! Just spectacular! *(pause)* Can ... *(pause)*

Young: Okay, Charlie.

England: Charlie, the DAC should be out of film; you can turn it off.

Duke: I can't reach it.

England: Okay, fine.

Duke: Gonna park at this heading?

Young: I don't ... No, I guess not. *(pause)* What I'd like to do is park where it's flat, and ... *(pause)*

Duke: Okay. Okay; where we came up over here, John, it won't be quite...They get a better view. Right here's where they get a great view of the interior [of North Ray]. Of the upper third of the wall. Okay. Tony, we're on the rim.

England: Beautiful.

Duke: There we go. If we go 360 and park right here, it'll be flat. *(pause)* Great, John. Super! *(excited)* Can't wait to get off. Got to get off.

England: We can't wait for you to either. You're ...

Duke: Okay, Tony. We're at ... *(stops to listen)*

England: ... about 17 minutes ahead of the timeline.

Duke: ... 360, 179. 5.5, 4.5, 60, 115, off-scale low, off-scale low, 130, 110; 225, 225 Forward Motors; 200, 200 Rear Motors.

They parked about 220 meters [722 feet] southwest of House Rock. The difference between the indicated and actual locations was an acceptable 250 meters [820 feet].

England: Okay. What was that Temperature on Battery 1 again?

Young: Okay. Primary [Drive Power] is going to Off.

Duke: *(answering Tony)* 100 and … About 110, I think.

England: Okay. Got it. And, Charlie, we'll need a [Hasselblad] frame count.

Duke: Okay. Pan … *(stops to listen)* Okay; stand by. I'm so dusty [I can't read the counter].

Young: Okay. Going to halfway between the Intermediate [PLSS cooling] and … *(pause)*

Duke: Gee, I don't know, Tony. I can't read it. Let John read it. John.

Young: What's that?

Duke: Read my frame count.

Young: Okay. Hold still.

Duke: Well, I wanted to get in the Sun so you could read it.

Young: 165. You better change that [Hasselblad film magazine] out.

Duke: Okay. Took 165 pictures coming up here, Tony.

England: Okay; and we concur on the changeout.

Duke: Okay. I'm going to put another black and white on. Kilo.

England: Okay. Kilo.

Young: Man, look at … *(stops to listen)* Okay, I'm walking down about a …

Duke: John, I'll get the TV for you.

England: Okay; and DAC off.

Young: Charlie, I'll get the TV.

Duke: *(answering Tony)* Yeah, okay.

John and Charlie spent nearly 2 hours on the rim of North Ray, carefully observing and describing the geology of the site and collecting representative samples. The white rocks near where they parked were breccias consisting of a lighter matrix that contained small, dark-colored inclusions. Later, when they made the run to House Rock, they found that it was also a breccia, this time with a dark matrix containing small, light-colored inclusions. Had there been any volcanic rocks, they would have recognized them, and there would have been fragments in the soil samples they collected. There were none. Like the crews of Apollo 15 and 17, they proved that well-trained non-professionals could do superb geologic field work.

The hills and plains of the Descartes site are underlain by anorthositic rocks, which are fragments of a disrupted highlands crust. There was no evidence that the site contained any volcanic rocks. Paul Spudis, a well-known selenologist at Johns

Figure 6.5. Charlie is pointing to a sample location on the face of Outhouse Rock – part of House Rock that probably broke off when the huge boulder landed. He has a pack of sample bags hooked to his finger because the bracket on the bottom of his camera that was designed to hold the bags has broken. The shadow cast by the sample bags is orange, which is normal when sunlight is transmitted through the translucent Teflon film from which the bags are made. In high-resolution versions of this picture, we can see that Charlie has magazine K (Kilo) on his Hasselblad and that his checklist is open to pages EVA-3 LMP-30 and EVA-3 LMP-31 (such high-resolution photography is accessible on the ALSJ site).

Hopkins Applied Physics Laboratory, has concluded that the Descartes Highlands were hummocky ejecta deposits from the impact that formed the Nectaris basin, an 860-kilometer (535-mile)-diameter crater, during the Moon's early history. Later, this already irregular terrain was further broken up and excavated by secondary impacts of debris from the Imbrium basin. The Cayley plains are also floors of large craters that were, in turn, cratered by impacts from debris out of the extremely large Imbrium crater; much of the early debris was excavated locally by secondary impacts. These highlands, consisting of debris from such complex, large-basin-forming impact sequences may be as much as 10 kilometers (6 miles) thick. No bedrock or volcanoes here.

The most common rocks collected by the crew are *cataclastic* – they had been crushed multiple times by impacts of bodies into the lunar surface. They were made even more complex when impacts melted surface material, creating glass and crystallized glass that mixed with the older rock fragments to form heterogeneous melt breccias. You might say the process is similar to preparing a batch of toffee into which you put every ingredient available in a well-stocked kitchen.

Eighty percent of the lunar surface consists of heavily cratered highlands. These regions have higher albedo (appear lighter) than the lower, smooth-mare lava flows. The differences in albedo are ascribed to a fundamental difference in the chemical and mineralogical character of the two types of terrain. The intensely cratered highlands are older and were subjected to the greater flux of impacting bodies that occurred very early in the Moon's history. Many of the largest craters, similar to those visible from the Earth, were later filled with darker colored lava flows. Despite their physical and mineralogical differences, samples from this region are similar in that they are chemically simple, with high aluminum and calcium.

Although Apollo 16 was flawlessly executed and scientifically successful, the broader scientific community was beginning to lose interest in lunar exploration by the time the mission results were published. Waning interest in the United States' lunar efforts was evident in many ways, and one example is very striking. Preliminary studies after each mission had always been published first in *Science*, one of the two premier interdisciplinary scientific journals. During the early missions, these reports had always dominated the issue and were in great demand; the Apollo 16 preliminary examination report, published in *Science* in 1973, was sandwiched between "The Saga of American Universities: The Role of Science," and "Population Cycles in Rodents."

The fact that the Descartes Highlands were not a volcanic field was a disappointment for many within the Apollo program. But not all preconceived expectations are fulfilled when mankind is exploring new worlds. This was good exploration, pure and simple, with goals based on careful preliminary interpretations of orbital photography. Those hoping for evidence of ancient volcanism on the Moon soon would be delighted when the Apollo 17 crew brought back samples from the Valley of Taurus-Littrow. Although Gene Cernan and Jack Schmitt didn't find the source vents for the many lava flows that spread across the floor of Mare Serenitatis, their discoveries during the second EVA produced surprises that changed the way we understand lunar volcanism.

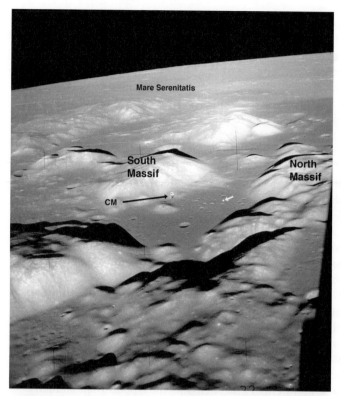

Mare Serenitatis

South Massif

North Massif

CM

Gene Cernan took this east-looking photo from orbit before the LM's descent to the lunar surface. On the final approach, Gene flew Challenger down over the Sculptured Hills (low, knobby hills just below center) and out over the valley floor before landing a little north of the Trident group of craters. The CM, flown by CMP Ron Evans, is barely visible at the arrow's point; the South Massif (which rises 2300 meters above the valley floor) in the background. The 500-meter-diameter Henry crater is the furthermost of the three largish craters at the foot of the North Massif to the right. The mountains that flank the landing site (white dot and arrow) have elevations of 1000 to 2300 meters above the valley floor.

North Massif

Sculptured Hills

scarp

Henry
Cochise
Shakespeare

Shorty Crater

LM

Light mantle

Taurus-Littrow Valley

Nansen Crater

South Massif

10 kilometers

This detail, from orbital photographs taken by the CM's panoramic camera, includes all the areas visited by the crew, including the light mantle, which is a deposit from a landslide that originated on the South Massif. Shorty is an impact crater that penetrated the landslide deposit and ejected material from the underlying mare lava flows and volcanic ash deposit(s). The summits of the North and South Massifs are 2000 and 2300 meters, respectively, above the valley floor.

7

The Volcanoes of Taurus-Littrow – Explosive Volcanism on the Moon

The well-known early selenologist G.K. Gilbert strongly proposed in his 1892 landmark paper "The Moon's Face," that impact processes were the primary source of lunar craters. Using the 67-centimeter telescope at the Naval Observatory in Washington D.C. for 18 days, he had gathered evidence for the most visible craters' origins, including the widespread "rays" of ejecta that can be seen today with even a good backyard telescope. He also gathered evidence to contrast the lunar craters with large volcanic craters on Earth. (Ironically, Gilbert also concluded that Meteor Crater, now a popular tourist locale in Arizona, was formed by a volcanic eruption rather than impact.) Many decades later, on the other side of the debate, California geologist Jack Green promoted the idea that most of the large lunar craters were volcanic and similar to large volcanic collapse craters (calderas) on Earth.

The hypotheses of both Gilbert and Green had supporters and detractors in the scientific community. And most geoscientists felt there was room for both processes in the shaping of the Moon's surface.

In the 1960s, author Heiken had NASA support in graduate school to study tuff rings, which are broad, low-rimmed volcanoes formed on Earth when rising magma mixes with surface or shallow ground water. From a great distance, these volcanoes are very similar in shape to impact craters. NASA's premise was that if some of the Moon's craters proved to be tuff rings, it would imply the presence of groundwater that could support a future lunar colony. Heiken began working for NASA at the LRL in 1969, 6 months before the Apollo 11 mission; one of his tasks was to be prepared to identify the products of explosive volcanic eruptions in the returned lunar samples. And, in fact, geologists saw particles that might have been volcanic ash in samples from many of the earlier Apollo missions, but there was no clear evidence to support this interpretation. That proof emerged during the exploration of the Taurus-Littrow region by the crew of Apollo 17.

Before Apollo, there was general agreement that the large, mare-basin craters were partly filled with lava flows. Optical properties (such as the lower albedo) of the dark crater fill are similar to those of lava flows on Earth. In addition, thin, widespread lava flows across the maria were visible on detailed photos taken by the Lunar Orbiter spacecraft and, later, by the Apollo CM cameras. Lava flows tend to cool quickly; if they were to flow any distance at all, enormous volumes must have been erupted very rapidly. Several of the youngest lava flows that cross the surface of Mare Imbrium are hundreds of kilometers long, and a single flow covers area equal to that of the State of Nebraska. It was also evident from orbital photos that such lava flows originated from vents near the mare basin margins.

On Earth, lava-flow vents are surrounded by cones or ramparts, formed by explosive volcanic activity; these explosions occur when gases come out of solution as magmas (molten

APOLLO 17 FACTLIST	
Crew	Commander, Gene Cernan; CMP, Ron Evans; LMP, Jack Schmitt
EVA CapCom	Bob Parker
Command Module	America
Lunar Module	Challenger
Launch from Earth	05:33 GMT 7 December 1972
Lunar landing	19:55 GMT 11 December 1972
Popular site designation	Taurus Littrow
Lunar stay	75 hours 00 minutes
Splashdown	19:25 GMT 19 December 1972
Command Module lunar orbits	75
Time outside the LM	3 EVAs, 22 hours 05 minutes
Maximum distance from the LM	7.6 kilometers
Distance walked or driven	35.0 kilometers
Lunar samples	110 kilograms
Lunar surface photographs	2237
Surface equipment	514 kilograms
Highlights	Finding and collecting the orange and black glass samples; first geologist astronaut on the Moon; refining procedures developed on earlier missions – a fitting finale to Apollo

rock) reach the surface. For lavas of basaltic composition – common over most of the Earth, but most easily seen in Hawaii or Idaho – the vent areas are marked by scoria or "cinder" cones, from which lava emerges and flows across the land. Lava fountains are another, more spectacular form of this activity, during which sprays of molten rock are erupted and rain down. The deposits include everything from delicate glass droplets to large blobs that spatter like a cowpat when they hit the surface.

Given this terrestrial background and looking at the Moon's images, volcanologists were wondering: Where are the lunar cinder cones? – Present, but few and far between, was the conclusion after studies of lunar orbital photography. However, there are many dark-haloed craters – usually consisting of a crater-like depression surrounded by dark deposits 2 to 10 kilometers in diameter – located along the margins of mare basins and along lineaments (linear structural features) or lunar rilles (sinuous canyons). These dark-haloed features are broad and flat and not cone-shaped as would be expected on Earth, and this observation confused most volcanologists.

In 1974, a solution to this dilemma came from the late Tom McGetchin (then Massachusetts Institute of Technology and later Los Alamos Scientific Laboratory). He and his colleagues used high-speed cameras and a variety of geophysical tools to conduct the first truly quantitative study of explosive volcanic eruptions on Earth. McGetchin's team studied the ballistic properties of molten particles that had been explosively erupted and then accumulated around the vent to form scoria cones. They used their results to calculate the particle trajectories that could be expected during eruptions of lunar volcanoes – where gravity is 1/6 that of Earth and there was no atmospheric drag on those particles (the eruption would have taken place in a vacuum). The McGetchin team's results predicted that such eruptions would have produced not cones, but broad, thin, circular deposits – just like the dark-haloed craters present along mare margins.

On the floor of the Valley of Taurus-Littrow, Shorty is a small crater that had been punched through the light-colored material left by an avalanche of debris from the South Massif. The crater is surrounded by dark ejecta. There was some hope that the dark-haloed Shorty, like many of the craters located west of the valley, might have been a source vent for some of the lava flows that had filled the valley and Mare Serenitatis – so a visit to Shorty Crater was planned for the second EVA). (Apollo 17 astronaut Jack Schmitt named Shorty Crater in honor of Richard Brautigan's collection of short stories, *Trout Fishing in America*, and other books that, in Jack's words, "bring the literature of youth and life from the foundations of Salinger into the present.")

Beginning their second EVA, Gene Cernan and Jack Schmitt drove to Station 2, which was 7 and 1/2 kilometers west of the LM, and thus, farther from their spacecraft than any of the earlier Apollo crews had ever ventured. At this stop at the base of the South Massif, they spent 1 hour and 10 minutes examining some large boulders, which had rolled down from outcrops high on the mountain

Throughout this second EVA, Houston kept a careful eye on the clock because of what was called the *walk-back constraint*. The length of time that the astronauts could stay outside the spacecraft was limited by the amount of oxygen and cooling water they carried in their backpacks. In the Rover, they could make 10 to 12 kilometers per hour (6.2 to 7.4 miles per hour) on reasonably level ground. From Station 2, they could have been back at the LM in under an hour. But what if the Rover failed? They would have to return to the spacecraft on foot – and that would take far longer.

Because of the Moon's weak gravity, the astronauts found that it was much easier to run than walk. Although none of the earlier crews had run more than 100 to 200 meters (90 to 180 yards) at any time during their EVAs, astronauts had done simulations of long-distance lunar running during training sessions on Earth. A suited astronaut was suspended in a contraption that made him feel like he was only 1/6 his actual weight; the test was to see how long he could run. Before Scott and Irwin flew Apollo 15, John Young and Charlie Duke – who wanted to convince NASA management to let them do a particularly long traverse on Apollo 16 – both did long sessions: John ran for 2 hours 43 minutes, covering nearly 18 kilometers at an average speed of 6.5 kilometers/hour, while Charlie ran for 3 hours 18 minutes, covering a similar distance at an average speed of 5.5 kilometers/hour. These simulations were performed on level ground and, as an extra precaution, EVA planning for the Rover missions was based on an assumed speed of 3.4 kilometers/hour. Assuming those conditions, a run from Station 2 back to the LM would have taken Gene and Jack about 2 hours 10 minutes. In order to allow time for packing up and getting back in the LM calmly and safely, NASA wanted them back at the LM no more than 6 hours 40 minutes after the start of the EVA. That meant that they would have to leave Station 2 before 4 hours 30 minutes of elapsed time. Gene and Jack left Station 2 when they were 3 hours and 10 minutes into the EVA – they had plenty of walk-back margin, but they had more stops to make and needed to move on.

Because of walk-back, it always made sense to go first to the farthest station planned for a traverse. After Station 2, Gene and Jack made their way generally northeastward, angling closer to the spacecraft as they continued geologic fieldwork. They made a brief stop at the top of the 80-meter-high Lincoln-Lee scarp for a gravimeter measurement and photographs taken with a telephoto lens, and then drove gingerly down the regolith-draped flank of the scarp for a sampling stop at its base.

During the long drive out from the LM to Station 2, the bracket holding Jack's camera to his suit chest had come loose and he'd had to hold onto the camera for most of the drive. Maintaining a grip against the internal pressure of the suit was hard work and, after an additional hour spent sampling and raking at Station 2, Jack's hands and arms were aching. By the time he and Gene got to Station 3 at the base of the Scarp, the fatigue began to show. While Gene drove a core tube into the surface, Jack tried to do some solo sampling on the rim of a 10-meter crater, but had trouble holding onto the scoop and bags. He spent a lot of time trying to pick up dropped items from the ground. He'd drop his scoop and, in the process of trying to retrieve it, would knock over a sample collection bag that he'd propped upright on the surface. As Jack floundered and flailed in frustration, CapCom Bob Parker suggested that Gene "give Twinkletoes a hand." As Gene and Jack were about to leave Station 3, Bob told them that the Houston Ballet had called Mission Control, requesting Jack's services in the coming season; the crater at Station 3 is now known throughout the NASA community as Ballet Crater.

SHORTY CRATER AND THE ORANGE SOIL

During the drive from Ballet Crater toward Shorty Crater, Jack was not only tired, but also bit discouraged. EVA-2 was not going smoothly. As Gene detoured around craters and navigated the hummocky terrain, Jack was simultaneously navigating and describing the geology. The Rover navigation system counted rotations of the four wheels to determine their distance from the LM and differences among the wheels' rotations to determine the direction to their home base (that is, when traveling in a straight line, all the wheels make the same number of rotations in a particular length of time, but when turning to the right or left, the wheels on the inner side of the curve rotate less frequently than those on the outer side). Although their destination was hidden by the rolling terrain, they knew they were 2.6 kilometers southwest of Shorty; with that information, Gene could steer the Rover on a rough heading that would point them toward Shorty, even though it wasn't immediately visible.

The scenery in the deep valley was spectacular. Their view of the landscape was sharp and crisp with no softening by an atmosphere. They could see tracks in the massif slopes, left by giant boulders that had broken away and rolled toward the valley floor, and to the west, the 80-meter-high fault scarp. Although draped with regolith, structures left from the violent early history of the Moon were easy to identify.

As we join the crew near Shorty Crater, the transcript voices are CDR Gene Cernan, LMP Jack Schmitt, CMP Ron Evans, and CapCom Bob Parker, a University of Wisconsin astronomer, who was selected for the second group of scientist astronauts. Cernan was LMP on Apollo 10, the final rehearsal before the first lunar landing, and was Alan Shepard's backup for Apollo 14. Jack Schmitt is a Caltech/ Harvard trained geologist who was working at the USGS Astrogeology Branch in Flagstaff when he was selected as a scientist-astronaut in 1965. He played an important role in devising the geology-training program that would properly prepare the pilot-astronauts for their work on the Moon. The Interviewer is author Eric Jones, whose mission reviews with the astronauts took place at various times from 1989 through 1991 in Houston, Albuquerque, and Santa Fe.

Excitement is evident in Gene and Jack's voices as they approach their goal of Shorty Crater. While Jack is describing geologic features on the distant massifs, Gene is "watching the road." The boulder-covered terrain levels out as they reach the crater rim and they discuss where they should park and begin their field descriptions and photography. Looking into the football-field-size crater elicits a breathless "Whoo, whoo, whoo!"

Adrenaline has kicked in and Jack no longer sounds tired. The dark-rimmed crater is rimmed with angular mounds and fractured boulders composed of rocks excavated from lava flows under the regolith. Here, it will be possible to sample the lava flows that poured into the valley from both local vents and those on the vast Mare Serenitatis. Jack is in his element, spouting rock descriptions and taking photographs as Gene finishes Rover housekeeping and dusting the TV camera.

At this moment, they spot something very odd:

Schmitt: Oh, hey!

Jack has just seen the orange soil. He is cautious, having been fooled by sunlight reflected off the LCRU at the Scarp stop.

Schmitt: Wait a minute ...

Cernan: What?

Schmitt: Where are the reflections? I've been fooled once. There is orange soil!!

Cernan: Well, don't move it until I see it.

Cernan – "Quite frankly, when Jack said he saw orange soil, I began to wonder if he hadn't been on the Moon too long. Until I saw it myself."

Schmitt: *(very excited)* It's all over!! Orange!!!

Cernan: Don't move it until I see it.

Schmitt: I stirred it up with my feet.

Cernan: *(excited, too)* Hey, it is!! I can see it from here!

Schmitt: It's orange!

Cernan: Wait a minute, let me put my visor up. It's still orange!

Cernan – "Like a pair of ordinary sunglasses, the visor attenuated the light but didn't really change the colors. If it was red, it was red; if it was blue, it was blue. But you could see the colors better with the visor up. I do that flying; I take my sunglasses off if I really want to see things clearly. So lifting up the visor was a natural thing for me to do."

Schmitt: Sure it is! Crazy!

Cernan: Orange!

Schmitt: I've got to dig a trench, Houston.

Parker: Copy that. I guess we'd better work fast.

Because of the amount of time Gene and Jack spent at Stations 2 and 3, they don't have much time before hitting the walk-back constraint. They are now 4 hours and 50 minutes into the EVA and running from Shorty to the LM will take about 1 hour and 20 minutes. They have only about 30 minutes left before they must leave.

Cernan: Hey, he's not going out of his wits. It really is.

Parker: Is it the same color as cheese?

This may be a serious question. Many American cheeses are orange.

Schmitt – "You'll have to ask Bob, but I think he was thinking of green cheese, as in 'The Moon's made out of green cheese.' My guess is that he wasn't convinced that we weren't trying to pull their leg again. And for good reason, because every once in a while we'd do that, either inadvertently or advertently."

In response to a question from author Jones, Bob remembers that it was a serious question – but with a joke thrown in.

Schmitt: It's almost the same color as the LMP decal on my camera.

Figure 7.1. Jack Schmitt is working with a double core tube at the Rover, which is parked on the southwest rim of Shorty Crater. Shorty Crater ejecta include streaks of dark grey, black, and orange debris. This orange material created quite a stir (both on the surface and in mission control), as colored deposits are unique in this mostly grey world. One of the first, quick interpretations was that the orange material was the result of hydrothermal alteration, which would imply the presence of water on the moon. Alas, the orange material is a very fine-grained volcanic ash – glass spheres from a lava fountain into a world with no water.

A B/W detail from the color photo Gene took at Station 1 shows an "LMP" decal on the side of Jack's camera. A similar decal can be seen on the top of the camera, just behind the lens. In a 2001 e-mail to author Jones, Jack wrote that the orange color of an f-stop decal on the Apollo 11 EVA magazine that is on display at the Smithsonian National Air and Space Museum "looks pretty close to what I remember, maybe not quite as bright or vivid."

Cernan: *(looking toward the orange soil near the boulder)* That is orange, Jack! *(pause)*

Cernan: Boy, this brush is getting harder to get on and off, too. But I sure don't want to lose it. Man, I may start putting that under my seat. *(pause)*

Schmitt – "The dust is going to get into anything that's not hermetically sealed. After a while, moving parts don't move any more with dust in them. We had a bracket [on the front of the Rover] for the dusting brush, and that bracket was now full of dust and he couldn't get the brush handle in it. One lock on the [tail]gate was no longer usable; and I was having an awfully hard time getting my scoop on and off."

Schmitt: Fantastic, sports fans. It's trench time! You can see this in your color television, I'll bet you.

Cernan: How can there be orange soil on the Moon?! *(pause)* Jack, that is really orange. It's been oxidized. Tell [CMP] Ron [Evans] to get the lunar sounder over here.

Schmitt: It looks just like an oxidized desert soil, that's exactly right.

Schmitt – "I don't know what Gene had in mind, but I suspect that the sounder was just an instrument in the CM that he remembered that you could point at something. But the sounder wouldn't have had anything to see here, because of its coarse resolution, if nothing else."

The orange/red color may have made Gene think about rust (oxidation) and, therefore, water and the sounder. Notes taken by author Heiken at the time of the mission indicate that after the discovery of the orange soil, Farouk El-Baz was quoted on television as saying that the color might be due to the presence of water at some time in lunar history. In fact, an article in the December 13, 1972 issue of the San Francisco Examiner states, 'Farouk El-Baz said the orange color most likely was the result of iron in the lunar soil rusted by the water vapor escaping from the vent.' Maybe Gene was thinking along the same lines.

Schmitt – "Of course, Ron started to see orange areas from orbit as soon as he heard about it. He saw orange in the southwestern part of the Serenitatis Basin, in what's called the Sulpicius Gallus region. And, when Gene and I got back into orbit, we were able to confirm and I mapped quite extensive regions of the orange and other colors. And, as we left the Moon, we were looking back at a full Moon and, for a little while, we had good color/tint definition."

At this moment, Evans was about 10 minutes from LOS on his 29th orbit. News of the discovery of the orange soil wasn't passed up to him until shortly after he reappeared around the Moon's eastern limb, and he had the following exchange with Houston.

CapCom: While you're doing your pre-sleep checklist, you might be interested that, at Shorty, the surface crew found some very, very orange soil, a great deal of it. Indicates strong oxidation and possibly indicates water and/or volcanics in the area. And they're really . . . Jack's kind of like a boy at Christmas time. I'll tell you, a little kid at Christmas time on that one.

Evans: *(laughing)* I bet he would be. Hey, that's a great find, by gosh!

CapCom: Yeah, that's the first time we find . . . It's orange. Boy, you can see it in the television; it's bright orange soil. No question about it.

Evans: I'll be darned.

CapCom: And, as luck would have it, they found it all and got working, and then they had to pull out of Shorty due to constraints, walk-back constraints in the area. You know, consumables versus walk-back.

Evans began his rest period about 10 minutes later and it wasn't until his wake-up call that he had a chance to look at Shorty Crater from orbit. He was convinced that there was an orange tint around the crater and in certain other areas of western Serenitatis. Throughout the mission Evans was conscious of variations in shades and tints and reported his impressions during his conversations with Houston. He remarked that "You know, to me the Moon's got a lot more color than I'd been led to believe. I kind of had the impression that everything was the same color. That's far from being true." However, the shades and tints are subtle and it would require a look at Shorty before he was convinced that he could see the orange from orbit.

Cernan – "When I heard that Ron had seen some orange patches from orbit, I was concerned that maybe the power of suggestion had taken over. You can stand on the Moon and suggest colors to yourself and you begin to see those colors. You could tint the bland gray color of the Moon blue or green or brown. By this time I was absolutely convinced that Jack and I could see the definitive color delineation between the orange and the gray (at Shorty), but I was concerned that Ron was being influenced by the power of suggestion. But, when we got back into orbit, I ate my thoughts because we really did see areas that you could definitively define as orange. Now, whether we would have seen those orange patches had we not seen orange soil on the surface, I don't know. If we had not found orange soil on the surface, we might have seen the orange areas from orbit and have just passed them off as reflections of some kind; because, who would have expected to find orange soil on the Moon? But, having known it was on the surface, we could feel comfortable about being able to say that we could see the orange from orbit."

Schmitt – "The reason that this was a little more exciting than it might have even been otherwise was that, in our thinking about the origins of Shorty, we really had two major possibilities, with a number of observations related to each one. Number One was what we really believed and what turned out to be true: that Shorty was an impact crater that had penetrated the light mantle and was exposing the dark-mantle and/or subfloor basalt beneath. The other alternative, just because of the lack of resolution and information from the pre-mission photographs, was that it was a dark-haloed volcanic crater; and, if that were true, one of the things that we had on our list to look for were signs of alteration, possible fluid-induced alteration of materials around the crater. And normally, on Earth, that alteration is colored, fumarolic alteration from oxidation. And so, when we saw orange, that was the immediate thought that everybody had. 'My God, there's a volcanic emanation here that's altered the soil.' Well, it turned out that that wasn't true, of course. It was volcanic material, but it was volcanic glass that had been spewed out of some fire-fountain-like eruptions 3.5 billion years ago that somehow had been protected from mixing with anything else, even though it was now at the surface. It had almost certainly been covered almost immediately by a lava flow, so that it was protected from meteor disruption and stirring. And then, when Shorty formed, somehow the pyroclastic ended up in the rim and a few other places in nearly pure form. There are some issues having to do with its origins that we can get to in a moment, but that's what caused all the excitement. We had sort of halfway predicted that we would find it there, even though it turned out not to be what we predicted we would find. They had a camera running in the Backroom, and what it shows is that, when that call came over, everybody sort of jumped up in the air and there was just total loss of control in the back science room."

Jack goes to his hands and knees to get the scoop and, to get up, pushes sharply back with his hands to get on his knees. He rotates backwards and, as his PLSS is now over his feet, rises easily. He may have put his feet in a small crater to get leverage, which would explain his kicking the scoop earlier. Once he was up, Jack goes about 10 feet farther west carrying the scoop and is ready to start a trench.

Cernan – "This is not meant as a criticism, but I think that Jack tended to fall more than the rest of us did. And it's maybe because he became more aggressive. And, thank God for 1/6 gravity. You would have dropped things anyway because of the lack of nimbleness and dexterity and you would have wanted to get down to pick things up and chip rocks and what have you. And 1/6 g made getting back up a lot easier than it would have been otherwise. You developed a personal technique – the way you skipped or hopped, the way you got on and off the Rover, the way you got up when you fell."

Cernan: Well, I'm going to clean their glasses so they know we're not crazy. Can you wait a minute on that pan you're taking?

Gene moves the TV camera so he can dust it.

Schmitt: *(thinking Gene is talking to him)* I already took it.

Cernan: No, I mean the television camera. *(to the TV operator, Ed Fendell, in Houston)* I'll put you back where I had you. *(pause)* Now, I'll put you right [back] where you finished your pan.

Gene repositions the TV and Jack starts the trench.

> **Schmitt** – "What we dug was half of a trench. That is, we'd only try to maintain one wall – the east-facing wall – and tear down the other one so that you could see in."

Schmitt: You know *(pause)* that orange is along a line, Geno, along the rim crest.

Cernan: What? Circumferential?

Schmitt: Yeah. Man, if there ever was a *(chuckles)* ... I'm not going to say it. *(and then does, anyway)* But if there ever was something that looked like a fumarole alteration, this is it.

Schmitt: Hey, I think we hit one of those things we've got to reconsider on, Houston.

Parker: Yeah. The problem is we're looking at PLSS constraints right now, as luck would have it, of course.

They will have to leave in about 25 minutes because of walk back. After watching Jack dig his trench, the TV camera is pointed down at the left-front wheel and is stuck in that position.

Cernan: What's wrong with the TV? Aren't you watching this? *(pause)*

Parker: It seems to have died slowly there. *(pause)*

Cernan: Well, stand by.

Gene raises the TV, and as the TV camera is turned toward Jack, the orange is faintly visible. Actually, this orange color would have seemed brighter on the monitors in Houston than it does in the TV recordings that are all we have now. Hasselblad images from a pan Gene will take later from the southeast rim of Shorty also show the orange much more clearly than the recorded TV images.

Schmitt: Okay, Bob, I've trenched across the trend of the yellow ... *(correcting himself)* or the orange. There is light gray material on either side.

Gene joins Jack and looks west into the trench.

Cernan: Oh, man, that's incredible!

Schmitt: Okay, Gene, we're going to have to ...

Cernan: That's incredible.

Figure 7.2. Jack dug a trench into an orange streak on the rim of Shorty Crater to see how deep the coloration might go. The boulder beyond (top left) is a block of lava shattered by the impact that excavated Shorty Crater. The gadget on a tripod is a *gnomon* that the astronauts placed on the ground before taking photos. The gnomon at the top of the rod is suspended on gimbals and, therefore, always points vertically; its shadow indicated the direction of down-Sun, indicating the orientation of the photo and the objects in it (the Moon's magnetic field is too weak for compasses to be useful). Both the gnomon's 62-centimeter length and the painted measurement on the support provided a scale for determining the size of objects in the photo. The grey-scale/ color chart attached to one of the tripod legs was used for color reference when film was processed after the mission.

Schmitt: You need to get a down-Sun color [photo] …

Cernan: That's incredible.

Schmitt: … as well as … I'll get my black-and-white.

Cernan: I'll get it. *(long pause)*

Schmitt – "I think there have been other instances, but this is a particularly good example of where Gene deferred to me in organizing how we would approach a geological problem, just as I would defer to him on the Rover stuff and things like that. It worked out pretty well; that had all been sorted out, almost without us knowing it, during the training cycle."

Cernan – "You can't specifically train for every moment of the mission, so the key is that you train for knowledge of each other and you train for teamwork. And, when you come up with something new and different, you just defer to each other's capabilities. In this particular case, Jack had pretty much surrounded the orange-soil project, if you will, and had in his mind what we needed to get. So we went ahead and did it."

Schmitt: We also got to get that rock up there.

This is the 2-meter-high boulder a few feet west of the trench that Jack thinks they ought to sample.

Cernan: Yeah, we'll get that. Okay, let's start sampling that trench. We've got to get ... That's ...

Schmitt: Okay.

Cernan: ... That's phenomenal. Look at where the contact between the gray and the ...

Schmitt: Yes. Right, and it's on both sides [of the circumferential band] ...

Cernan: Before you disturb it, let me just get a couple of close-ups of that.

Gene does a deep knee bend to get the close-ups from a few feet southeast of the trench.

Cernan – "You can see here how I could bend down without kneeling – without getting to a static position – to get some close-up pictures. In 1/6 g you can go down slow enough and you can waver in that almost-kneeling position – uncomfortable and hard to sustain – long enough to get a couple of pictures at a 125th of a second."

Schmitt: Hey, can you get a down-Sun? I think your color will be best down-Sun.

Cernan: Okay.

Schmitt: Hey, you want any of this bagged in the can, Bob? Canned in the bag ... or whatever it is?

Parker: Stand by. They [meaning the geologists in the Science Backroom] are debating that right now.

Cernan: *(garbled)*

Parker: Roger. Let's get the short can for some of that and ...

Schmitt: Okay, the color is ... *(listens to Bob)* Okay, let us get the ...

Cernan: Better sample it first, then we'll get it. *(pause)*

They move in to sample: Jack manipulates the scoop, and Gene opens a bag.

Schmitt: It's quite indurated.

Jack means that the soil is cohesive. He goes to one knee to scoop soil from the bottom of the trench, but he spills some of the soil getting up.

Cernan: Aah ... Go back and get that one. Go get a new chunk!

Schmitt: I'll go get a new chunk.

> **Schmitt** – "It came out in chunks. It was cohesive. It eventually fell apart as they worked with it in Houston, but it was cohesive enough that when we put it in the bags it was in chunks."

Cernan: Give me that, and get a new one. Give me that. Get some more.

About a cup full of the orange soil stays in the scoop, and Jack pours it into the bag.

Schmitt: I'm going to slow down here.

Cernan: Yep, just take it easy.

Schmitt: I can't see into this [trench].

Jack is standing east of the trench, casting a shadow on it.

Cernan: I can't see when your shadow is there.

Schmitt: Can you get around on the other side?

Cernan: Yep.

Schmitt: Because I can't see to sample. *(pause)*

Jack goes to the south end of the trench, Gene to the north. Jack drops to one knee to sample and then kicks surface material into the trench as he gets up.

> **Schmitt:** Oh. Well.

Jack does a deep knee bend and gets a sample this time.

> **Schmitt:** Yeah, that's it. *(pause)*
>
> **Cernan:** See if you can get a sample right across that contact too.

Jack pours the sample into the waiting bag.

> **Schmitt:** I will. Okay, bag that one.
>
> **Cernan:** Bag 509 has got the orange material from, oh, about 2 to 3 inches down.
>
> **Parker:** Copy that. *(long pause)* Okay, we're suggesting Intermediate [cooling] for you, Jack.

Jack is working hard as he gets another sample.

> **Schmitt:** Okay, the light gray, which is on either side ... We sampled the ...

Jack pours sample material into a bag Gene is holding open for him.

> **Schmitt:** Want me to get some more?
>
> **Cernan:** Yeah, a little more.

Jack collects more soil by scooping sideways, knees bent.

> **Schmitt:** All of this is getting mixed a little bit with about a half-centimeter-thick light-gray or medium-gray covering over the whole area.
>
> **Cernan:** Bob, the gray material that's adjacent to the red material is in – what would I say – [bag] 510.
>
> **Parker:** Copy that.
>
> **Cernan:** I had it, and I can't see it now.
>
> **Schmitt:** And the LMP is Intermediate.

Parker: Copy that.

Cernan: 510, Bob.

Parker: Copy that.

Cernan: And that orange band is about a meter wide, I think.

Schmitt: About a meter.

Schmitt – "Post-mission analysis showed that the orange soil turned out to be very small, spherical beads of glass, rich in titanium and low in silica and with the intensity of the orange color apparently related to the titanium/iron ratio."

Cernan: You can't get to the bottom of it though, can you?

Schmitt: I haven't been able to yet.

Parker: Okay ...

Schmitt: *(moving to the north end of the trench)* Just to be sure, why don't we sample this side of it, too.

Cernan: Then I'm going to go get the can.

Schmitt: Okay ... One ...

Cernan: If I can remember where we put it. Bob, where did we put the small can?

Gene opens a bag and Jack pours.

Schmitt: It's in bag 7 under my seat.

Cernan: Okay. *(commenting on the amount of soil Jack has poured in the sample bag)* That's good.

The short can or SESC was 13 centimeters tall and had a 6-centimeter inside diameter. Like both the long can (CVSC, or core sample vacuum container used to transport one of the Station 3 drive tubes) and the rock boxes (SRCs), the short can had a knife-edged indium seal to maintain a vacuum and to protect the sample from terrestrial and spacecraft gases. To increase the chances of getting a good seal, both the knife edge and the indium were protected with Teflon covers that Gene would remove just before he closed and latched the lid. In addition, Gene put the short can into the rock box at the end of the EVA, which doubled the chances that at least one seal – on the short can itself or on the rock box would be good. Some of the knife-edge seals on previous missions were compromised by dust despite all the precautions.

Schmitt: [Bag] 511 has the gray from the other side of the orange band.

Cernan: And the other side happens to be the crater side.

Schmitt: That's right. North side.

Cernan: Okay. Why don't you look around a minute, and I'll get that can.

Schmitt: Okay. I'm going to see if this goes on down here as a zone. *(pause)*

Schmitt: It looks like it's an ellipsoidal area if my footprints are any indication.

Parker: 17, Houston. We'd like to get the double core here instead of the small can. Double core, please, instead of the small can.

Cernan: Okay.

Schmitt: Did you want it in the orange?

A chorus of voices in Houston said "Yes!"

> **Schmitt** – "The reason I asked the question was not so much because I disagreed but, as I did several times during the EVAs, I wanted to make the Backroom think a little bit, just to make sure they thought through some of the options they had. Based on the information we had up to this point, it looked like something vertical [the orange-glass appeared to be a vertical structure] in the rim of the crater; and, fortunately, we did what they suggested and didn't change [the coring location] because, otherwise, we might not have hit the underlying black soil."

More than 20 years after the fact, Gene questioned whether Jack was quite so logical at the time.

Parker: Roger. That's affirm. We can put cores in gray soil all the time.

Schmitt: Well, it's a vertical stratigraphy. Do you want to go sideways a little with it? *(pause)* Or you just want to get it as deep as you can, huh?

What Jack means by "vertical stratigraphy" is that the orange soil deposit appears to have vertical contacts with the adjacent grey material.

Parker: I expect we want to get ... Let's go as deep as we can in the orange, please, there, Jack. And the one problem at this station, Jack, is not that ...

Schmitt: *(responding to Bob's "deep as we can")* All right.

Jack goes over to the Rover.

Parker: ... we can decide priorities between this station or any other station. It's the fact that we're running up against the walk-back constraints here in just a very few minutes, about two-zero [20] minutes.

Schmitt: Okay.

Parker: Roger. That's some crater!

Schmitt: *(to Gene)* Got your hammer?

Cernan: Yeah. I've practiced too long on taking stereo pans of craters, without getting one here.

Schmitt: I got mine from right down there, Gene. So ...

Cernan: What is that right there?

Schmitt: What?

Cernan: That right there.

Schmitt: I don't see. Oh, it's a piece of glass, probably.

Cernan: Boy, it sure is.

Schmitt: Hey, how about right up here [just west of the trench, for the core]? *(pause)* You know, we just about got *(laughs)* to the upper edge of this little ellipsoid zone. I think we're going to have to ... We've messed up most of it. Let's try right over here.

Cernan: I've got a little piece of glass in my pocket.

Schmitt: Bob, the upper portion of the core is going to be a little bit disturbed, because we've walked around the area so much.

Parker: Okay. Copy that.

Cernan: That was a little piece of black glass ...

Schmitt: *(to Gene)* You may want to get uphill ...

Cernan: ... solid black glass.

Schmitt: Okay. That ... Did you get a ... Hold it, and I'll get a shot.

Cernan: Take your picture.

This photo, like the others Jack has taken since finishing his pan, is badly overexposed, probably because he doesn't have the f-stop set correctly.

Schmitt: That's about as far as I could shove it [the core tube] in [by hand].

Parker: Okay. And, 17, while you're doing that, was the gray mantle over the top of this, or was this showing all the way through to the surface?

Schmitt: No, it was over the top. About a half a centimeter over the top.

Schmitt: He's getting about 3 centimeters a whack.

Cernan: I'll tell you, it's a lot harder going in than that double core was back there [at Station 3].

Schmitt: Yeah, it's pretty hard. *(to Bob)* It acts like it's inherently cohesive. It breaks up in angular fragments.

Schmitt: The central portion of the zone actually has a crimson hue, or a red hue. Outside of that it's orange. And outside of that, it's gray.

Cernan: *(breathing heavily)* Wait a minute, Jack.

Schmitt: That's all right, take it easy. I'd offer to hit it, but I don't think I can, my hands are so tired.

Cernan: I'm going up to Max [cooling] here for just a minute or two.

Cernan: Okay, let me hit some more. Ready?

Schmitt: Yeah, go ahead.

Jack holds the core tube while Gene hammers it. Gene is very anxious about that. He stutters a couple of phrases before he got the thought out.

Cernan: Why don't you ... I'm afraid ... Jack, it's stable enough. Why don't you get out of the way? I'm afraid if I leave go of this thing, you'll get it in the head.

Jack moves out of the way to the south. Gene raises the hammer to level with the top of his helmet and swings it down to hit the top of the extension handle at about waist height. He uses the flat of the hammer to have a better chance of connecting.

Cernan – "You had to hold the hammer tightly with your fingers. It was like gripping your golf club too tightly: all your muscles tighten up and your swing goes to hell and you get tired. And to make it harder, the core wasn't going down all that readily. It was just a heavy bit of work, particularly clutching the hammer."

Schmitt: Okay. Have at it. *(to Bob)* He's still getting a centimeter a whack, poor guy. Let's see. I didn't get a locator, I better get a locator. *(pause)* Oh, it's in the [pan] ... No, it isn't. *(pause)*

Cernan: The only thing I question is our ability to get it out. *(pause)*

Gene stops to rest while Jack takes a picture toward the Rover as a locator. Gene then hits the core a half a dozen times more and stops again. Jack's visor is halfway up.

Cernan: *(breathing heavily)* Man, that's really hit bottom.

Schmitt: Has it?

Cernan: Yeah.

Cernan: ... I really wonder about getting it out.

Schmitt: Well, we'll give it the old college try.

Cernan: Yeah, we ought to be good at getting cores out by now.

Schmitt: It'll come out.

Cernan: Whew! Okay.

Schmitt: It wouldn't dare not come out. Wait a minute ...

Cernan: Is there enough to hold on to?

Schmitt: Whoops. Which side you got?

Cernan: I was just getting this out for you. Let me. I can get this side better.

Schmitt: Okay.

Cernan: You ready?

Schmitt: Okay.

Cernan: Go. Okay, pull slowly.

The core comes out about a foot with apparent ease.

Cernan: Slowly so I can cap it all right. Let me get a cap.

Schmitt: Okay.

Cernan: Okay. Hold it. Hold it. Let me get a cap. Turn.

Schmitt: All right, get the cap.

Cernan: Okay. Now, wait a minute.

Schmitt: Are you ready?

Cernan: Okay, very slow. Even the [outside of the] core tube is red!

Schmitt: Look at that!

Cernan: Even the core is red! The bottom one's black and orange, and the top one's gray and orange!

Schmitt: The fact is, the bottom of the core is very black compared to anything we've seen.

Cernan: Hey, we must have gone through the red soil because it's filled, but it's filled with a black material.

Schmitt – "Post mission analysis showed that the black soil was mostly beads of devitrified [crystallized] glass of the same composition as the orange, but with a different cooling history."

Schmitt: Let me see, Gene.

Cernan: Dark gray, almost a very, very fine grained ...

Cernan: *(capping the bottom of the core)* But it, it, it ...

Schmitt: Let me ... God, it is black isn't it?

Cernan: Yeah. I've got to get it so I can get the ... Boy, it is black and is it contrasted to that orange stuff.

Schmitt: Very black. Well, not very black. It's a good dark gray.

Cernan: Very dark bluish gray.

Schmitt: Yeah ... contrast.

Cernan: Okay, turn that thing so I can push this cap a little bit. Just turn.

Schmitt: Which way?

Cernan: Either way. Just turn the whole tube.

Cernan: ... some more. I don't want this cap to come off. Okay. I'm going to Intermediate cooling. Okay. Now you don't have any caps, so let's take this back to the Rover.

Cernan – "You can see in the TV how much inertia I got into my arm to get it back there [to swing the hammer]. I didn't move back slowly at all. I already had my hand chest high and I sort of raised my hand and then moved back very rapidly. You just couldn't go back slowly and sneak up on it. Other times, when my arm was down more, I'd have to do the double pump; but, here, my hand was already up there. You adapted mentally and physiologically to find the easy way to do things like this and, rather than bust my butt trying to stretch and push that suit to go back, I'd just bring my hand up chest high and, with the strength I'd need if I was swinging the hammer, just swing my arm back there. Then, once your hand was back there, the suit was stiff enough that, before it rebounded, you could reach the control and not have to forcibly hold your arm back there."

Parker: ... And, 17, for your thought ... We *have* to be leaving here ... Not "*like* [to be leaving here]" [there are no ifs, ands, or buts about this departure time] ... we *have* to be leaving here in 14 minutes. On the move, because of walk-back constraints.

Parker: And we'd like to get a quick sample of the basalt up there on the rim, and Gene's stereo pan, and then press on. And I emphasize that it's walk-back constraint we're up against in 14 minutes. 13 now.

Schmitt: Okay, Bob, I'll sample it by hand. But it'll be documented. And I'll get it in a bag in a minute since I don't have any.

Parker: Why don't you leave the core there, Gene, and you can take your stereo pan while Jack's getting that sample. And then you can get together and ram the core home.

> **Schmitt** – "When we knew we only had a little bit of time, we got things done a lot better than when we did a station for no specific purpose, like Station 3. Station 3 was a disaster. We got a lot done, but it was a very poor use of time. This one was both productive and fun."

Cernan: *(at the Rover, disassembling the core)* Okay. Bob, the bottom of the upper core is also dark.

Parker: Copy that. *(pause)* Sounds [like the orange layer is] a little thin. *(long pause)*

Jack leans on the boulder to examine it, then pulls off a hand-size sample. He then moves southeast to take some "after" pictures.

> **Schmitt** – "The boulder had been intensely shocked. It was falling apart."

Cernan: And, like you might expect, the top of the bottom core is dark, too!

Schmitt: *(garbled; pause)* If I ever saw a classic alteration halo around a volcanic crater, this is it. It's ellipsoidal; it appears to be zoned. There's one sample we didn't get. We didn't get the more yellowy stuff, we got the center portion ...

Cernan: Let me get those caps, Jack.

Schmitt: Okay.

Cernan: I'm going upslope. I'm circum ... I'm on the circum ... Oh, you know, on the rim. And I'm up. Oh, that ought to be a beautiful shot, if I could see what my settings are.

Schmitt: Okay, the lower core is "chucky-jam" full. *(a favorite expression Jack adopted from Stanford professor, R. H. Jahns)* I don't think I've budged that thing.

Parker: Okay. Copy that. *(pause)*

Parker: Okay, and, Jack, I copied – aside from three trench samples – I copied one single bag of basalt samples. Is that correct?

The TV picture shakes as Jack works at the Rover.

> **Schmitt:** That's right. 512.
>
> **Parker:** Copy that.
>
> **Cernan:** *(looking west)* Hey, Bob, from where I am, [I can see a place] about 100 meters around the west side of the rim of this crater [where] the mantle on the inside of the rim turns from this gray material we've been sampling, in here, to a very dark gray material. And there's a lot of orange stuff that goes down – radially down – into the pit of the crater.

> **Schmitt** – "A first-rate observation."
>
> **Cernan** – *(tongue slightly in cheek)* "I really do appreciate getting an A-plus on the quiz. I really wondered; it's taken a long time [nearly 19 years] to find out how I did."

> **Cernan:** I got to take a couple of more pictures at that contact slope over there. You can't see it from where you are, Jack, but I guess we got to leave. Otherwise it would be nice to sample that dark stuff up on top.

Gene moves about 10 feet west. As he moves, he kicks flat sheet-like sprays of dust a few inches off the ground. The sheets reflect sunlight into the TV lens, creating a short-lived patch of light.

> **Parker:** We need you guys rolling in 7 minutes.
>
> **Cernan:** *(gesturing toward the west)* That stuff [inside the crater rim] – and you're looking at me with the camera – that stuff is up toward that boulder. About as far away from that boulder on the other side as we are on this side. And we want a hack at that boulder, too. Jack, let's see if we can't get that boulder, anyway. *(pause)* But I don't have any film.
>
> **Parker:** Guys, we don't have that much time.
>
> **Cernan:** I know, Bob, I know. *(pause)* There's a lot of little pieces – not a lot – but enough that I've seen five or six of them. Little pieces of obsidian-like glass. I got one in my pocket. Unbagged. Undocumented. *(pause)* This boulder that you were looking at with the TV, I'm going to take a sample. Undocumented.
>
> **Schmitt:** I got it [meaning the boulder samples]! I got it!
>
> **Cernan:** Oh, you got it?
>
> **Schmitt:** Yeah.
>
> **Parker:** Yeah.

Figure 7.3. In this portion of a panorama of Shorty Crater, taken by Gene Cernan, the North Massif is on the horizon at the right. The 110-meter-diameter crater is quite fresh. Fortuitously, it excavated a sequence of lavas and deposits of volcanic ash and regolith. By understanding the physical processes that take place during such an impact, geologists were able to help the astronauts select samples that had been excavated from well below the lunar surface. Excellent photography contributed a great deal to the geologists' later ability to interpret the crater-forming impact and the underlying structure. The clarity of these photos required a steady hand, which was not an easy task for the excited astronauts at Shorty Crater.

Schmitt: Let's go.

Cernan: *(to Bob)* Okay. As you look at the inner rim – as it goes down *(cough)* to the right – you see a lot of boulders, a lot of rocks – that are protruding out. Where that rock pattern thins out, just beyond that is an orange – a visible orange – radial pattern, and then beyond that is a definite change in albedo where you get the gray material, and a definite change in the number of rocks on the slope.

Cernan: ... and that particular material ... *(hearing Bob again)* That par ... *(stopping a last time for Bob)* Let me finish, Bob! That particular rim material there continues around to the due north, and then there's a drastic change again where you see the inner rim completely terraced with this boulder fill.

Parker: Okay, copy that, Gene ...

Cernan: And I can't bet on it, but I can see it ...

Parker: ... and you can talk about it when you get home.

A SIGNIFICANT FIND – VOLCANIC ASH BUT NO WATER

The discovery of orange deposits was truly exciting for the entire Apollo team. The crew was exhilarated – literally hopping around on the lunar surface – and the scientists in the Backroom were like rambunctious schoolboys. On Earth, a rim of orange, yellow, or red deposits around a crater margin would imply that those deposits had been altered as volcanic gases were released slowly as the volcano cooled. If this had been the case for Shorty Crater, there also would have been clues about the volatile phases released during volcanic eruptions on the lunar surface. But this was not the case for Shorty's intriguing deposits – this crater was not on Earth.

Shorty Crater is definitely an impact crater. However, it excavated a deposit of lunar volcanic ash and the sequence of black and orange materials were beautifully preserved – albeit inverted – because the Moon has no atmosphere. Before being ripped out of its original position during the impact, the deposit may have been part of a volcanic ash sequence deposited during an eruption. The deposit's optical properties are akin to those of the dark deposits that rim Mare Serenitatis and the floor of the Taurus-Littrow valley; this similarity helped to illuminate their origins.

The orange deposits consist of mostly beautiful, transparent glass spheres. They are fairly uniform in size, with an average diameter of 40 micrometers, or slightly less than the thickness of a human hair. Some spheres contain crystals and a few have preserved gas bubbles; similar glass spheres are found in the most violent of Hawaiian lava fountains. The lunar ash differs in composition from its Hawaiian counterparts – as a result, the lunar magma had a very low viscosity, erupting in lava fountains that produced a spray of tiny molten droplets. The droplets chilled quickly to form glass spheres. What gas phases drove these eruptions? Perhaps carbon monoxide, plus sulfurous gases. Many of the glass spheres are partly coated with materials that condensed from these volcanic gases, which later analyses indicate had contained sulfur, zinc, cadmium, and lead.

The black deposits collected by coring the deposit at Shorty Crater have a composition identical to that of the orange glass. The particles are not spherical, but are composite grains made up of many droplets stuck together. They are also partly crystalline, with feathery crystals of iron oxide and silicate minerals separated by areas of orange glass. Geoscientists have inferred that these deposits were also part of a lava fountain, but rather than being blasted free and quickly chilling like the orange spheres, these composite particles were trapped near the center of a lava fountain where falling droplets impacted still-molten droplets being erupted. Minerals grew within the melt before the droplets solidified.

Among all the lunar samples, the orange and black droplets stood out as products of volcanic action because of their physical and chemical homogeneity. Glasses of impact origin are a mess: they are crisscrossed with bits of rock and minerals and display bands with complex chemical compositions. The orange glasses from Apollo 17, green glass spheres collected at the Apollo 15 site, and many of the clear glass spheres gathered from all of the mare landing sites have provided evidence of the variety of magmas erupted from along the mare margins 3 to 4 billion years ago. These samples from the Moon's interior are pristine: there has been no alteration by water and therefore – even after billions of years – are fresher than 1-day-old volcanic ash from an eruption in Hawaii.

Planetary exploration works in strange ways. Using remotely sensed data and terrestrial analogs, we can prepare initial interpretations of what we think occurred on the surface of another planet. But we don't really understand the origins of geologic features and deposits until we visit. Perhaps, in part, because of the excitement of discovery and, almost certainly, influenced by preconceptions, the explorers who visited Shorty Crater in the Valley of Taurus-Littrow got the story right even though they got some of the plot wrong. Although they were incorrect in their on-the-spot interpretation that the orange color was result of hydrothermal alteration, they were correct in recognizing that these were indeed the products of explosive volcanic

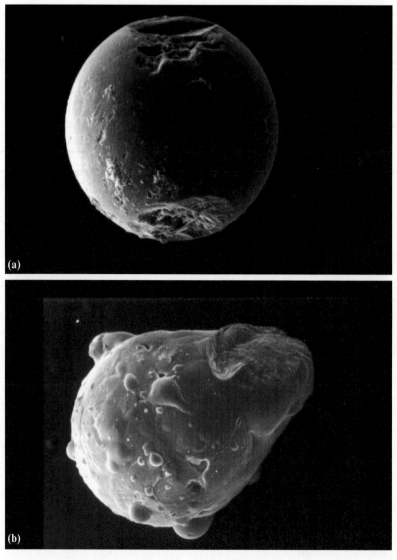

Figure 7.4. Droplets from the "orange" and "black" soils collected on the rim of Shorty Crater have been interpreted as spray from lava fountains of melts that had very low viscosities. (a) This 0.2-mm-diameter sphere, when viewed in an optical microscope, is clear orange glass. The small spalls on the sphere's surface may have been caused by impacts of micrometeorites. (b) A lumpy grain, formed when a larger, chilled droplet was impacted by smaller droplets from the same lava fountain, has the same composition as the droplet in (a), but the black droplet is partly crystalline. In the field, these deposits appeared black and Jack and Gene called them "the black glass," even though they aren't entirely glassy. By studying these deposits, which were likely erupted from volcanic vents, scientists began to understand the explosive activity that accompanied the filling of the maria with lava flows. (*scanning electron microscope images from JSC in Houston.*)

eruptions. Their carefully collected, detailed samples later revealed a complete storyline for volcanism on the Moon.

The orange and black deposits sampled at Shorty Crater during Apollo 17's second EVA provided information about explosive lava fountains – events accompanying the eruption of massive lava flows that filled the large impact basin of Mare Serenitatis. The next EVA was planned to sample boulders that rolled down the North Massif. The North and South Massifs, which now rise high above the Taurus-Littrow Valley floor, were the deposits left around the rim of Mare Serenitatis by a massive impact. Samples from the massifs have provided information on the composition of the Moon's crust and the event that formed the Serenitatis basin.

Gene Cernan is preparing the Rover for a drive to the North Massif (at left-horizon). This was the last trip across the lunar surface during the Apollo program. The crew crossed the valley floor north of the LM and then began their ascent up the lower slopes of the North Massif to some house-size boulders that had rolled down the side of the 2000-meter-high mountain. The Sculptured Hills form the right horizon and Wessex Cleft is above the U.S. flag.

8

Boulder Rolling – the Last Apollo EVA

The Moon's early history was one of unimaginable violence. Ripped from the Earth during a collision with a Mars-size planet 4.5 billion years ago, its childhood was not a pleasant one. After the molten mass began to cool, the surface began to chill to form a solid crust. Constant bombardment by meteors and the turbulent mixing with molten rock rising from the interior combined to create a remarkably complex, but steadily evolving early crust, which is now exposed in the lunar highlands.

Collecting clues that would help scientists understand the early violent history of the Earth's constant companion was one of the chief goals for the Apollo 15, 16, and 17 crews. Apollo 15 explored along the margin of Mare Imbrium, Apollo 16 in the highlands, and Apollo 17 along the margin of Mare Serenitatis.

The Apollo 17 landing site is in the Valley of Taurus-Littrow, a narrow defile flanked by massifs of more than 2000 meters (6600 feet). The valley is an open fracture that cuts through the eastern rim of the Mare Serenitatis (Sea of Serenity) basin. The basin's origin, about 3.8 billion years ago, was far from serene, belying its name. The basin was excavated when a large body collided with the Moon at a velocity of over 3 kilometers per second (more than 6700 miles per hour). The impact shock at the lunar surface – producing pressures millions of times greater than air pressure on Earth – fractured, melted, and even vaporized rocks. This impact fractured rocks to a depth of nearly 25 kilometers (16 miles) and brought great chunks to the Moon's surface. Debris from the crater was deposited around the basin margins; those deposits are now exposed in the high massifs. It was this margin area, where the basin and valley meet, that Apollo 17 crew would explore.

Slopes of the high massifs above the Taurus-Littrow valley range from 20 to 30 degrees – steep enough to shed much of the ejecta from later impacts and other debris that covers the Moon's surface, but far too steep for either the Rover or astronauts on foot. Whether working in mountain regions on Earth or in the lunar highlands, field geologists face the same difficulty: "How do we get a sample from the exposures high on the walls above the valley floor?" The answer is to look where Nature had indulged

Figure 8.1. This photo was taken from the Rover during the Apollo 17 traverse to the North Massif during EVA-3. Directly ahead is the large boulder called "Turning Point Rock," where the crew changed direction to reach another large boulder (visible at right center) on the lower slope of the mountains. Both boulders were dislodged from outcrops high on the North Massif (likely by impacts) and rolled downslope, leaving ragged tracks in the regolith. The area highlighted shows the feature known simply as the "dark boulder" as well as the track it gouged out as it bounced, rolled, or slid to its current position. The Rover didn't have enough power to safely climb the steep slopes of the massifs up to the visible outcrops near the summit, so Gene and Jack took samples instead from the boulders that had rolled into reach.

in some "boulder rolling." In this case, several large chunks from the massif outcrops had been broken off from outcrops by impacts or perhaps moonquakes and had rolled or slid down to a break in the slope just above the valley floor. Most of the boulders came to a halt when the slope angle decreased to less than 11 degrees about 50 meters (160 feet) above the valley floor.

High-resolution photographs of the site, taken from the command modules of earlier Apollo missions, showed several boulders resting at the base of the North Massif that are over 10 meters (33 feet) long and had originated high on the massif. We

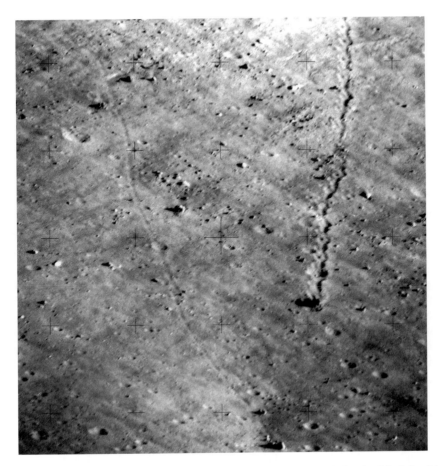

Figure 8.2. Deep tracks on the slopes of the North Massif were made by the "dark boulder," which may be as much as 10 meters long. A fainter track, made by a smaller boulder, crosses from upper left to lower right.

knew where they had come from because they left deeply gouged tracks in the regolith before bouncing, rolling, or sliding to a stop. Some of these boulder tracks are up to 2 kilometers (1.2 miles) long. The boulder chosen for sampling (at station 6) by the crew of Apollo 17 had rolled cross-slope for a distance of about 1200 meters (3900 feet), with a vertical drop of more than 500 meters (1,600 feet). Thanks to gravity, the crew was able to sample a piece of outcrop that had originated high on the massif.

This mission segment covers much of the third and last Apollo 17 EVA, which was rewarding but not easy. The transcript begins with Gene Cernan and Jack Schmitt getting ready to leave the LM for their third traverse across the lunar surface. It would be not only their last moonwalk but also the last of Apollo; after the excitement of discovering the orange soil at Shorty, they were eager to make the most of the opportunity. The capcom is Bob Parker.

PREPARING FOR THE TRAVERSE

> **Cernan:** Okay, Bob. I'm going down the ladder.
>
> **Parker:** Roger, Geno. *(long pause)*
>
> **Cernan:** "Godspeed the crew of Apollo 17."
>
> **Schmitt:** Good.
>
> **Parker:** Amen there, Gene. Amen.

Cernan – "It had become a little prayer at the start of those EVAs. I just shouted to the world, "God, bless us. Give us a shot at it." I'm beginning to think that there wasn't anything printed or written anywhere on that ladder. And if Jack doesn't recall it, it wasn't there. I think I just decided to say that when we started every EVA. Sort of like blessing the shrimp fleet."

Gene first "reads" the message at the end of EVA-1. At that time, Jack said, "Who signed it? I forgot to read it." Gene replied, "I'm not going to tell you, but I like the message. Probably shouldn't tell you." The only sure way of knowing if the message was on the ladder or strut is to look at photos of the LM that were taken during the mission. At the start of this third EVA, Jack will take a color picture of the front of the LM, so we have been able to examine blow-ups of the hatch and ladder area. Although the blow-ups are quite grainy, there is nothing visible that looks anything like a note. A white spot just below the hatch is part of the normal LM hardware. At the end of this third day on the surface, Jack will take some pictures of Gene and the Apollo plaque, one of which shows considerable detail on the ladder and strut up as high as the second step below the porch. There appears to be nothing that might be a message. The mystery remains.

> **Cernan:** Okay, Bob, I'm on the [foot]pad. And it's about 4:30 [on] a Wednesday afternoon [Houston time], as I step out on to the plains of Taurus-Littrow. Beautiful valley. The first thing I'll do is I'll turn the TGE [Traverse Gravimeter] on, and I'll give you a reading.
>
> **Parker:** Okay; we're ready.

Gravimeters are commonly used for oil exploration and other applications on Earth. The Apollo TGE was a portable device that could be stowed on the back of the Rover; it was used to make accurate measurements of the strength and direction of the Moon's gravity at each of their stops. Small differences from place to place provided information about subsurface geologic structure. At Taurus-Littrow, the valley was formed when two massive blocks – roughly forming the South and North Massifs – were thrust upward during the Serenitatis impact; a central block fell back and formed the valley floor. Later, volcanic eruptions filled Serenitatis with lava, and these flows also flooded the valley floor. Analysis of the TGE measurements would indicate the depth of the lava fill and provide insights not only the mare filling, but also the forces that moved the blocks during the Serenitatis impact. To make the measurement, Gene sets the TGE on the ground as upright as possible, pushes a "Start" button, and waits about 30 seconds until the TGE flashes a light to indicate that the measurement is complete.

Cernan: It [meaning the TGE light]'s on and [I'm pushing the] Read [button]. Bob, it reads 222, 262, 207, 222, 262, 207.

Parker: Okay, I copy that, Geno. *(pause)*

Cernan: *(to himself)* Okay, get the visor down, Geno. Get the visor down . . . Holy Smoley. Think it'd be better to leave it up. Beautiful out here today, Bob! We can look to the east for a change – a little bit, anyway.

Gene's next job is to change the battery that powers the Rover's LCRU.

Cernan: Well, let's see if I can change this little baby now. Supposed to be simple. *(pause)* Bob, we have no use for the old [LCRU] battery, right?

Gene is asking if he can have some fun and see how far he can throw the battery.

Parker: That's affirm. *(long pause)*

With a premium on weight and space, constant housekeeping was necessary. The Apollo crews left dead batteries, used equipment, miscellaneous packaging, etc., on the Moon so they could make best use of the remaining LM propellant to get themselves and the priceless samples back to lunar orbit.

Schmitt – "During Apollo, the answer to Gene's question was very simple: you weren't going to go back there any time soon. But at a lunar base, any used material will have to be inventoried, catalogued, and stored in some kind of recoverable way because the value of practically any material will, because of transportation costs, be very, very high – at least until the Moon is self-sufficient and is producing it's own stuff. And that won't be for a long time. You're just never going to know when you're going to want to get something and, with a good inventory system, you can always ask, 'Do I have that or something like it for refurbishment or reuse.' "

There is some light static for a few minutes, which ends when Gene turns on the LCRU just before they drive off on the Rover.

Schmitt: Okay. I'm on the porch and the hatch is closed.

Cernan: *(to himself)* Oh! Don't bump into that.

Schmitt: Are you talking to me or you?

Cernan: I'm talking to me. *(long pause)*

Schmitt: Okay, that sounds familiar and looks familiar – the old plains.

Cernan: The Valley of the Taurus-Littrow.

Parker: And, 17, if you guys are interested, your shadows will be 8 feet long tonight.

Schmitt: How many meters is that, Bob?

Cernan: I'll draw it out ... I'll step it out for ya. You can measure it. *(laughter)*

Schmitt: Well, I don't know. Should I take my gloves off? I mean my cover gloves.

Cernan: Why don't you leave them on for a while and see where we're going. See what the boulder field looks like up there [at Station 6].

Schmitt: Well, I know what it's going to look like.

Cernan: No, you don't.

Schmitt: The point is: my hands will be much better off without them.

Cernan: Take them off, then. *(long pause)*

Although the cover gloves were invaluable in protecting the space suit gloves, they did add bulk to the hands, lessen flexibility, and hasten fatigue. Jack had difficulty holding onto equipment toward the end of the earlier EVAs because his hands had become tired. A few minutes later, Gene removes his cover gloves and puts them in a pocket on his left thigh so that he can bring them back to Earth. He later commented to author Jones, "They were pretty badly abraded, but not cut or torn."

At the end of EVA-2, Gene had discovered that two latches on the back of the Rover designed to secure the SCBs (sample collection bags) had been fouled with dust and would no longer work. Houston had been thinking about the problem during the rest period, trying to find a way Gene could use the latches – particularly to secure what they called the Big Bag, which was rather like an SCB on steroids: its top and bottom had the same dimensions as an SCB but it was twice as long.

Parker: Okay. A couple of things on that, Geno. You might try tapping the thing [bag latch] to see if that loosens the dust. There's also the hook business on the inside of the pallet that you could hook it [the big bag] on. Caution: if you open the pallet, be careful not to knock the clamps off the fender. But you can also reach over the pallet to put the big bag on.

Cernan: Okay, Bob. I brushed it and tapped it [the bag latch] yesterday. I'm not sure we're going to have much luck with them.

Parker: Say again there, Gene.

Cernan: *(trying to hang the big bag on the gate)* I brushed them and tapped them yesterday.

Parker: Okay; copy that. You might want to put the big bag on the inside of the pallet there, if you can't operate them.

Cernan: Okay.

Usually, Gene opens and closes only the gate, leaving the geopallet – to which the gate is attached – fixed in place. Apparently, there is a place to mount a bag on the forward surface of the geopallet, far enough inboard that Gene can't reach it from the side of the Rover. Instead, Bob is suggesting that Gene open the pallet. Because the pallet pivots on the geopost at the right rear, Bob is urging that they watch to make sure the pallet assembly doesn't hit the replacement fender or the clamps holding it in place.

Several minutes later, Gene discovers another dust-related problem.

Cernan: Jack, I'm also going to keep this [dustbrush] in there [under the CDR seat] ... Because it's too hard to get [the dustbrush off the front end]. We'll find a place for that in there.

Gene is having trouble getting the brush in and out of the bracket on the top of the LCRU.

Schmitt: *(not convinced that this is a good thing to do)* Well ... Okay.

Schmitt – "I may have been thinking that we didn't want any more dust under the seat."

Cernan: It's just too hard to get off the front end. Okay, let's see. "Big bag to gate, dustbrush to [LCRU] ..." Let me get that big bag on the inside of the gate if I can. Inside the gate or the pallet, Bob?

While Gene continues to work around the Rover, mounting the traverse maps on a holder in front of Jack's seat, loading film magazines in each of the cameras, and mounting packs of individual sample bags to a bracket at the bottom of each camera, Jack is scheduled to take a pan. He's a little late getting to that because the Velcro holding down the cover of the SEP receiver, which is mounted on the Rover behind his seat, has become fouled with dust. Houston asks him to use some duct tape they have under Gene's seat for such contingencies to tape the cover shut. While he works on that, Bob has another request – and a "weather" report.

Parker: Okay. And after you take the pan, we'd like you to retrieve the cosmic ray experiment. They're expecting a little solar storm, and before the rain gets on the cosmic ray experiment, they'd like to retrieve it. We'll leave it in the ETB during the traverse.

Schmitt: Okay, [I'll do it] after the pan. All right.

Solar flares are common events (several per day when the Sun is active) and most of them are weak – even on the Moon, which doesn't have a strong magnetic field like the Earth's to provide a high level of protection. If this expected solar flare is weak enough that the ETB offers adequate protection for the cosmic ray experiment, it does not represent a risk to the suited astronauts.

Solar storms severe enough to be very hazardous to anyone on the lunar surface occur only about once or twice per decade. During Apollo, one such event occurred during the first week of August 1972, about 4 months after Apollo 16 and about 4 months before Apollo 17. Scientists estimate that, had an Apollo crew been on the Moon at the time, they would have received a radiation dose of about 400 rem, which is high enough to produce nausea, vomiting, and diarrhea. There also would have been a significant risk of death unless the crew received intensive treatment within about 2 weeks. Large solar flares and storms are usually associated with high levels of sunspot activity around the time of the maximum in the 11-year solar cycle. Because all of the Apollo events occurred during the broad peak of the solar cycle, the risk of a large flare or storm during a 10-day Apollo mission was roughly 1 in 100; compared with other risks inherent in a program like Apollo, it was one that NASA management and the crews were willing to accept.

One final task for Gene and Jack before they jump on the Rover is to mount an SCB on the side of each other's PLSS. Even this seemingly simple task is sometimes a challenge.

Schmitt: Okay, here, let me get this on you first since I got ...

Cernan: Okay.

Schmitt: And I'm going to ask you to turn 180 degrees because you're up on a hill. I'll never be able to do it.

Cernan: How's that? I'm down in a hole now.

Schmitt: That's beautiful. *(pause)* Okay, just a minute. Can't get this fixed. *(long pause)* Tallest man on the Moon right now.

Schmitt – "Gene is several inches taller than I am and reaching up to that level was tough in the suit."

An alternate interpretation is that, with Gene in a hole and Jack on a little rise, Jack is above Gene slightly and is laying claim to the title. Certainly, this would be in character. The TV camera angle makes this a difficult call.

> **Schmitt:** Okay, that's done.
>
> **Cernan:** Okay?
>
> **Schmitt:** Just a second. Let me close the cover. *(pause)* Not a very good cover. Okay.

THE DRIVE TO TURNING POINT ROCK

About 50 minutes after they opened the hatch, Gene and Jack are both on the Rover and ready for the 3-kilometer (1.8-mile) drive north past Henry Crater to a large boulder called Turning Point Rock. There, as planned, they will turn to the northeast and drive about 400 meters (0.25 of a mile) to another large boulder at the Station 6 site, where they will stop for about 1 hour's worth of geology. They are starting the drive from a spot close to the SEP transmitter, which they deployed about 100 meters [330 feet] east of the LM as they completed EVA-1. They also had laid out the SEP transmitting antenna, which consisted of four 35-meter-long wires that extended along the ground to the north, east, south, and west of the transmitter. With both the transmitter and receiver on, Gene will tell Houston as he starts driving north along that arm and when he reaches the end of it. This information will help the SEP experiment time-calibrate their data.

> **Cernan:** That's exactly . . . I just came right over. Okay, we're starting Bob. Mark it.
>
> **Parker:** Copy that. *(long pause)*
>
> **Cernan:** We can't go too far in this heading. We've got a big hole up here.
>
> **Parker:** Okay.

This is the large, unnamed feature north of the SEP.

> **Cernan:** Like a big one. *(pause)* Wonder if that's Rudolph?

Gene is referring to a Poppie-size crater on the northwest rim of the unnamed crater. It is not Rudolph.

Schmitt: Well, let's see, this is east ... Looks awf[ul] ... It's a double crater but it's much bigger than I thought Rudolph would be.

Parker: No, if you're where you think you are, you're east of Rudolph quite a ways.

Cernan: Gee, I think you ought to know where we are by now, Bob.

Parker: Roger that.

It is a little curious that no one has yet told them that – as determined by Rover readouts at the various stops they made during the first two traverses – they landed just north of Poppie Crater (the original plans were to land west of Poppie). Bob will, however, tell them in just a few moments.

Schmitt: Maybe that's Lewis and Clark.

Jack is referring to the famous Americans Meriwether Lewis and William Clark, who explored the drainages of the Missouri and Columbia Rivers in the years immediately following the 1803 Louisiana Purchase. The pair of craters named for them is about a kilometer NW of the current Rover position.

Parker: After you give me a mark there, I'll talk to you about it.

Cernan: I'm sorry, Bob. I guess you didn't hear it. We're passed the end of the [SEP] antenna and we're headed northeast.

Parker: Okay, I ...

Cernan: That screw you up?

Parker: Did you give me a mark when you started or a mark when you passed the antenna?

Cernan: I gave you a mark when I started and it took about 20 seconds to get to the end [of the north antenna arm].

Parker: Okay, copy that.

Cernan: Is that good enough or do you want me to go back?

Parker: No. No. Press on.

As Gene drives, Jack is observing geologic features for the benefit of the Science Backroom – and for Gene, who is watching the road and giving Houston readouts from the Rover's navigation systems.

Cernan: Well, I tell you, it's not exactly the greatest place to navigate through.

Schmitt: I think you ought to bear left, don't you?

Cernan: Yeah. That's where I'm going here. I just want to get across this . . . around these boulders.

Schmitt: There's a crater we're just passing at 207/0.4 about 20 meters in diameter, with the pyroxene gabbro blocks on the rim, a few of them. It's not an exceptionally blocky rim crater, but we are in an area where the block population is up to about 5 percent [coverage of the surface], in contrast to most of the area we traversed yesterday.

Cernan: I tell you, going is a little bit rough; there's a population of blocks as Jack said and an awful lot of small craters.

Schmitt: Range is quite a bit higher than we were used to yesterday. *(pause)* Oops, there's one.

Cernan: Yep.

Schmitt: Snuck up on you. *(pause)* *(to Bob)* And they all – although not exceptionally blocky rim – they all have a slightly, maybe 2 or 3 or 5 percent more blocks in their walls and on their rim than does the normal terrain.

Parker: Roger, Jack. Copy that.

Schmitt: Still no obvious structure within the dark mantling material itself.

Cernan: Bob, you said 185/1.5?

Parker: That's affirm.

Schmitt: What do you want? For the Rover [sample]?

Cernan: Yeah, for a sample.

Schmitt: Oh, they changed it on us. *(pause)* Okay. Still seeing the little pit-bottom craters with the glass in them. I've forgotten the acronym already, Bob, I'm sorry. *(pause)*

Jack is referring to an exchange during the inbound drive from Shorty, when Bob invented the acronym GLPBC for the glass-lined, pit-bottom craters.

Schmitt: And you asked me for an LMP frame count a while back and I believe it was 5. That was at the SEP.

Parker: That was after the SEP photos, right?

Schmitt: That's affirm. *(pause)* Negative; that was *before* the SEP photos.

Parker: Copy that. *(pause)*

Schmitt: Okay, Bob, looking up at the North Massif, we see the scattered, strewn field of boulders that generally seem to start, more or less, from a line of large boulders, which might indicate some structure. And those lines are roughly horizontal across the face that we're looking at. The boulder tracks are irregular in shape, obviously downhill, but you'll see in the pictures that they are *(pause)* curved in places. But they're all – that I see – tend to be aggregates of little craters where the boulder was obviously tumbling and bouncing a little bit. *(pause)* We're out in [a] population of fragments now in the immediate area, at ... Is that 188?

Cernan: 188/0.9.

Schmitt: [The population's] generally about 1 percent between craters. But at the crater rims, it's up to about 5 percent.

Parker: Okay. Copy that, Jack. And how far down the North Massif ...

Schmitt: And these craters ...

Parker: ... is the line of boulders?

Schmitt: *(listens to Bob)* Oh, there are several of them, Bob. What I'm talking about is about 100-meter-long lines where the boulder trains initiate. And there's one [that] looks like about halfway, maybe two-thirds of the way down in perspective. Another one that's probably about halfway. They're just sort of scattered around on the Massif. *(pause)* I think we're getting close to ... Well, we couldn't be.

Cernan: I've got to move over here a little.

Schmitt: That must be Jones [Crater, named for Scottish-born John Paul Jones, a naval hero of the American Revolution].

Cernan: Where are you looking?

Schmitt: Off to the right.

Cernan: Yeah, our heading that they're sending us down here, it really should put us to west of Jones. So that's about right. *(pause)* A lot of static in the background today.

Parker: Yeah, I think we are talking to you guys through the LM right now; and how about a speed reading?

Cernan: Okay. We're at 12 clicks [kilometers per hour] and we're full bore.

Parker: Copy that. *(long pause)*

Schmitt: Okay.

Cernan: 187/1.1.

Parker: Copy that. *(pause)*

Schmitt: Bob, I wish I could give you more on that structure in there [that is, structure in the North Massif], but I think those lines of boulder sources are about all we can see right now. Talked about the lineaments yesterday and they're not nearly as obvious today in the higher Sun. Looking up Wessex Cleft, even with the Sun in the flat area there, it looks darker than the North Massif side. But again, the Sun angle may be fooling us, but I recall it was darker on the [overhead] photos.

Cernan: The old man wrinkled face on the ... [meaning wrinkled appearance like the face of a very old man]

Schmitt: Sculptured Hills.

Cernan: ... Sculptured Hills, though, is evident as soon as you come out of [look to the right of] the Wessex Cleft.

Jack named Wessex Cleft in honor of the novels of Thomas Hardy and to Wessex, England, as a symbol of the "intangible strength that the land transmits to its people the world over." Wessex Cleft, at the center of the frontispiece photograph for this chapter, separates the North Massif to the west (left) and the Sculptured Hills to the east (right).

Schmitt: Yeah. And they look like there are boulders up on the side of Sculptured Hills; except that they aren't nearly as big as those on the North Massif. The areas where the boulder source is look like they're made up of boulders no bigger than a meter maybe; whereas, the North Massif boulders are up to several meters. Those boulder sources all seem to be up within a third of the height of the Sculptured Hills, just east of the Wessex Cleft. Here is a boulder track that crossed the slope. See that, Geno?

Cernan: Yeah. Yeah. I sure do now!

Schmitt: It looks like it goes, rather than perpendicular to the contours, it probably is crossing them in a fairly straight line on an angle of 60 degrees, maybe.

Cernan: Back to the east.

Schmitt: Yeah, to the east. That one may be fairly near ...

This cross-slope track was made by the Station 6 boulder; they are currently about 2 kilometers from the base of the mountain.

Cernan: Jack, see that big boulder with that big track? It looks like it's an elongated, rolled-up boulder. Look at that.

Schmitt: Yes, it does. Looks like it may be broken now.

Cernan: Okay. Here we are: 1.5 and 185.

Parker: Okay; copy that.

Cernan – "In listening to Jack's descriptions in here while I'm driving, he's doing an excellent job. For someone who can't remember a great deal about it – about the craters and boulders – it brings a very descriptive picture to my mind. Jack, being the geologist, was better equipped to describe this stuff. But, boy, as you can tell from some of the comments, driving takes your full concentration. Every once in a while I could make a comment about something; but, once you start looking around – and you're moving at 6 or 8 or 10 kilometers per hour – peripherally you don't have the contrast to see obstacles and holes and craters. They don't stand out. They blend into the countryside and you've got to be looking for them or you don't see them. So you learn very quickly that you don't go rubbernecking as you drive. Everywhere you go – and it doesn't just have to be while you're in a boulder field – there's a hole or a pit or a rock or something. And, to some degree that may limit my recall because I was narrowly focused on a path and on a direction and on a destination. Jack, on the other hand, could back off and look at the big picture. And that was a good situation. You had two people, with one looking at the big picture and one getting you there. And it explains why I don't remember Turning Point Rock when I do remember taking the pan at Tracy's Rock [also known as Split Boulder at Station 6]. There, I remember how steep that hill was; I remember having to crawl up that hill; and I remember having to watch my footing as I was taking the pan. That, I remember. I was taking in the big picture then. But I don't have any impression of driving up to Turning Point Rock or to Tracy's Rock. I was probably intent on driving and I don't have any recollection about any of this until I got off the Rover. The only exception is that I do recall the boulder tracks because I could drive and take a moment to look at those."

About halfway to Turning Point Rock, Gene stops the Rover so that, as planned, Jack can use a long-handled sampling tool to collect a small amount of soil without leaving his Rover seat. After that is done, Gene asks for readings on the bearing and range to the LM he should expect to see on the Rover navigation system when he gets to Turning Point Rock.

Cernan: Okay, Bob, I'd like a bearing and range.

Parker: Okay. Bearing and range for the large block, just beyond . . . Let's see; it's just beyond the crater Henry. The large block there near the break of the slope, which is our next aiming point. The bearing and range there is 188 and 2.8.

Cernan: 188 and 2.8. Roger.

They have collected the sample at a spot from which the LM is 1.5 kilometers (0.94 miles) from them on a bearing of 185, which is slightly west of south. As an experienced naval aviator, Gene knows that he has to drive about 1.3 kilometers (0.81 miles) on a heading slightly east of north to get to Turning Point Rock. Because the boulder is about the size of the LM, they will see it long before the reach it.

Parker: Okay. And, Jack, what do you see in the way of boulders coming down the base of the Sculptured Hills, in terms of sampling opportunities at Station 8 [on the lower slopes of the Sculptured Hills] and in terms of any boulder tracks that might lead down to boulders that might just possibly be accessible at Station 8?

Schmitt: Watch it, Gene. *(to Bob)* Boulder tracks are not obvious on Sculptured Hills at all. It looks like there are fragments over there that would have had their sources higher up the slope. I think we can get boulders there.

Not long after the brief sampling stop, Gene's attention is drawn away from the road by a dark boulder and its track up on the flank of the North Massif.

Cernan: Okay, thank you. See that big boulder, Jack, with those tracks?

Schmitt: Yeah, it looks like …

Cernan: That's [a] funny looking boulder.

Schmitt: It looks like it may have stopped rolling because it broke up. *(pause)* Looks broken to me now. *(pause)*

Cernan: Boy, they've got the low-gain right on 'em. But, I tell you, we still got static.

The comm has gotten a little noisy and Gene has checked to make sure the Rover's low-gain antenna is pointed at Earth.

Schmitt: I don't have any, Gene. You may …

Cernan: Well, I sure do.

Schmitt: I don't like the sound of your bounces. Okay, you've got yourself in some holes here. *(pause)* You've never … I've read you all along, though, so there's no problem. Okay, there's a big crater. I haven't recognized Jones yet. *(pause)* Looks like you're getting up on the rim of Henry here.

Cernan: *(mulling it over)* Well … No, Henry should be to … I should be well west of Henry, I think. I wouldn't be surprised if Henry isn't right over that little rise on the right.

Schmitt: Bob, the surface structure hasn't changed … *(correcting himself)* texture [hasn't changed]. We're on a little bit of a rise in here now and still about 1 percent of the surface …

Cernan: Here's Henry right there, Jack.

Schmitt: There's Henry!

Parker: Okay, how about a range and bearing.

Schmitt: I thought you were close to Henry.

Cernan: Yeah. *(responding to Bob)* 188/1.8.

Schmitt: And we're just southwest of Henry.

Parker: Okay. Copy that.

Schmitt: On the rim. Old Prince Henry the Navigator!

Jack named the crater for Prince Henry the Navigator, most famous for the voyages of discovery that he organized and financed, which eventually led to the rounding of Africa and the establishment of sea routes to the Indies. The crater has a diameter of about 0.5 kilometers (0.3 miles).

Cernan: *(to himself)* Watch that foot.

Schmitt: It's called a wheel, I think. *(pause)* And Henry looks much like Horatio did. Has boulders on its inner wall [but] not as many. They look light colored: a light-albedo, gabbroic appearance. There may be some right down there, though, that are fine grained; they look a little greyer.

Gene and Jack drove past the 0.4-kilometer (0.24-mile)-diameter Horatio Crater during EVA-2. Jack named it in honor of author C. S. Forester and his series of novels about his fictional naval hero, Horatio Hornblower.

Cernan: Jack, there's our target … There's … Either one of … That's one right down there on *(garbled)* break in slope.

Schmitt: See the one we've got over has a boulder track. That's the one, that crossed slope.

Cernan: Yeah, if we could get up …

Schmitt: Can we get up there?

Cernan: It's awful high. We'll see.

Schmitt: That's the one. That's Station 6, and that was the … [identifying the rock] the Turning Boulder.

Cernan: Yeah, that's it.

Schmitt: The one right there. *(pause)*

Cernan: Station 6: we can probably get up there.

Schmitt: I think we can; it doesn't look too bad. *(to Bob)* The break in slope, right now, doesn't show anything obvious, except that's where the boulders start.

Parker: Okay, we hope that's fairly obvious.

Schmitt: And on up the hill you have ... *(stops to listen to Bob)* But as I was saying, Henry just looks like a somewhat more mantled Horatio. *(laughs, then makes an aside)* [This dialogue is] getting to be ridiculous.

Cernan: Say, Bob, I'm navigating – headed northwest now – to get around the western rim of Henry.

Parker: Okay ...

Schmitt: And on that west rim, we've got about 10-percent boulder cover.

Parker: Okay. And a reminder, Jack, to keep taking your Rover photos.

Schmitt: Yes, sir! And by "boulder," I generally mean "fragment," Bob, in this case. When I say 10 percent, I'm looking at stuff greater than about a centimeter in diameter. *(pause)* I'll try to say "fragment" from now on and be more precise. *(pause)* Okay. Here's a little area where there's ... This is the one part of the rim of Henry I see that has fairly large fragments, or boulders, on them up to 2 or 3 meters. But, again, they all appear to be buried. There are very few – except small ones – sitting out on the surface.

Cernan: And, you know, the fragment population out here only goes out maybe 200 meters, I expect.

Gene means that blocks can be seen out to about 200 meters (600 feet) beyond the rim.

Schmitt: Okay. Now this [particular] concentration of boulders [we are driving past] is because of a 50-meter crater in the rim of Henry.

Parker: Okay, that sounds like Locke [Crater].

Locke is 100-meter crater on the northwest rim of Henry. At the moment, Gene and Jack are driving by a smaller crater about 200 meters south of Locke and right on the west rim crest of Henry. Jack's size estimate is quite accurate.

Schmitt: I think that was one that we ... *(hears Bob)*

Cernan: Take a picture in here, Jack.

Schmitt: *(responding to Bob)* No. Locke, I can see. *(to Gene)* I'm getting the picture.

Cernan: Okay, that's not ... Locke's right ahead of us.

Schmitt: This is one [of] about 50 meters [diameter] right on the rim crest of Henry, the due-west rim. Now Locke is just ahead of us. It also has boulders in its walls but has relatively few on the rim.

Parker: Okay. Copy that.

Schmitt: Characteristic of both Henry, Locke, and Horatio is essentially no change in the average frequency of boulders on the rim. The increase comes in the wall.

Cernan: We're at 184/2.3. We're just about between Henry and ...

Schmitt: Locke.

Cernan: Locke. Yeah; right between them.

Parker: Okay. I copy that. And you guys are heading for that big boulder, which must be just dead ahead of you there, about half a kilometer.

Schmitt: Well, Gene's sort of headed for Station 6 now.

TURNING POINT ROCK

Cernan: I'm going to take a tour around that boulder [meaning Turning Point Rock] and give them a fix on it.

Schmitt: Okay. Go ahead.

Turning Point Rock, visible in overhead photographs taken from both the Apollo 15 and Apollo 17 CMs, is about 400 meters NNE of Locke. The navigation readout there will provide a more precise indication of the LM's location – derived from both the well-defined location of the rock and the fact that they haven't driven far enough for any substantial drift in the Rover's navigation system.

Parker: Yeah, and that would be a good mark to give us a range and bearing on, since it's a pretty discrete point.

Cernan: Yeah, we are.

Schmitt: Bob, the boulder concentrations in the wall of Henry have their upslope start [that is, upper limit] at about, oh, I would guess an average of 30 meters down from the rim crest. The rim crest of Henry is not very well defined, but it's there. And from that initiation of boulders, they stream down the slope to the break-in-slope down at the floor. Still no obvious change in the dark mantle, as we're just to the east of Locke now. *(pause)* There's a 30-meter crater, fairly subdued but still quite deep ... [That is, with a] subdued rim. Again, it looks as if it were mantled; [and it] has no significant increase in blocks on its rim. That crater, in any other place, would have been a very blocky-rim crater. It's maybe 30 meters [across] by 5 meters deep. Man, that is a big rock up there! *(pause)*

Schmitt: Turning Point Rock is a split rock; [it] has [what] looks like a northwest-southeast overhang, with another block just this side of it, just to the south of that overhang. It's a pyramid shape in cross section – triangular shape in cross section – and it looks like it is pretty well fractured, although not pervasively like the rock at Shorty was.

Cernan: Okay, Jack, I know I can get up to Station 6.

Schmitt: Yeah.

Cernan: I can drive up there.

Schmitt: Yeah. Now, Bob, Station 6 rock – one of them – is from that boulder track that runs obliquely across the contours.

Parker: Okay. I copy that, Jack. Sounds like good news.

Schmitt: And the pictures ought to pin down at least the [upper] end of the boulder track pretty well.

Cernan: Boy, this is a big rock, Jack. Whew.

Schmitt: As I saw it, the boulder track stopped about halfway up the slope of the North Massif. *(to Gene)* That [Turning Point Rock] is a big rock.

Cernan: We're at Turning Point Rock. And it looks like it's – I don't know if it's mantled on top, but it's certainly filleted. There's a lot of the dark mantle up and on some of the shallower slopes of the boulder. And it's on a little mound itself, as if much of it might be covered up.

Schmitt: Yup. Okay. It looks like a breccia from here.

Cernan: Can you get a sample of it right here? You see these little chips?

Schmitt: Yeah, I probably can.

Cernan: Okay, Bob. I'm 3 meters from Turning Point Rock on the east side, and I'm reading 186 and 2.8.

Parker: Roger. Copy that. Sensational.

Cernan: Ahh! *(garbled)* that over.

Schmitt: Okay. Can you drive up ...

Cernan: Yep.

Schmitt: ... to the ... right there. Let's see ... No, I can get them. The thing is, I don't know what it is.

Cernan: Well, but it's part of these fragments around here. *(pause)* I guess Turning Point Rock is *(doing a visual measurement of the rock)* 1, 2, 3, 4, 5, 6 ... 6 meters high anyway. It's a . . . Well, I'd say it's a very rough, subrounded type of rock. By the face . . . Let me get this, Jack. Okay.

Gene's estimate of Turning Point Rock's height is quite accurate; the length of its shadow, as shown in orbital photography, is comparable to the LM's, indicating that it is also about 7 meters (23 feet) tall.

Cernan – "That's big! I'm obviously impressed with how big it is. Jack says 'Gee, that's a big rock;' and all of a sudden I'm counting 'One, two, three . . .' All of a sudden I'm eighteen feet high! That's a big rock. And I hate to admit it, but I just cannot picture that rock in my mind. Now, the color pictures I took do bring back some memories. And it may be that I remember because we'd stopped to take the pictures."

Cernan: Man, we're on a little rise looking at this boulder [from the north]. That's incredible. *(pause)* Okay. We're on the roll, Bob.

Parker: Copy that.

Schmitt: *(to Gene)* You know that . . . *(to Bob)* Bob, my guess is, right now, is that Turning Point Rock is a big piece of subfloor gabbro.

Parker: Okay. I gather you changed your opinion.

Schmitt: What looked like fragments are just big spalls where the zap pits have cleaned off the rock.

Schmitt – "From my description, it sounds like Turning Point Rock was the same kind of rock as the north half of the Split Boulder (at Station 6). That boulder was two kinds of rock: a vesicular breccia that had intruded a blue-grey breccia. My first instinct here was to call it a vesicular breccia and then I backed off from that."

Parker: Okay. I copy that. And, guys, you might be happy to know that we think we've finally found the LM, because we were calling that for 188 and 2.8, and you got there at 186 and 2.8.

Cernan: That's not bad. *(pause)*

Schmitt: Okay.

Because Turning Point Rock is almost due north of the landing site, the bearing gives a very good east-west position for the SEP transmitter, which turns out to be only 260 meters (850 feet) east of the planned deployment spot. The 2-degree difference in bearing between Houston's expectation and the navigation system readout is equivalent to a 100-meter east-west position difference at a 3-kilometer range.

Figure 8.3. Turning Point Rock is a 6-meter-high boulder on the lower slopes of North Massif. At this point in the drive, the slopes are becoming noticeably steeper, testing the Rover's capabilities. In this view to the southeast, part of Henry Crater can be seen beyond Turning Point Rock, and the blocky ejecta from Locke Crater on the northwest rim of Henry is visible just above center on the right.

> **Cernan** – "260 meters out of 250,000 miles ain't bad! They ought to give you that much leeway."

By the time of Apollo 17, the Rover navigation system has more than proved its worth. Although there had been a bit of uncertainty about locations in the early stages of the first traverse on each of the three Rover missions, once a crew reached a well-defined location, the Rover navigation readouts yielded an accurate LM location; that, in turn, made it possible for the crews to find other planned locations without difficulty. Post-mission examination of the high-resolution photographs taken from the CM showed just how superbly the Rover navigation system had performed.

ONWARD TO STATION 6

> **Cernan:** It [the Station 6 boulder]'s the split one up there, Jack. I've had my eye on it. *(pause)* Get some more pictures *(garbled)*.

Schmitt: I've lost ...

Cernan: There's some big boulders down here. Got it.

Schmitt: I sort of lost track of Station 6.

Cernan: Nah. I got it. I've had my eye on that boulder. You can't see the track from here. I'll bet you can. I can see it now. We'll see it. We'll be looking right up it; looking right up the old boulder track. *(pause)* Man, I tell you, this navigating through here is not ...

Schmitt: Okay. We're in a region where, really, the general fragment population is no different. We're up off the break in slope, although you wouldn't notice it.

They are now on the lower slopes of the North Massif, with presumably little or none of the basaltic lava beneath them.

Schmitt: But we are, quite a ways. But the fragment population is not much different than out on the plains. The big difference is that there are these scattered blocks that are from 1 meter to probably 10 meters ... *(correcting himself)* No, 5 meters in diameter. Hard to say, maybe 8.

Cernan: See that track coming down? We'll be looking right up that track.

Schmitt: Yeah, yeah, you got it. I didn't realize you were *(chuckling)* that far upslope.

Cernan: Yeah, we're way upslope!

Schmitt: Yeah. You did it.

Cernan: Not very uncomfortable for me on this side. *(laughing)* How do you feel?

Schmitt: Oh, I feel fine. I just ... Until I looked down here and saw the slope we're on.

Cernan: Yeah, I know it.

They are angling upslope toward Station 6, with Jack's side down. During an interview with author Jones, Gene commented on his particular hillside driving technique.

Cernan – "You had to be careful on the side of a hill, because if you hit a bump with an uphill wheel, you could lift the thing off the ground and possibly become unstable and tip over. Of course, the situation was more obvious if you were in the downhill seat and *(tongue-in-cheek)* I tried to keep Jack on the downside. It was much more comfortable on the uphill side; and that's a commander's prerogative when he's driving the Rover.

"Here we have the responsibilities of command, again. Always watching out for your crew. And Jack deserved every minute of it. And I'll tell you one thing, I got myself on the downslope a couple of times and in 1/6 gravity ... I won't say it's uncomfortable; but the Rover is so free to move, so free to bounce, that it seemed if you so much leaned an arm out you'd tip over. The slopes got very noticeable. Now, on a 45-degree slope, sometimes, you'd think you were going straight up. I compare it to bombing runs. You could be in a 60-degree run and you'd absolutely know that you were going straight down at 90 degrees. You always tend to overestimate your angle. If you were looking for a 30-degree bombing run, you might be at 15 and think you were at 30. And it was the same way here. A 15- or 20-degree slope can be awfully steep when you're on it."

Schmitt: And I can't see any obvious change in albedo, like we could see with the light mantle yesterday. *(pause) (to Gene)* You got a ... Don't ... There you got her: a nice, nice place. Oh, oh, you don't want to go over that way.

Cernan: I can make it. I want to park right ...

Schmitt: *(garbled)*

Parker: And, 17, you want to park at a heading of 107; we're going to open the battery covers and let them cool at this station. So a heading of 107.

Cernan: *(pause)* 107, huh? Okay. I'll get it up here. *(pause)*

Schmitt: Hey, that's going to be moderately level right there [on the west side of the boulder].

Cernan: Yeah.

Schmitt: Trouble is, they're [Houston is] looking into the shady side of the block.

Cernan: Well, if I park on the other side, they won't be able to ... I can go right upslope a little bit.

Schmitt: That's all right. We can work in there. No, that's all right.

Schmitt – "I didn't know what I was going to see on the other side of the block at this point, so I told them that we could work in the shade. You got quite a bit of diffuse light, particularly on a down-Sun face because the Sun backscatters directly on it. You can see pretty well, as some of the photos show."

Cernan: Yeah, I can't go up there. Let me just ... This is going to have to be good. I can't go up there. *(pause)*

Schmitt: Yeah, I think you're all right.

Cernan: That's not very level, but ...

Schmitt: Oh, not too ... Not too hard. Watch that turn. *(pause)*

Cernan: That's not very level, but we're not going to get much more level than that.

Schmitt: No, that's good.

Cernan: Let me … They wanted 107. *(pause)* That's the best I can do. That's not very level for the gravimeter, but … Let me see if I can get comm.

The Station 6 boulder actually comprises five large pieces, which are fragments of a large rock that tumbled down the North Massif long ago.

Cernan: Hey, Bob, how do you read?

Parker: Loud and clear, 17. How do you read?

Cernan: Okay. We're parked on a heading of 107. *(Jack is laughing)* Are you happy with that?

Parker: Roger. Sounds great.

WORKING ON A SLOPE

Schmitt: *(laughing) You parked on a slope, too.*

Cernan: There's no level … There's no level spot to park, here, though.

Schmitt: You want some help getting off? *(laughs)*

Because of the slope, neither Gene – who has to get off the Rover on the uphill side – or Jack – who has to get off on the downhill side – is having an easy time dismounting.

Cernan: *(laughing)* I've got to go uphill!

Schmitt: I just about ended up down at the bottom of the hill.

Cernan: Okay; *(static)* 192, 3.8, 3.1; 88 and 80; *(static ends)* 108 and 0 on the batteries. The forward motors are 220 and about 270, and the rears are off-scale low and 220.

Schmitt: You want me to block the wheels? *(both laugh)* You got the brake on, I hope.

Cernan: You betcha! *(pause)* I don't know if I can lean uphill enough [to get off]! *(hearty, playful laughter)* I can't. Holy Smoley! Boy, are we on a slope!

Schmitt: You okay?

Cernan: Yeah. Let me get this thing set again.

Schmitt: I don't think you can get a [gravimeter reading] …

Cernan: Boy, are we on a slope!

Cernan – "Even the pan I took from up above Tracy's Rock doesn't really show you the slope; but the comments do. And we were parked cross slope, pointed roughly east, because they wanted us pointed 107. I had to get off uphill, and it was really pretty hard to get off. Jack said he almost rolled to the bottom of the hill. It was almost like parking in San Francisco, except we were parked side slope. Now, when we got back on, it was just the opposite: it was very difficult for him to get on and it was very easy for me. We might have wanted to park uphill or downhill, but we parked at their heading for battery cooling."

On Apollo 15, at Station 6a, Dave Scott and Jim Irwin also parked on a steep slope. In both cases, the Rover was parked perpendicular to the local downhill direction but, on Apollo 15, the soil was particularly loose and Scott had Irwin anchor himself on the downhill side to keep the Rover from sliding.

Cernan – "Working on hillsides wasn't something that you couldn't handle. Let's face it, you're on the Moon, it's a new environment, and all of a sudden you're on a hill. It wasn't something that you didn't expect. It's like you get on an airplane and you go to San Francisco and you sort of know what to expect. But you go and park on one of those hills and you still say "Wow!" In one sense, we were trying to relate to the people who were listening what we were confronted with, because they couldn't see any of it."

Schmitt: Okay. I'm going to stay out from between the rocks. It's a beautiful east-west split rock. It's even got a north overhang that we can work with. *(pause)* And let me see what it [the boulder] is! We're right at station 6. You wouldn't believe it.

Cernan: I would. Oh, man, what a slope!

Schmitt: And this boulder's got its own little track! Right up the hill, cross contour. It's a chain-of-craters track, and it looks like it stops *(static)* off where it started. It starts in, what looks to be, a lighter-colored linear zone. Trying to give you perspective, it's probably only about a third of the way up the North Massif. *(pause)*

Cernan: Bob, are you reading us?

Parker: Read you loud and clear; and we've got a picture.

Cernan: Oh, man, I tell you, are we parked on a slope! I don't know whether your TGE's going to hack it.

Parker: Okay. It'll take up to 15 degrees.

Schmitt: Bob, this is a . . .

Cernan: *(working behind Jack's seat)* Well, it's [the TGE] going to have it [that is, a slope of at least 15 degrees].

Schmitt: It's a coarsely vesicular, crystalline rock . . . finely crystalline. Looks like, probably, an anorthositic gabbro. [I'm] trying to see the zap pits for glass color; I don't have a good one yet. *(pause)*

Gene backs a step or two from the Rover and has to lean about 20 degrees into the hill to stay erect.

Cernan: Oh, man, is it hard to get around here.

Schmitt: Bob, it looks like the glass is fairly light colored. It's not white. *(pause)* Well, no; it's black. It's anorthositic gabbro, rather than gabbroic anorthosite [in composition], I think. Yeah, that's black glass in the pits.

> **Schmitt** – "I had worked out a rule of thumb that the glass color was indicative of composition. The glass in anorthosite feldspar was translucent white; and if you have any amount of mafic minerals at all, you're going to get darker color to black. I think that, right here, I'm mouthing off, getting my bearings, and don't really have things sorted out as yet."

Mafic minerals are usually dark colored and rich in magnesium and iron.

Parker: Okay. And, Gene, did you happen to notice the temperature on the SEP when you dusted it?

Cernan: I didn't dust it yet.

Parker: Copy that.

Schmitt: Bob, some of the vesicles are ... They're flattened. All of them are flattened. There's a strong foliation of vesicles in the rock. Most of them are flattened, and they are up to 15 or 20 centimeters in diameter and about 5 to 6 centimeters thick ... or wide.

> **Schmitt** – "The flattened vesicles indicate that the rock was flowing while it was molten, but was still too stiff to allow the vesicles to take a spherical shape."

Parker: Outstanding.

Schmitt: And there's some beautiful north overhangs all around the block. Well, *(laughs)* on the north side of the block.

Parker: Okay. That's the best place to have north overhang; and I guess that means one of you guys might grab the SEC – the small can – before you leave the Rover.

A north overhang offers a chance of sampling soil that has been continuously shielded from the Sun for as long as the rock has been in place. Later, detailed analysis will compare the effects of solar wind particles on soil grains from the shielded sample and those from a sample of exposed regolith collected nearby; geoscientists use the data to estimate how long it had been since the shielded sample was last exposed to Solar radiation. The results indicated that the Station 6 boulder rolled, bounced, and slid down the mountain about 22 million years ago.

Cernan: Okay, Bob. It's going to take me a while to dust. I tell you.

Schmitt: Okay, . . .

Cernan: Hard to get around here.

Parker: Roger on that.

Schmitt: Bob, let's get it straight. You want the north overhang sample in the SEC – or the short can? *(pause)*

Parker: Miracle of miracles. They don't want the short can. I'm not sure I understand that, Jack, but they don't want the short can here, they say. *(pause)* I guess they're looking for volcanics today [at Station 9].

Schmitt: Okay, we'll put them in bags.

Cernan: Oh, man . . .

Parker: They're looking for volcanics today, Jack.

Schmitt: Oh, they are, huh? We found those yesterday [at Station 4].

Parker: Well, they're hoping [to find volcanic material] again at Station 9 [and want to save the sealed "short can" for that possibility].

> **Schmitt** – "Because of Shorty, they were thinking that another fairly dark, fresh-looking crater (Van Serg) might have some volcanic material around it. Of course, when we got there, it was just another impact crater into a very thick regolith with some dark-mantle material in its wall. As I recall – and we'll see – we never used the short can there. I don't recall."

Jack is examining the west face of the north boulder.

> **Schmitt** – "I was close to it to see better in the shadow, to keep my eyes out of the Sun."

Schmitt: Now, that foliation I mentioned does not go all the way through the rock. There are variations in texture. One zone was strongly foliated. There's another; it almost looks like a large – it is – a large inclusion of nonvesicular rock within the vesicular rock. There may be some auto-brecciation involved in the formation of this thing. It really looks, mineralogically, like the light-colored [breccia] samples from the South Massif. But I tell you, that's only because it's light colored, and I ... I can't give you anymore than that right now, until we get a fresh surface.

Jack has moved around to examine the southwest face of the north boulder, which is in shadow because of an overhang. There is a spot of sunlight reflected onto the rock off of Jack's visor. He now moves south. Gene is doing a bit of housekeeping at the Rover, primarily dusting off temperature-sensitive surfaces after the drive.

Cernan: Man, I never ... *(laughing)* You can't believe how tough it is getting around this Rover, on this slope!!

Jack moves to the west face again.

Schmitt: I think I'll ...

Cernan: Man, that ... I think we're probably pitched 20 and rolled 20!

Schmitt: I think I'll get over here and get a pan while we're waiting to sample. *(pause)*

Gene is dusting the left-side battery covers; Jack moves north west of the Rover to take the pan.

Cernan: Oh, I got to dust those radiators. I can't leave them like that. I tell you, this is not a very good [easy] place [in which] to dust them, though. Let me try one time. *(a bit exasperated)* Oh, boy!

Schmitt: Be careful, Geno. Need some help?

Cernan: Nope. I need a little finesse, though. *(pause)*

Jack has some trouble moving uphill.

Cernan: It's one thing to reach over here and do this on level ground. *(pause)* I don't know if I can do that without falling on the battery.

Schmitt: Well, I found a place to stand where I can take a pan. *(pause)*

Cernan: Bob, I'm going to have to give you a good battery brushing at the next site. I can't get ... I can get half of them, but I can't get the other half. It's too slopey.

Parker: Okay. We copy that.

Figure 8.4. Gene Cernan is removing the TGE from the back of the Rover, which has been parked on a slope at the Station 6 boulder cluster. This cluster apparently was originally a single boulder that rolled downhill from an outcrop about 500 meters above the valley floor, but appears to have broken apart when it encountered a decrease in the slope.

While Jack starts his pan, Gene deploys the TGE, eventually using his boots to create a level place to set it down. Houston has been watching Jack and it's been obvious to everyone watching the TV for the last few minutes that Jack has his gold-plated visor up. On Earth, the atmosphere protects us from most of the Sun's ultraviolet radiation, which can cause sunburn or – after very long exposures – eye damage or skin cancers. At high-elevation locations on Earth, there is necessarily less protection and on the Moon, there is none at all. Jack raised his gold visor when he was examining the shadowed side of the boulder, but he won't leave it up long; he avoids looking directly at the Sun and will suffer no ill effects.

> **Parker:** Hey, Jack. And we see your gold visor is up. You may want to put it down out here in the Sun.

Jack climbs out of the crater and moves cross slope, angling uphill and to the east.

> **Schmitt:** Well, I think I might ... I can't see with it down; it's scratched! Bob, I'll use it. I think I can monitor that one [meaning the use of the visor and his exposure to direct sunlight]. *(pause)*
>
> **Schmitt:** Hey, I'm standing on a boulder track. *(to Gene)* How does that make you feel?
>
> **Cernan:** That makes me feel like I'm coming over to do some sampling. *(pause)* Think how it would have been if you were standing there before that boulder came by.
>
> **Schmitt:** I'd rather not think about it.

Gene moves uphill and east.

> **Cernan** – "It's not just walking uphill that's difficult, it's walking sidehill, too. In 1/6 gravity your stability changes and you need a new approach. You have to work at it a little harder."

Jack is standing in the middle of the boulder track, manipulating the scoop and a sample bag. In contrast with the problems he had at Station 3 on EVA-2 the day before, Jack isn't having any particular difficulty doing solo sampling and seems to pour soil into the bag with relative ease.

> **Schmitt** – *(laughing)* "My hands must not have been tired. And you learn how."

SAMPLING SPLIT ROCK

> **Cernan:** Okay, let's go. You got a spot picked while you're here?
>
> **Schmitt:** Well, the big thing is, let's get the boulder and then get in that east-west split. Bob, I got an undocumented sample from the middle of the boulder track.
>
> **Parker:** Copy that. Soil sample?
>
> **Cernan:** Whew!

Jack leans his stomach on the scoop handle as he takes a breather. Gene is now at the shadowed, northwest corner of Fragment 2, the boulder immediately north of the east-west split.

Schmitt: Soil sample. *(pause)* Gene, if you hit them off in there, it's going to be awful hard to find them, that's the problem.

Jack realizes that pieces Gene hammers off the rock will be hard to find if they fall in that shadowed area.

Cernan: Did you pick a spot – a good spot – while you were over here?

Jack has joined Gene at the north boulder; Jack puts his gold visor up, again.

Schmitt: No, I didn't. I just was looking at it. I think we need to get in the light, though.

Cernan: I can see with my gold visor up.

Schmitt: Let me put a sample in your bag [SCB].

Cernan: Okay. Go ahead.

Jack stands uphill of Gene and puts the sample in Gene's SCB.

Schmitt: It's bag ... *(quietly, to himself)* Shoot. *(to Bob)* It's 534.

Cernan: This boulder looks fairly uniform from top to bottom.

Schmitt: We've got to get a reference sample of this soil.

Cernan: Let's get where we can get that 90-degree picture, too. So we really ought to get on the Sun side.

Cernan: *(pointing at a spot about shoulder height on the boulder)* Let me get that slab right there, though, to start with. I can get that one off.

Schmitt: Well, there's no ... Let's go over on the Sun side because we can't really photograph it.

Cernan: Okay. I got to get out of here first. Let's go through the split [between the boulders].

Gene moves south; his gold visor is up also. Then, he pulls it down.

Schmitt: Well, okay. Be careful, though. *(pause)* Why don't we sample the split first so we don't ...

> **Cernan:** Look at that overhang. Man, I tell you, if you get your shovel down there, you'd have a *(garbled)*.

Jack follows Gene to the split.

> **Schmitt:** Yes, let's sample in the split first so that we don't get it too messed up. And then we can sample some of this stuff. Not ... *(gesturing south)* We want this overhang over here, Geno – the north-facing one.

Gene has been looking at the west face of the north boulder and Jack wants him to work the north face of the south boulder.

> **Cernan:** Right here?
>
> **Schmitt:** Yeah. I got to sneak by over there. Whoops! Don't shuffle too much dirt in there.
>
> **Cernan:** Okay. [Are] you by [past] me so I can set the gnomon down?
>
> **Schmitt:** Not quite. Don't think I can make it – without hitting you. I can't.

At the surface, these two large boulder fragments are separated by about 3 or 4 feet. Jack tries to pass Gene to the north. Gene takes a couple of small steps closer to the south boulder, rests his hand on it, and creates a space wide enough for Jack to pass.

> **Cernan:** Okay. Now try it.
>
> **Schmitt:** Okay. *(pause)*

Cernan – "The video of our work at this station really shows what it takes to work on a slope in the confines of a pressure suit. You can see that our movements are totally different than they were on the flat and level ground. If you start stepping downhill, your center of gravity shifts. You could hear me talking about leaning over the Rover to dust things. If you do it from the topside, you're going to fall down over the Rover. If you do it from the bottom side, you can't quite reach that far. And, if you do it crosshill, you've got a similar problem. I think this whole station is a good example of working in a sloped environment in 1/6 gravity. You don't have boulder fields to trip over, but you've got an undulating, sloped surface you're trying to work on. I haven't seen this [film] in a long time. You look at the TV picture and the bodies are all cocked at 15 or 20 degrees. It's totally different from anything else we've seen. Now, granted, the Rover is on a slope as well and that probably accentuates it. But the combination of the TV and our comments really tell the story."

Cernan: Ready?

Schmitt: Okay.

Cernan: Let me set the gnomon down ...

Schmitt: Set it down just outside the shadow there. Right ... Whoa. Right there. That's good. There's still some good clean ground there. Okay.

Cernan: *(dropping the gnomon)* Okay. I can get back far enough to take these pictures. I want to go get a stereo pan around the corner anyway. Let's see if I can't start here with about [f-stop] 5.6. I'm so close. *(pause)*

Schmitt: I'll tell ya ... I'll get a ...

Cernan: I must have a boulder behind me.

Schmitt: I'll get it. Let me ...

Cernan: I'm going to go around the cor[ner] ... I got it now. *(pause)*

Schmitt: Okay. You got a bag?

Cernan: All set.

Gene briefly raises his gold visor and then puts it down again.

Schmitt: Okay. I'm going to get the shadowed material. *(pause)*

Jack has moved a little east and now leans way under the north-facing overhang, supporting himself with his left hand on the boulder while he uses the scoop in his right hand and leans far forward.

> **Schmitt** – "I was trying to get as far under the boulder as I could."

Cernan: *(holding the sample bag)* It's in bag 312, Bob.

Parker: Copy 312.

Schmitt: And it's ... It's from ... I think you saw where I got it. It's about a half a meter back of the limit of the overhang. *(to Gene)* Put it [the sample bag] down. Put it [lower] down.

Gene is holding the bag while Jack tries to pour in soil from the scoop, a difficult operation because of the stiff suit arms. Jack wants him to hold the bag lower to make the pouring easier.

> **Cernan** – "I wasn't holding it in a way to make it difficult for him, it was just suit restriction. You seek out the most convenient and least restrictive positions while

reaching out far enough and low enough. From his point of view, it would have been most convenient if I got down on my knees and held the bag, but only if I could have stayed there for 10 minutes. But you take a sample and then you move 10 feet; you take another sample and then you move again. So, unless he's going to go back and forth, you have to move together and you both have to be on your feet. It's a compromise."

Cernan: Okay. Can you reach it?

Schmitt: I will in a minute. You can turn it a little bit towards me. *(pause)*

Jack hops uphill to the east of Gene, while Gene turns to keep the sample bag in position.

Schmitt – "I'm surprised that, with me moving like that, the soil stayed in the scoop. And I was reaching upslope some, so it was a tough maneuver."

Schmitt: Okay; 312. *(pause; presenting his SCB)* And the soil outside the overhang will be next.

Cernan: Okay. Go get it. *(pause)*

Schmitt – "There is more TV coverage at this station than I thought. I had forgotten that we had sampled together for a while. What I remember is being alone, working on the contact between the two breccias (the vesicular and nonvesicular materials Jack describes later) while Gene was taking some pans."

Jack scoops a sample just outside the shadow, leaning on the boulder again with his left hand. The scoop mark on the ground is very reflective.

Schmitt: And the first one is from the upper 2 centimeters.

Cernan: Bag 313.

Parker: Copy 313. *(pause)*

Schmitt: *(pouring)* And the second one is from probably 2 centimeters down to about 8 [centimeters].

Jack collects the second sample, then stands still while Gene puts the bagged first sample in his SCB.

Parker: Copy that.

> **Schmitt:** Bob, it looks like the fragment – or the boulder – just to the south of us has some inclusions in it – light-colored inclusions.

Jack pours the soil into the bag and then turns so that his SCB is toward Gene, who puts the sample bag in it.

> **Cernan:** Bag 472 on that.
>
> **Parker:** Copy 472 on that. You mean the south half of the split boulder?
>
> **Schmitt:** *(gesturing)* Yeah. I haven't seen inclusions in the other half.
>
> **Cernan:** Okay?
>
> **Schmitt:** Okay. Now we need boulder stuff. You happy with that, Houston? *(to Gene)* Let's get *(garbled, overlapping conversation)*.
>
> **Cernan:** *(garbled)*
>
> **Schmitt:** *(to Gene)* Got your hammer?
>
> **Parker:** Yeah, we're happy with that for an east/west split.
>
> **Cernan:** *(garbled) (pause)*

Near the boulder, the gnomon staff is inclined about 20 degrees, pointing north and indicating just how much of a slope they must contend with. Gene is hammering on a protrusion on the south boulder at about shoulder height. He hits it a dozen times.

> **Schmitt:** It's a little hard, huh? *(pause)* I think ...
>
> **Cernan:** *(no longer hammering)* I've got to find a corner I can get at.
>
> **Schmitt:** Yeah.
>
> **Cernan:** *(moving north, away from the boulder)* Let me get an "after" picture down in this hole [that is, where they got the soil sample].
>
> **Schmitt:** Oh, that's right. You almost stepped on the ... I forgot the "after," too. *(pause)* Hey, there are chips up here on top, also. That's been spalled off.
>
> **Cernan:** Yeah.
>
> **Schmitt:** We can get some of those, but ...
>
> **Cernan:** Looks like somebody's been chipping up there.
>
> **Schmitt:** Looks like there's been a geologist here before us.
>
> **Cernan:** Let me get the gnomon. I think I can get some of these pieces over here. I want to get that 90-degree angular flight line around this boulder, too.

In order to reach down for the gnomon, Gene comes forward on his toes to keep his PLSS centered as he bends his knees.

> **Cernan** – "It was just easier to bend my knees like a guy with a bad back might do. It was just easier to bend my knees to let my arms down low enough. You learn very, very quickly. The human being is very adaptable. You don't have to think about those things; you just do it. Somehow, that computer up above your neck just picks out the easiest approach."

They move off to the east behind the boulder, stepping gingerly and apparently downhill.

MEETING THE OBJECTIVES – A WELL-SCHOOLED TEAM

Schmitt: Bob, the more I look at this thing ... Now, here's the piece that fell off. Here's the piece that was knocked off up there.

Cernan: Yeah.

Schmitt: Look at that.

Cernan: We ought to bring a big piece of that home. That's obvious. It's obvious ...

Schmitt: How about this one up here? Take your picture. I think we can just lift that off. See that?

Cernan: Stand by. *(pause)*

Schmitt: I'd better get ...

Cernan: I'll get a locator [photo] from here.

Schmitt: Okay. I was going to get my down-Sun, but I'm afraid I'll ...

Cernan: You may be down-Sun if you do.

Schmitt: Yeah, we'll get some. [Can] you get it?

Cernan: Yes. Will it come off?

Schmitt: Let me see. *(pause)* Yeah.

Cernan: Just throw it in my bag. It's broken, but it's "in place." That's a nice, big piece, too. It's about the size of a ...

Schmitt: *(laughing)* Why don't you put it in mine. I can't get up to you. [Jack's downslope and Gene can more easily put samples in Jack's SCB than Jack can move upslope and put it in Gene's SCB.]

Cernan: *(garbled)* okay?

Schmitt: Got it?

Cernan: Yeah, I got it. *(backing up to take a series of flight line photos)* Don't move.

Schmitt: Okay, Bob, there's a big spall lying on the ground here that has been knocked off up there, from right on top of the boulder. And, I tell you, the more I look at the south half of this boulder, the more heterogeneous in texture it looks. It looks as if it may be either a recrystallized breccia of some kind, or you had a gabbroic anorthosite magma catch up an awful lot of inclusions. I guess I prefer the latter explanation *(pause)* because of the extreme vesicularity of the rock.

Parker: Okay; very interesting.

Schmitt: Now, a few of the inclusions are. Well, they're all subrounded to rounded, and a few of them are very light colored. I'm going to try to get ...

Cernan: I'm coming around the corner with a flightline stereo.

A flightline stereo is a series of pictures taken by stepping just a little to the side between frames but without turning the camera as would be done with an ordinary stereopair.

Schmitt: Are you going to do it now? Okay. Well, you know, I ought to get one shot back here with a black and white. I'll get this half black and white.

Cernan: Okay, if we could get ...

Schmitt: I think we ought to pick up a piece of that spall there by the gnomon ...

Cernan: I can break it off. *(garbled)* top.

Schmitt: There's one right by the gnomon we can just pick up. It's a finer grained vesicular rock than ... Wait ... Where ... Jeez ...

Gene moves behind the south boulder. Jack follows, using the scoop as a cane in his right, uphill hand. His PLSS remains visible until Fendell pans away. Otherwise they are both out of view for quite a while.

Cernan: Oh, oh, oh, oh ...

Schmitt: I thought I was going to get this half. Okay.

Cernan: I don't care. I started down, Jack.

Schmitt: Well, they like to have some of it in black and white, you know.

Cernan: I'll get that rock. *(pause)*

Schmitt: I forgot to look at the objectives for the station. I hope we're meeting them.

The fact that neither of them has even glanced at their cuff checklists at this station – and rarely do so at most of the other stations either – is an indication of just how well prepared the J-mission crews were for their field work. None of the Apollo 15, 16, and 17 astronauts depended very much on the geologic portions of their cuff checklist because they understood the geologic objectives and also were prepared to make real-time decisions (with input from Houston) when they got to a station and realized what was going to be important. When one compares the mode of operation of the crews on the last three missions with that of the earlier crews, the difference is striking. Of course, on the first three flights – particularly Apollos 11 and 12 – geology was understandably of less importance than the main task of flying the LM. The J-mission crews also had the advantage of training as backup crews on earlier landing missions; consequently, they could devote as much as 40 percent of their own mission training time to geology and other scientific tasks. In addition, with three 7-hour EVAs at their disposal (compared with two 4-hour EVAs) and as well as the mobility of the Rover to give them a greater number and variety of geology stops to explore, real-time decisions made more sense.

The mission planners acknowledged this evolution by including only generic tasks in the later checklists. Nonetheless, here at Station 6, Jack and Gene are operating more independently than any other crew had during all of the Apollo surface operations. Houston is making almost no suggestions, confident that with 45 minutes budgeted for this station, Jack and Gene won't miss anything important. This independent thinking is evident later in the transcript when Jack takes a moment to get his thoughts organized about what they want to do at this station, Bob summarizes the objectives listed on the cuff checklist, and Jack ignores him. A geologist himself, Jack probably needs no instruction about sampling!

Cernan: Well, we want to get 500's of that boulder track. I know I want to get that. *(pause)*

Schmitt: Okay. A piece of that spalled rock that was sitting by the gnomon ... Ooh, watch out gnomon. How about that? ... [The sample] is in bag 535.

Cernan: You got one in there already?

Schmitt: Yup.

Parker: Okay. We copy that one, Jack.

Cernan: You won't be able to reach ... You won't be able to reach my bag.

Schmitt: No, but you can put it in mine. *(pause)* Can you reach it? *(pause)*

Cernan: Working at it.

Schmitt: Bob, one of the light-colored inclusions looks like it may be anorthositic ... *(correcting himself)* gabbroic anorthosite ... Let me get my terms straight. The host rock has dark-enough zap pits that it's probably gab – ah –

anorthositic gabbro, if I didn't say that. Some of the light-colored inclusions have slightly lighter colored glass, and they may be the gabbroic anorthosite.

Parker: Okay, I copy that, Jack.

Schmitt: *(to Gene)* Inclusions like this one and that one.

Cernan: Yeah. Some of those inclusions get to be bigger than the size of a baseball. There's one here and a couple up there.

Schmitt: Let me borrow your hammer.

Cernan: Yeah. Jack, try a little higher. See that one right on the ... Right there. Right ... *(pause)* Well, that's a hard rock.

Schmitt: *(unable to break it loose)* No.

Cernan: That's a hard rock.

Schmitt: You might be able to do it; I can't. *(pause)*

Cernan: I can't get down there. *(pause)* Okay, we need some of the soil outside the shadow here.

Schmitt: Yep. How about over where your bag went? Let's move around here ... I think there is some ... Oops! Get on this slope over here. Okay. How about out over here? Are we supposed to get a ... *(glancing at his checklist)* Where are we here?

Cernan: I don't know. I'd like to get ... *(pause)* Well, when you face uphill, your camera faces down.

Gene has to lean forward to maintain his balance and, as a result, his camera is pointing into the hill. In order to get a picture, he must either lean back or take the camera off the RCU bracket.

Schmitt: We want to get a rake on the rim of that little crater down there, I guess. And ...

Parker: Okay, 17. Roger. You were asking about objectives. Of course the primary objective is documented samples of the blocks; and then, also, we'd like to get some of the rake and soil sample out in the surface, namely, the rim crater there, if that's available. And one of the things, of course, we're looking for is the variety of rocks here, if there's more than just the one boulder. You can sample the boulder for a while, but we would be interested in seeing if there is more than just the single type of rock. Probably, also, samples from both sides ...

Schmitt: *(trying to interrupt Bob's instructions)* Okay ...

Parker: ... both halves of the rock ...

Schmitt: *(to Gene, speaking under Bob)* Let's get working.

Parker: ... What we said this morning in terms of combining Stations 6 and 7 to 1 hour and 20 minutes ...

Schmitt: Come on up here, Geno.

Cernan: *(to Jack)* Okay.

Schmitt: If you can.

Parker: And so it's sort of your option as to how much time you spend here and how much you go on to Station 7 and spend. If you feel that it's worthwhile, we could spend essentially all that 1 hour and 20 minutes at this station. But if we did that, we'd like to get a fair variety of blocks, if they're available.

Schmitt: *(responding to Bob, matter-of-factly)* Okay.

> **Schmitt** – "It's fairly clear from the tape that, during this period when Bob was talking to us, it was the wrong time because he made it difficult for us to talk to each other and we were trying to get some sampling done. And that's probably why I was fairly short, responding with just an 'Okay.' Now, I suspect we continued to work while he was talking; I don't think we stood around. Of course, I'm not sure when would have been a good time to interrupt us. *(laughing)* Even though we asked the question, we probably didn't want an answer."

> **Cernan** – "I want to emphasize the point. We never stood around and listened. We never stopped doing what we were doing. We could listen while we were taking a pan or getting a sample or whatever it was."

Schmitt: *(to Gene)* Geno, we can sample some of the light-colored group ... As a matter of fact, this block looks different.

Cernan: Well, so does that big one ...

Schmitt: It's grayer.

Cernan: ... That's why I've been photographing it.

Schmitt: What it is, I think ...

Cernan: ... It's a blue-gray.

Schmitt: ... it's a big blue-gray rock, itself is crystalline, I believe. The inclusions are much more sharply defined, and it's nonvesicular; and it's included, or at least it's in contact with the very vesicular anorthositic gabbro. Right up there. See that?

Cernan: Yeah, a whole big one. I just ...

Schmitt: Did you get some pictures of it?

Cernan: As I bounced around there, I took pictures of it.

Schmitt: Look, we can get some of that light-colored stuff in there, along with the blue-gray.

Cernan: We ought to get as big a piece of that inclusion as we can. There's …

Schmitt: See it up in there.

Cernan: Yep. *(pause)* I think we're out of line-of-sight with them. We're behind a boulder.

Schmitt: Yeah, sorry about that. But …

Parker: We can hear you loud and clear. We're just looking at rocks right now.

Schmitt: *(garbled)*

Cernan: Okay, Bob, the boulder downslope [that is, Fragment 2] is more of a light-gray, vesicular boulder. The one Jack just talked about [Fragment 1] with some of the larger white inclusions is less vesicular, and it's more of blue-gray rock. And if I don't fall on my tail here, I'll get …

Schmitt: The "locator" is of Henry.

Being high up the slope, Gene and Jack have a good overview and can easily distinguish craters on the valley floor.

Parker: Copy that. *(pause)*

Cernan: Okay, let me try and get up there. *(pause)* Henry? We must be high enough to see something. I haven't even looked back.

Schmitt: Let me get a close-up before you start pounding.

Cernan: I might go from this angle too. *(pause)* That will give them something. (pause) A little different up in there too, Jack.

Schmitt: Yup.

Cernan: We ought to try and sample that. *(pause)*

Schmitt: Okay. Let's get the … You want me to get my scoop under there? Probably will fall out.

The scoop head is only 11.4 centimeters wide, 15.2 centimeters long, and 5.1 centimeters deep (4.5 by 6.0 by 2 inches), so the chances of a fragment coming off the boulder and actually staying in the scoop is small. The scoop was designed for trenching and collecting soil and small rocks that lay on the surface.

Cernan: Okay. [Let's] get as many of these pieces as we can. I don't know how many are going to come out. *(long pause)*

Schmitt: *(hearty laugh)*

Cernan: Outstanding! Outstanding! *(laughing)* This whole thing will come out here in a minute. *(pause)*

Cernan: I'll watch it. I'll watch it.

Schmitt: Got it?

Cernan: Move your arm up or down. *(pause)* Okay. I got it in case we don't get another one. *(long pause)* Hey, we're getting good at that.

Schmitt: Yeah. Can't hold that much longer.

Cernan: Yeah. Let me get up on this ... Up here. Oh.

Schmitt: Why don't we get a bag out. Let me put these in a bag.

Cernan: That's why I'm getting up here so I can ...

Schmitt: Oh, okay.

Cernan: ... just get my balance. Bob, 556 is one of the light-colored inclusions in the blue-gray rock.

Schmitt: It's chips.

Cernan: Chips of it.

Parker: Okay. Copy that.

Schmitt: I think we lost that other one. That's good enough.

Cernan: I got it; I know where it is.

Schmitt: That's all right. It's not a lot of sample, but it's representative, I think. It looks a lot like that sugary rock I sampled yesterday [at Station 2], doesn't it? Found in the ... That we sampled in the ...

Cernan: Yeah, it's pretty easy to break up; it's really not very coherent [means "cohesive"] at all.

Jack is referring to the last boulder (Boulder 3) they sampled at Station 2. It was a blue-gray-matrix breccia that contained light-gray clasts of up to 10 centimeters (4 inches). Gene and Jack had commented from the LM after landing that the outcrops at the top of the South Massif appeared to be a blue-gray color; the boulders they sampled later at Station 2 appeared to be the same material. Jack subsequently told author Jones that "one of these fragments [from boulder 3] turned out to be the oldest dated rock ever sampled on the Moon. It was dated at 4.6 billion years, plus or minus 0.1 billion years."

Schmitt: You know, I thought last night, Bob, that I should use the word "aplitic" for a texture that we saw in that inclusion yesterday on the South Massif.

Schmitt – "A good layman's translation of 'aplitic' is a coarse, sugary texture – like raw sugar."

Cernan: If I could keep from falling on my tail.

Schmitt: Can you get a ...

Cernan: I want to ... *(pause)*

Schmitt: *(garbled)*. *(pause)* Okay, you going to get some of that?

Cernan: Yeah, that's a different kind; that's a more beat up inclusion of some sort. *(pause)* Oh, there's a nice piece coming out. Oh, wait a minute ... Don't lose it.

Schmitt: I got it. I've got it.

Cernan: Got it.

Schmitt: Okay. *(pause)*

Cernan: Okay. We have another inclusion that, on the surface, has a more reddish-brown texture [means "color"]. Interior looks pretty much the same; it's a very light gray.

Schmitt: This looks like a piece of breccia. Looks like a fragment breccia that got caught up in this thing.

Cernan: Yeah, well, the whole thing is obviously a breccia. I'd sure like to get that ...

Schmitt: Well, I'd say ... I'm not sure [that] it's obviously a breccia. I think it may be an igneous rock with breccia inclusions.

Cernan: Well, okay, but look at all these things ...

Schmitt: Which is sort of in the same class.

Cernan: [That] sort of makes a breccia ...

Schmitt: Well ...

Cernan: ... out of the big rock.

Schmitt: Okay.

Cernan: Except you can ...

Schmitt: I can't get in there, Geno, you'll have to.

Cernan: Okay.

Schmitt: *(trying to fit a sample in Gene's SCB)* No way ...

Cernan: Let me ...

Schmitt: Watch it. Hold still. Oops. *(pause)* I think it's easier for you [to put the sample bag in Jack's SCB]. Did I give them a number on that? No.

Parker: Negative.

Cernan: It's 536.

Schmitt: Squash it ... Cramp it a little bit, if you can; a little more. *(pause)*

> **Schmitt** – "I was trying to get a sample in Gene's SCB and he must have been upslope of me. So I probably handed it to him so that he could put it in my SCB."

Parker: Do you guys have a feeling that the two halves of the big boulder are different rocks? Or is it the same rock split?

Schmitt: No, they were all one boulder, I think. They are just two major rock types in wherever they came from. And I tried to describe that to you. We have the contact in the central boulder [Fragment 2]. There're really three big boulders [Fragments 1, 2, and 4 and 5]. The central boulder had the contact between the light-gray rock – or the blue-gray rock – and the vesicular anorthositic gabbro.

> **Schmitt** – "The boulder didn't split along the contact, largely because the vesicular breccia (which is referred to here as 'vesicular anorthositic gabbro') was intruded as partially molten material. It was hot enough to metamorphose about 1 meter of the blue-gray breccia it intruded. You could tell that because there were small vesicles – which indicate some partial melting – in that contact zone, which then died out as you got away from the contact."

Parker: Okay. And you guys have that pretty well photo-documented, right?

Schmitt: Yeah, it's in pretty good shape. We're working on it still.

Parker: Okay. Copy that.

Schmitt: Try going on the side there, Geno.

Cernan: Just went from the side, Jack.

Schmitt: That's enough. You got a piece of the ...

Cernan: *(garbled)* host rock.

Schmitt: I think you can get this one up here, can't you?

Cernan: I wanted that one 'cause it had that inclusion wrapped in it. *(pause)* Let me go to high [cooling] here for a minute. *(pause)*

The combination of a high Sun angle and the physical exertion of working on the hillside makes this one of the few times that either of them uses the high cooling setting.

Cernan: Which one are you talking about? This one here?

Schmitt: Yeah, I just ... *(pause)* It's about to come. *(pause)* Oh, oh, oh, okay. I've got it. I've got it.

Cernan: Okay.

Schmitt: Okay. I need this in a bag.

Cernan: They're both host rocks; we can put them in the same bag.

Schmitt: No, let's don't. No, they're different places. 537 ...

Cernan: *(garbled)*

Schmitt: ... is a chip of the blue-gray rock; and the blue-gray host rock ... *(pause)* And let me get that other one. *(pause)* Ahh!

Cernan: Oop. Be careful. *(pause)* Pick the rock up while you're there. It's right at your hand.

Jack has probably fallen.

Schmitt: I will. *(pause)* Okay.

Cernan: [Let me put] my little hammer somewhere.

Schmitt: Okay. And 538 is another sample of that material ... a little dustier [sample]. *(guffaws)*

Parker: Okay. We copy that.

Schmitt: That's the blue-gray, Bob, with the inclusions in it. Now the blue-gray, the more you looked at it, it looks like a ...

Cernan: Give me your left ... *(correcting himself)* Your right hand.

Schmitt: Huh?

Cernan: Give me your right hand. Turn it over. Turn it over. Turn it over.

Schmitt: Well, I did. How do you want it over?

Cernan: You kept turning it over in the same direction. Like that, so I can fix that.

Schmitt – "I don't know what this was, unless my watch band was coming loose or something like that."

Cernan: Okay. Now give me your bag, and I'll get it in there. *(pause)*

Schmitt: *(to Bob)* The blue-gray rock, on closer examination, looks like a partially recrystallized fragment breccia. It's very hard. *(pause)*

Cernan: And I'm going to ...

Schmitt: Are you going to get the "afters" [photos showing the site after the sample is removed] in there?

Cernan: Yeah, I'll get them. I want to do a little bit better documentation on this thing. *(calling)* Bob ...

Schmitt: I'm going to go over and look at that contact.

Cernan: ... I got a few close-up stereos of the inclusion that we tried to sample, and I'm going to see if I can't give you a little flightline stereo around this thing. If I can stay on my feet. *(no answer; long pause)* Do you read me, Jack, okay?

Schmitt: Yeah, I hear you.

Parker: And Houston reads you loud and clear, also. *(pause)*

Cernan: *(to Bob)* You can see where we've been pounding on this rock. We didn't succeed in getting samples everywhere. *(pause)* And I'm giving you a "90-degree corner."

Jack is looking at the southeast corner of the north boulder, almost "treading" sand to keep his balance on the slope.

Schmitt: Bob, it looks to me like there are inclusions of blue-gray in the gabbro ... *(correcting himself)* in the anorthositic gabbro.

Parker: Positively outstanding.

Cernan: Are you saying you think this whole big ... You think this whole big blue-gray thing is an inclusion?

Schmitt: Yes, sir. And there's some little ones over here.

Cernan: Yeah, but then within the blue-gray, we've got all these other fragments.

Someone in Houston notices that the bottom of Jack's SCB seems to be loose.

Schmitt: Well, that's right. It's just several generations of activity; and it looks like the gabbro, though, picked up the fragmental breccia as inclusions. *(pause)* Bob, It really looks that way right now. There's a small one here in the ...

> **Parker:** Okay, Charlie [Duke] is here [in Mission Control] mumbling something about it looking just like House Rock.

John Young and Charlie Duke visited House Rock at the rim of North Ray Crater on EVA-3 of Apollo 16. It was the largest lunar boulder sampled by any of the crews: literally the size of a good-size house, about 12 meters high, 25 meters long, and 20 meters thick. This Station 6 boulder consists of five large fragments of what was once a single boulder of about 6 by 10 by 18 meters. For comparison, Turning Point Rock is about 6 to 7 meters high.

> **Schmitt:** It's very crystalline. I'll tell you, it's not a breccia, not like House Rock. Not to take anything away from House Rock, though.
>
> **Cernan:** Hey, Bob, there's a lot of mantling on a very shallow slope of a fracture here on one of the upslope blocks [Fragment 1]. I would assume it's just part of the talus picked up as it's rolled down. But if it's worth sampling, you might think about it. *(pause)*

Bob waits to get a recommendation from the Backroom but, getting no prompt answer, makes the decision himself to have Gene grab a sample.

> **Parker:** Okay, Gene, if you can get that fairly readily, why don't you ... You can perhaps just scoop it up with the bag.
>
> **Cernan:** That's exactly what I can do.
>
> **Parker:** If you can get up to the rock there.
>
> **Cernan:** It will be in my flight line stereo, and it's going to be bag 557. And I'll take an "after" and show you where it came from.
>
> **Parker:** Okay. Copy that.

TRACY'S ROCK

Out of sight of the TV camera, Gene steps to the northeast corner of Fragment 1, puts his left hand on the dust-covered shelf to support himself, and reaches out as far as possible toward the center of the dusted area. He sweeps the bag twice from his right to his left, leaving furrows and, at the left-hand end of each furrow, a small mound of the cleared dust. Later Gene will take a panorama from higher up the hill, which will also show the marks left in the dirt where he took the sample.

> **Cernan** – "I haven't seen the rock from this perspective in nearly 19 years. My handprint really shows you how big the rock is and, in [photo] 21482, you can see

across to the South Massif and the Scarp. The Scarp looks small in this photo, but I remember how big it was [80 meters] because we went up it."

The patch of dirt on the north face of the boulder is the subject of a painting by Apollo 12 astronaut-turned-lunar-artist Alan Bean. Those who know this story call the Station 6 boulder "Tracy's Rock," for Gene's daughter, who was 9 years old at the time of the mission.

Cernan – "It was on this part of the rock that Al Bean wrote "Tracy," my daughter's name, in one of his paintings. After we came home I started to see a picture of the boulder in lots of places. It was the picture of Jack going past the corner of the rock in one of the pictures from my pan. It became very popular. One day Al Bean came by and said he was doing a painting of it. And it was a big painting, 6 feet by 3 feet or something like that. Al likes to have stories in his paintings and he wanted to talk about it. So we talked about the slope and how hard it was to climb up there, and I said that, if I'd known the picture was going to get so much notoriety, I wished I would have done something that I hadn't even thought of at the time. And that is to have printed Tracy's name in the dust. Al's daughter Amy and Tracy had grown up together and he asked 'How would you have done it'. So I wrote it out on a piece of paper and, some time later, he called and wanted me to come over and look at what he'd done. He had erased the place where I'd taken the sample and had put in Tracy's name, instead. And in the little story he put with the picture, he said that he'd done it to save me the trouble of going back to do it myself, and to save the taxpayers the expense of sending me back."

Jack leaps upward about a foot, kicking his heels back briefly before he lands.

Schmitt – "I was having trouble getting up to exactly where I wanted to be to look at the boulder."

Cernan – "I can't see why a geologist would jump up and down in the air like this except to show his enthusiasm over finding a new and different rock. And, all joking aside, it really looks like he's trying to get closer to the rock and the slope made it hard to move."

Cernan: This is the easiest part of the rock in the world to work. Man, here's a big white clast. There's one on top about a foot and a half across, and here's one. [It] must be 2 feet across . . . 3 feet. And that's in the blue-gray. *(pause)* Feel like a kid playing in a sandbox. *(pause)*

Schmitt: *(balancing on his downhill leg)* Well, Bob, I think I've done the best I can. I'd say that they're pretty clearly inclusions of blue-gray in the anorthositic gabbro here near the contact.

Jack walks slowly uphill, leaning into it, knees dramatically bent. His SCB is clearly loose. Bob makes a misidentification.

> **Parker:** Okay. And Gene, your bag is hanging by one hook there. Be careful, if you can ... *(correcting himself)* Or, LMP ...
>
> **Cernan:** Okay. I gave you 557, I believe. Didn't I?
>
> **Parker:** Roger. We have that one. *(pause)* And whoever is giving us 557 ...
>
> **Cernan:** Okay, I'll have Jack fix my bag.
>
> **Parker:** Yup.

While facing more or less north, leaning into the hill and using the scoop in his left hand as a cane, Jack leans far enough forward that he can grab a fist-size rock off the ground.

> **Schmitt:** Okay, Bob, by accident ... I didn't think I could do it, but I got a sample of the inclusion. And it's in bag 539.
>
> **Cernan:** Hey, Jack, that's your bag [SCB] that's hanging by one hook. Let me go get it.
>
> **Schmitt:** Oh, they're talking to me, huh?
>
> **Cernan:** Yeah, because I didn't ...
>
> **Parker:** Yup.
>
> **Cernan:** I didn't think they could see me. I'm way up on top! *(pause)*

While Jack examines the bagged sample, Gene hops into view from behind Fragment 2. Jack seals the sample bag by grasping both ends of the metal sealing strip, flipping the bag over it, and then folding in the ends of the strips.

> **Schmitt:** And it's blue-gray with light colored ...
>
> **Cernan:** *(presenting his SCB)* Put these in my bag.
>
> **Schmitt:** ... inclusions in it. *(to Gene)* All right. *(to Bob)* But the whole thing [Fragment 2] seems to be pretty well altered or metamorphosed ... compared to the major rock [Fragment 1] we just sampled ... [compared] to the other blue-gray rock. *(pause)* This bag [Gene's SCB cover] is terrible. I can't ... It won't latch.
>
> **Cernan:** Man, there's a dark hole in there where you ...
>
> **Schmitt:** Don't let me ... *(pause)* Ah! I'm not ...

Jack is having a difficult time reaching Gene's SCB. Because of the slope, if Jack is to reach the top of Gene's SCB, they have to be standing on virtually the same hill contour. Gene can stand facing uphill, leaning into the hill, but Jack has to face him and lean sideways into the slope. Jack's right leg is fully extended downhill, and his left leg is bent at the knee.

> **Schmitt** – "Plus, he's taller than I am. That was always a problem."

Cernan: Here's another bag to put in there before you go away.

Schmitt: *(taking a full sample bag from Gene)* Oh, okay. *(pause, grunting)* It won't latch.

Cernan: Well ...

Schmitt: Not at this angle.

Cernan: Just put the thumb on one side, and I'll ...

Schmitt: It's bent or something. There, that's pretty good.

Jack presents his SCB by turning to face uphill and using the scoop as a cane.

Cernan: Now let me fix your bag. *(pause)*

WHAT A VIEW! – A PANORAMA OF THE VALLEY OF TAURUS-LITTROW

Cernan: Bob, we don't move around from here too much. I tell you, these slopes are something else.

Parker: Yeah. We agree with that, from what we see on the television. So use your judgment, and get them [panoramic photos] where it's the best place.

Cernan: Well, you might take a look at me walking up. But I don't think I can get to the top. *(pause)*

Cernan: I just got to get a place I can get a pan from. Right here. Right in this little hole. *(pause)* *(gesturing)* Okay, now I left the gnomon down there.

Schmitt: Okay. I'll have to go get it. *(pause)* I think we'll setup right here near the Rover.

Cernan: I think I'll go and save some water, back on Intermediate. Okay.

Parker: Copy that.

Figure 8.5. At Station 6, located on an 11-degree slope about 50 meters above the valley floor, driving and even standing up was a challenge. Beyond the boulder are the Valley of Taurus Littrow and the South Massif. The massifs appear light gray, an indicator of their composition of lunar crustal materials and welded impact breccias. The valley floor is darker because of the underlying dark-grey to-black lavas and volcanic ash deposits. The distance from this station to the point where the slope of the South Massif meets the darker material is about 10 kilometers – which demonstrates how deceptive lunar distances can appear.

Gene switches his cooling and then starts the pan. Jack crosses the TV picture, headed east to get the gnomon. This is the reason he appears in Gene's pan.

Cernan: Hope my lens is clean. *(pause)* Bob, from up here, the light mantle is not evident until you see the angular reflection up on the scarp. Very thin, light patches might be evident out on the valley, but not nearly as pronounced as I might have thought from this altitude. *(long pause)*

Cernan – "You really had to lean to get the horizon in the pan. And I don't think I was able to get the uphill pictures."

Cernan: Oh, and there's Challenger! Holy Smoley! *(pause)* You know, Jack, when we finish with Station 8, we will have covered this whole valley from corner to corner!

Schmitt: That was the idea.

Cernan: Yeah, but I didn't think we'd ever really quite get to that far corner. Not [Station] 2, but this other one [Station 8]. And we're going to make it!

Cernan – "I wish I'd climbed to the top of the mountain to get even a better view. But, even where we were it was a much better panorama than we got from the Scarp. We couldn't see much from the Scarp. Over here, we could see everything: Challenger, the Scarp, the Sculptured Hills, and everything. I do remember this. It was fantastic – like coming over a hill and getting a look into a great big valley somewhere in the middle of Idaho."

Schmitt – "In order to find a level place, Gene went to a crater and stood, not in the bottom, but on the downslope wall which would be nearly level."

Schmitt: *(having taken another opportunity to examine the boulders)* Bob, that blue-gray rock near the contact with the anorthositic gabbro does get some vesicles in it. I think they'll show up in Gene's pictures. *(pause)*

Parker: Okay. We have that too, Jack. *(pause)*

Schmitt – "This examination of the boulder shows the evolution of my thinking as I was getting new information. The final impression that I had – which was borne out [later] by the samples – was of a blue-gray breccia intruded by – or enclosed by, we don't know which – vesicular, tan-gray material, which also turned out to be a breccia. Now, when I first approached the boulder, the vesicles and crystals I could see in the matrix around them led me down the path of thinking that I was looking at a truly igneous rock. And it had not occurred to me yet that I might be looking at an impact breccia that got hot enough to melt. And it took the whole time of examining it – until right at the last – before I came to the correct conclusion. At the last, I saw the vesicles in the contact zone, which pretty well confirmed it all. It took a while to work that through and to get away from thinking that I was looking at an igneous, anorthositic gabbro that had caught up pieces of other things. Normally, a field geologist would sort all those multiple working hypotheses as you go along and then eventually come to a conclusion. And that's what you write down. But here, what we see is a large part of my thinking process being verbalized. If we had had to stop after the first few minutes, we would have had all the wrong answers."

"My guess is that the vesicular breccia was produced in the Serenitatis impact, that it was injected downward into the crust and then migrated back up along fractures in the ejecta. And, in that context, the blue-gray breccia may be pre-Serenitatis – who

Figure 8.6. Jack Schmitt is dwarfed by one of the largest boulders at Station 6. The boulders are heterogeneous; the interwoven fragmental and melt rocks may have been formed as ejecta from the catastrophic impact that formed the Imbrium basin about 3.8 billion years ago. The boulders probably rolled downhill to their present location about 22 million years ago.

knows, perhaps Tranquillitatis ejecta or something like that. We'd seen the same blue-gray rock at Station 2. There was tan-gray breccia at Station 2 but that wasn't vesicular. Probably, the first place we saw the tan, vesicular breccia was at Turning Point Rock and, again, I was thinking there in terms of it being igneous."

As he is coming down the hill after taking the pan, Gene trips and falls to his hands and knees, apparently without hitting the camera on the ground. His path to the Rover took him into the boulder track and back out; as he came out, he caught his trailing left foot, started to tilt sideways over his right foot, and lost his balance.

Cernan – "I had some speed coming down, even though I was coming cross slope. It was so hard to get up there and so easy to come down, so I was moving. I was coming cross slope because, to come straight down, I probably would have gone head over heels. You couldn't come straight down that hill very fast. It's like skiing. If you go straight down, you're really going to motor. So you've got to shallow up the slope. But I got off balance, got on one leg, and fell into the hill. When I was watching this just now, I thought I was going to roll; but I didn't. I kept my balance pretty well.

Schmitt: *(having seen the fall)* [Are you] okay?

Cernan: Yup. *(pause)* *(calling to Bob)* Remind me to dust my camera, too, will you?

Schmitt: Don't forget to dust your camera.

Schmitt – *(laughing)* "That's a bad habit of mine. When people ask me to remind them to do something, I'll say it right then so I'm not held responsible for remembering."

Gene shifts position so that his feet are downhill. He then pushes back with his hands to get up on his knees and, as his PLSS rotates past the point of his knees, he rises to his feet.

Parker: Okay. We'll keep track of that for you, Gene.

Gene continues on his way, using the same gait at the same speed.

Schmitt – "By this third EVA we were very confident. We didn't worry about falling. Maybe we should have, but we didn't. What I want to know is how come he didn't get kidded about that fall like I got kidded about the fall at Ballet. That's just as spectacular."

GETTING READY TO LEAVE STATION 6

At 2 hours and 15 minutes into the EVA, the crew is 3.1 kilometers from the LM. At Station 6 they had been stunned by the view; they comprehensively described and sampled the large boulders that rolled down the North Massif and fulfilled their goal of collecting core, soil, and rake samples that will help geologists back home sort out the dramatic history of Split Rock. They also have taken some of the most spectacular pictures from the Apollo missions, including Jack's telephoto shots of the LM. Now, just 1 hour after they arrived at Station 6, Gene and Jack are both back at the Rover. Gene has now hammered a core tube into the surface and, as soon as he gets it capped, he'll take a final reading of the TGE before stowing the instrument on the back of the Rover. After that's done, they'll be ready to jump in the seats and go.

Schmitt: Oh, man!

Cernan: *(reaching the gate)* Yes, sir, we got a couple of dented tires!

Schmitt: Okay. My hands have had it.

Parker: Okay; good enough.

Schmitt: You aren't going to get anything else out of me if I keep taking pictures.

Parker: And, Gene, what's a "dented tire"?

Cernan: A dented tire is a little, oh, a little golf-ball size or smaller indentation in the [wire] mesh. How does that sound to you? Doesn't hurt anything.

Parker: That sounds like a dented tire; that's how it sounds.

Schmitt – "I think we're both sounding tired after working on this slope. There isn't as much joking as when we first got there. We're quieter."

Schmitt: *(to Gene)* What did you do with that [duct] tape?

Parker: Let's worry about it at Station 7, if we're going to worry about it. Press on.

Schmitt: Okay.

Cernan: Yeah. Let's forget it now. It's too hard to work on here, and it's not going to take just a minute. It's going to take too much time.

Schmitt: I'm not sure I can get back on here.

Parked near the boulder, the Rover is tipped about 20 degrees – with Jack's seat on the downward side. Jack looks downslope to the south, assessing the situation and speculating about whether Gene should drive to a more level spot before he tries to get on.

Cernan: Well, let me give you a hand.

Schmitt: We need any a ... We don't need any ...

Cernan: No.

Schmitt: Nothing. As a matter of fact ...

Cernan: I can drive, Jack.

Schmitt: Why don't you drive down and get so you're not ... You can get on ...

Cernan: You can go downhill very easy.

Schmitt: Yeah.

> **Cernan:** Okay. Let me get the TV; The battery covers are closed.
>
> **Schmitt:** Let me carry ...
>
> **Cernan:** Why don't you just go down there.
>
> **Schmitt:** ... I'll carry the Rover sampler, just in case.

Jack extracts the sampler, which is stowed vertically next to the control panel.

> **Cernan:** Got it? Okay. I'll get that out of your way, too.
>
> **Schmitt:** Okay. I'll head down to that ... Actually, I'll sidehill over to those boulders right over there and then see if that's any change [in rock type].

Jack starts downhill; Gene leans over the right-side battery covers.

> **Cernan:** Okay. You might, if you get another sample, a large sample, you might grab it, and we'll throw it in the footpan here.

Gene wants to get a large rock that they can use in the closeout ceremonies back at the LM. The plan is to split the rock into dozens of pieces for distributions to museums around the world.

> **Cernan:** And I'll see if I can't find a level spot to ...
>
> **Schmitt:** I sort of ought to have my scoop, too.
>
> **Cernan:** ... help you get on. No, don't take too much; just take that [LRV Sampler]. That's all you need.

Gene goes around the front of the Rover; Jack has only gone a few feet to the southeast.

> **Schmitt:** How about letting me have your hammer, then?
>
> **Parker:** Okay; and, 17, can you verify that the gnomon is back in the Rover?
>
> **Cernan:** Gnomon is on the Rover. The TGE is on the Rover.

Schmitt – "One of the things that bothered me physically was doing a lot of sidehill traversing at Station 6. On going east on the south-facing slope for some distance, the inside ligament on my left leg began to feel fatigued because the right leg was so much down and the left leg was so much up. Most of the stress was on the left leg, and it is the same ligament that I had damaged skiing years before in '58 or '59; and I suspect that that was the reason for that. It was a pretty bad pull and, after it recovered, I never felt it again except for going sidehill right in this part of the traverse. I started to tire and I could feel that ligament. It wasn't bad, but I could tell it wasn't normal."

Cernan – "About 5 or 6 weeks before the flight, we were playing softball and I pulled my planteres [a small calf muscle associated with the Achilles tendon]. I was running between second and third base and I thought somebody had hit me in the bottom of the right leg with a machete. And I just doubled up. I was on crutches for about a couple, 3 or 4 days. If it was ripped, I was done (that is, off the flight). It was strained or pulled but not ripped, I guess, and I had some great concern as to whether the doctors would consider it recovered enough for me to fly. But I guarantee you that, less than a week after I was off the crutches I was in a suit doing EVA training. Hurting like hell. And I really wondered how it would be on the Moon. When we did go, I could feel it the whole time I was doing some of these things, but never to the extent that it either hobbled me or cut down my mobility. It just hurt. It had healed very, very quickly and I'm not so sure … And it sounds a little crazy; but I think part of your physical well being depends on your mental well being. And I'm not so sure that, mentally, I didn't want to fly that flight so bad that it either healed faster or I didn't know it hurt. There were times during the surface activities that I did feel it; but I don't remember feeling a thing during the stuff we've just gone through at Station 6. Also, it's possible that, subconsciously, the injury may have influenced my choice of hopping and skipping rather than Jack's long, loping stride. When you were hopping, you had to plan on maintaining a special kind of balance each time you came down, but you had two feet supporting you all the time. I'm not sure, but maybe I did hop to make my leg feel more comfortable."

Schmitt: The rake.

Cernan: The rake is on the Rover. The scoop's on the Rover. We got the … You put the core under your [seat] pan, right?

Schmitt: Yep, that's right.

Cernan: Okay. I'm going to power up and see if I can't come down and get you. *(pause)* It's fun walking downhill! Boy, that boulder track is impressive.

Parker: Okay; and, 17, when you get moving …

Cernan: [The track is] very symmetrical.

Parker: … we want to get, and I quote, "a maximum variety of hand samples with a minimum amount of documentation, in a minimum amount of time at Station 7." It's just an attempt to see what kind of variety we can get along the face of the front. Over.

Cernan: Roger. *(pause; static)*

Cernan – "I'd like to reminisce with Jack – a little bit – because, for what it's worth, we're sort of authenticating history, 19 years later. It wasn't until Eric and I sat down in my office and watched the films and listened to our voices that I recalled the steepness of that slope at Tracy's Rock. Because Al Bean painted the picture, I've seen the hill a million times; but, in the years since we were there, I've never fully appreciated that you and I were walking at an angle on the side of the hill and that getting on and off the Rover was typical of the problems it caused."

Schmitt – "I think it was one of the USGS guys who said that it was a 20-degree slope where we were parked. And there was a break there. The boulder probably hit and broke when it hit that break in slope and, above us, it was probably 25 or 26 degrees."

Cernan – "I recall climbing up there to do that pan, and being way above you. And I think this is important so that people can know what your ability is to maneuver around on slopes. And I don't think anyone else ever really encountered . . . I don't know what they did on Apollo 14 when they went up Cone Crater . . ."

Schmitt – "Apollo 14 didn't have a 20-degree slope. They had enough to make them work [about 10 degrees], but they didn't have as good a suit. They didn't have the new waist joint."

Cernan – "I just wanted to mention this. It was the one thing that impressed the devil out of me when we reviewed it. The still pictures and a lot of the TV don't show it, but it was really steep."

Schmitt – *(laughing)* "That's because the camera was always tilted 20 degrees."

Cernan: Okay. Well, I'm not sure I can get on without ending up in your seat.

Gene's side of the Rover is the upslope side; as he literally jumps up and pulls himself sideways onto his seat, he'll need to be careful not overshoot his mark and, perhaps, end up with his head on Jack's seat.

Schmitt: Need some help?

Cernan: No *(garbled)*.

Schmitt: I shouldn't have left.

Cernan: No, no. I don't need any help. I'll get on.

Cernan – "For Jack to have gotten on that downhill seat, even if he'd jumped up there he probably would have fallen back off. For me, it was much easier to get on the uphill side but it was steep enough that it really did feel like I was going to end up in Jack's seat."

Schmitt – "It wasn't worth the effort for me to try to get on. By this time, we didn't have any problem with walking some distance."

Cernan – "The only thing about it is that Jack really did enjoy riding on the downhill side."

Schmitt – *(laughing)* "That might have influenced my decision as well."

Cernan: *(static clears)* I probably ought to turn my water off of Max if that's where it is. It's cold. I don't want to run out today. *(long pause)* Well, the roll indicator says 15 degrees; and the pitch indicator says about 12 [giving a total slope of nearly 20 degrees]. I don't know if I believe all that. Bob, you with us?

Parker: Go ahead. Right. We're with you.

Cernan: Okay. I'm rolling.

Finally, nearly 5 hours after they started the drive to Split Rock at the start of the traverse, Gene and Jack were back at the LM with their geologic treasures from Station 6 and three other very productive stops on the way. Soon, they would re-join Ron Evans in orbit and, 1 day later, would be on their way home. It might be a long time before anyone returned to the Moon, but the crew had collected a fine group of lunar materials for scientists to use as they probed the Moon's history.

THE STORY OF SPLIT ROCK

The cluster of boulders at Station 6 was most likely a single boulder before it broke apart and came to rest at a change in slope. The pieces and the variety of rock types in them can be put together like a puzzle. The pieces are large, ranging in size from a Hummer SUV to a suburban bungalow. Before it broke apart, the boulder that broke off an outcrop high on the North Massif was about 18 by 10 by 6 meters (59 by 33 by 20 feet). It wouldn't have been healthy to get in its way as it rolled because this boulder weighed about 125,000 tons (if weighed on Earth).

Jack and Gene successfully sampled all of the rock types visible in these boulders and took many photographs to document each sample. After the samples and the film were returned to Earth, it was the job of the Lunar Sample Preliminary Examination Team to quickly characterize each sample and to link them to the appropriate boulder photos. Author Heiken recalls, "We were struggling with the quality of photographic prints until Jack Schmitt asked, 'Why don't you simply use the first-generation color transparencies?' This simple solution had never occurred to us because the NASA photo lab guarded the film as national treasures – which they were and still are. Carrying enlarged prints and pens, Jack and I donned nylon suits, hats, and clean cotton gloves and entered the inner sanctum of the photo lab. There, working on a light table with magnifying lenses, we were able to work with the original Hasselblad transparencies. It was a successful week. Jack described not only what he saw on the film but also what he had observed at Station 6. I looked at the transparencies as well and, after discussing what we saw, transferred our observations to the large prints. It served as a wonderful framework within which we could interpret the later laboratory analysis of the rock samples."

The samples were a challenge for the geologists, many whom had a lot of experience looking at deposits from large impact craters on Earth. All of the rocks at Station 6 are impact breccias composed of angular rock fragments bonded together by everything from glass to finely crystalline rock. These angular rock fragments – all rock types that might be expected in the highlands and the Moon's crust – were deposited after the cataclysmic Serenitatis impact, along with rock melted by the impact(s). The layering seen in photographs of the boulders at Station 6 is related to variations between rock units, depending on how many rock fragments were encased in the glassy impact melt. The speed at which the melt was quenched (cooled) to glass depended on these rock fragments: if there were many, they acted as "cold sinks," quickly cooling and bonding the fragments with the quenched glass; if there were fewer fragments, the melts had more time for cooling and crystallization and the result was a finely crystalline rock. This difference in timing to reach thermal equilibrium is only about 100 seconds – a few seconds and you have a glassy rock, but more than 100 seconds and the product is a crystalline rock.

By using various techniques for determining the age of rocks, scientists have determined that the impact debris deposit surrounding the Moon's Serenitatis basin is between 3 and 4 billion years old and that the debris sampled by the astronauts is most likely representative of the age of the basin-forming impact. Then, by analyzing physical effects that occur on the surfaces of both the rock and regolith after exposure to radiation from space, it is possible to conclude that the boulder rolled downhill about 22 million years ago, a mere moment in the Moon's history.

Studies of the Moon are crucial to our understanding of the Solar System's early history. We also are discovering valuable clues about the variations of meteor and radiation flux through time and can project them into the future as we design craft and habitats for space explorers. This is impossible to do using information gained solely from Earth-based observations, in part because we are shielded by an atmosphere that screens out much of the radiation and most of the smaller bits of rock. In addition, plate movements have erased much of the Earth's early history. Less dynamic in that sense (with a rigid crust and no plate movements), the Moon serves well as a detector for both radiation and the debris that has traveled through the Earth's neighborhood. The Apollo Program has provided us with invaluable first-hand descriptions and photos as well as geologic samples and geophysical data from stations established at several lunar sites.

Apollo gave us new perspectives on our place in the universe. The scientific ground truth gained by our exploration of the Moon during Apollo became the cornerstone of our growing understanding of the Moon, its history, and the Solar system as a whole. If the earlier Gemini and Apollo missions proved that man could survive during long spaceflights, Apollo missions 11, 12, 14, 15, 16, and 17 increasingly demonstrated that, with appropriate technology, human beings could adapt to the lunar environment. Ever since our remote ancestors began to spread out of Africa, we have used an ever-expanding kit of tools and skills so that we could live and work in virtually every environment on Earth. Apollo provided our first lessons in living and working on the Moon

Gene Cernan (left) and Jack Schmitt captured their shared sense of jubilation in the photos they took when they were back in the cabin after their final EVA. Theirs was the longest and most productive of the Apollo missions and, if much of that success was built on lessons learned from the prior missions, the combination of their talents and cross-training was also very important. Gene, a naval aviator with an engineering background, learned to be a capable field geologist and was able to recognize key features of the lunar site geology. Jack was a professional geologist who had become a capable co-pilot, knowledgeable about the LM systems, the Rover, and other lunar surface equipment. Their skills meshed well and, together, they made the most of this final – at least during Apollo – opportunity for lunar exploration.

One of the clearest photos of the full Moon was taken at the Lick Observatory. The six Apollo landing sites are indicated with dots, each much larger than the area explored at that location during Apollo. There is a lot of territory yet to be explored – including the Moon's entire far side!

9

Lessons from Apollo for Future Operations on the Moon

In 1910, the great polar explorer Roald Amundsen arrived in Antarctica far better prepared than anyone ever had been before. He had experienced conditions at the edge of the continent once previously, when sailing as first mate on an 1897 Belgian expedition that was poorly organized and – like so many prior expeditions to both the Arctic and Antarctic – nearly ended in tragedy. Determined to make his mark in the annals of polar exploration and to avoid the mistakes of the Belgian expedition, Amundsen spent the next decade learning how to live and work in polar conditions and how to put together an expedition that could achieve a well-defined goal. He spent time in the Arctic learning about such practical issues as suitable clothing and how to handle sled dogs from the peoples who live there. Although he was well prepared when he returned in 1910, Amundsen spent his first southern summer establishing depots for the following year's attempt at the pole and, in the process, evaluating his plans, his equipment, the members of his team, and himself. He also had with him a small library containing the available accounts of previous polar expeditions and read these repeatedly to reduce the chance that he had overlooked or misunderstood some important piece of information. As a consequence of this intense preparation, Amundsen made the trip to and from the Pole look easy.

The Apollo astronauts made their lunar surface explorations look easy, too. However, unlike Amundsen, Captain James Cook, and other explorers, Apollo crews had unprecedented support: thousands of people who designed and built the equipment they used on the lunar surface, organized the training exercises that let them make the best possible use of their brief time on the Moon, or devised simulations that let them become familiar with some aspects of the lunar environment. In addition and very importantly, television allowed later Apollo teams to watch the earlier crews in action giving them confidence in what they planned to do during their own missions and to get a feel for what worked and what didn't work. In some ways, post-mission debriefing sessions and sample analysis were the equivalent to the explorers' daily logs; they allowed the astronauts and support personnel to fine-tune the training and

Figure 9.1. This group portrait of Apollo astronauts was taken during a June 1964 geology field trip at the Philmont Scout Ranch in northeastern New Mexico. At this early stage of Apollo, NASA had a great deal to learn about preparing professional pilots to be field geologists but, in the final analysis, most of those who walked on the Moon did outstanding fieldwork. From left to right, Pete Conrad, Buzz Aldrin, Dick Gordon, Ted Freeman, Charlie Bassett, Walt Cunningham, Neil Armstrong, Donn Eisele, Rusty Schweickart (behind Eisele), Jim Lovell, Mike Collins (partly hidden behind Lovell), Elliot See, Gene Cernan (behind See), Ed White, Roger Chaffee, Gordon Cooper, C.C. Williams (behind Cooper), Bill Anders, Dave Scott, Al Bean. In 2005, planetary scientist Jim Scotti noted, "It's a bit sad looking at this image to note that 6 of the 20 gentlemen in it didn't even live to see the first moon landing; and 6 of the 20 left footprints on the Moon. They had an equal chance of dying before the first landing or walking on the Moon." As of early 2007, we have lost three of the moonwalkers and, even for the survivors, memories of the details are fading. Although many of the astronauts wish that time had been allowed immediately after the missions to focus on lessons learned, Apollo remains the best-documented exploration program in history.

Figure 9.2. Dave Scott and Jim Irwin have arrived at Apollo 15's Station 2 on the flank of Mt. Hadley Delta, just down slope from St. George Crater. Dave is setting the gnomon down next to the only sizeable boulder they'd spotted during the approach: a 1-meter, glass-coated breccia. The quality of the TV images – although not as detailed as this Hasselblad image – and the fact that the TV could be operated remotely from Earth, allowed both mission controllers and the scientific support teams to be active participants in the exploration, providing better interaction with the crew than on prior missions

simulations, adapt equipment to the field environment, and improve maps and visual aids – vastly increasing capabilities and productivity from mission to mission.

APPLYING THE LESSONS

In the years since the end of Apollo, NASA has learned – from Shuttle, Mir, and International Space Station operations – a great deal about living outside the Earth's environment and about the support structure needed for such exploration. New methods for task scheduling, ground support, training, and the overall conduct of longer duration operations has been accompanied by advances in spacesuit construction, food supplies, crew comfort, and medical treatment. We have learned a great

Figure 9.3. Apollo 15 EVA CapCom Joe Allen gestures as he talks with back-up Commander Dick Gordon and Deke Slayton in the Mission Operations Control Room in Houston. The live TV image in the background shows Dave Scott and Jim Irwin sampling at the rim of Hadley Rille. The crews were necessarily the center of attention then and now; however, a large part of Apollo's success was due to the crucial real-time partnership between the crews and mission support personnel in the Control Room, various NASA facilities, and tracking stations around the world.

deal about both the physiological and psychological adaptation necessary for living and working in space. Equipment too has evolved along with digital technology and lighter materials that have greater tensile strength. How some of these improved features will fare in the lunar environment remains to be seen. Apollo still represents our only experience of working in 1/6 gravity and a far-from-pristine environment.

During Apollo, NASA put little effort or thought into how they could use what was being learned about lunar surface operations for possible applications in future space programs. In those early years, the strongest focus was, quite necessarily, on achieving the first landing. By the time that goal had been reached, it was becoming more and more obvious that Apollo would not continue beyond Apollo 20 at most; by the time came to prepare for the Rover missions, it wasn't certain that even Apollo 17 would be flown. Out of operational necessity – and because political support for further lunar missions was subsiding – many lessons learned from mission to mission were important only to the extent that valuable crew time on the remaining missions could be put to its best use, wasted effort could be avoided, and confidence could be gained by adding new capabilities. This was an eminently practical approach at the time. None of the crews spent more than 3 days on the Moon or ventured farther from the LM than they could walk in 3 to 4 hours if the Rover failed. The critical lunar

	Apollo 11	Apollo 12	Apollo 14	Apollo 15	Apollo16	Apollo 17	All Six Missions
Commander	Neil Armstrong	Pete Conrad	Al Shepard	Dave Scott	John Young	Gene Cernan	
CMP	Mike Collins	Dick Gordon	Stu Roosa	Al Worden	TK Mattingly	Ron Evans	
LMP	Buzz Aldrin	Al Bean	Ed Mitchell	Jim Irwin	Charlie Duke	Jack Schmitt	
EVA CapCom	Bruce McCandless	Ed Gibson	Fred Haise	Joe Allen	Tony England	Bob Parker	
CM	Columbia	Yankee Clipper	Kitty Hawk	Endeavour	Casper	America	
LM	Eagle	Intrepid	Antares	Falcon	Orion	Challenger	
Launch Date	16 Jul 1969	14 Nov 1969	31 Jan 1971	26 Jul 1971	16 Apr 1972	7 Dec 1972	
Time	13:32 GMT	16:22 GMT	21:03 GMT	13:34 GMT	17:54 GMT	05:33 GMT	
Lunar Landing	20 Jul 1969	19 Nov 1969	5 Feb 1971	30 Jul 1971	21 Apr 1972	11 Dec 1972	
	20:18 GMT	06:55 GMT	09:18 GMT	22:16 GMT	02:24 GMT	19:55 GMT	
Lunar Stay	21 hr 36 min	31 hr 31 min	33 hr 31 min	66 hr 55 min	71 hr 02 min	75 hr 00 min	280 hr 35 min
Splashdown Date	24 Jul 1969	24 Nov 1969	9 Feb 1971	7 Aug 1971	27 Apr 1972	19 Dec 1972	
Time	16:51 GMT	20:58 GMT	21:05 GMT	20:46 GMT	19:45 GMT	19:25 GMT	
Lunar Orbits	30	45	34	74	64	75	322
EVAs	1 EVA	2 EVAs	2 EVAs	3 EVAs	3 EVAs	3 EVAs	
Time Outside LM	2 hr 24 min	7 hr 29 min	9 hr 23 min	18 hr 33 min	20 hr 12 min	22 hr 05 min	80 hr 6 min
Maximum Distance from LM (km)	0.06 (Armstrong)	0.45	1.4	4.8	4.6	7.6	7.6
Distance Walked or Driven (km)	0.12 (Armstrong)	1.3	3.0	27.9	27.0	35.0	89.9
Lunar Samples (kg)	21	34	42	77	96	110	381
Lunar Surface Photographs	339	583	401	1151	1787	2237	6498
Surface Equipment (kg)	102	166	209	550	563	514	2104

surface gear – the LM itself, the suit and backpack, the Lunar Rover, and the communications gear – were designed, built, and tested to ensure a high degree of reliability for only that week's use. Few individual pieces of equipment flew again on a later mission. And, other than the rock boxes and some items brought back as personal souvenirs, virtually all of the lunar surface gear was left behind to save room and weight on the return trip.

As we begin to think about more ambitious, future lunar operations, with crews spending weeks or months at a time on the Moon, detailed consideration of how best to live and work on the Moon becomes more important. Equipment will have to be more durable; the crews will be required to do more servicing and repair work than was possible during Apollo; and more attention must be paid to work schedules, daily debriefings, and day-to-day planning within the particular context of lunar surface operations.

The authors stress that what follows here is a mere introduction to the general subject of lessons learned from Apollo: it is not intended to be complete or authoritative. But future exploration programs will certainly discover that the issues discussed briefly below – and the many others not mentioned – will require the same composite of expertise, careful analysis, innovation, ingenuity, and dedication that was the hallmark of the Apollo program in its time.

A NEW ENVIRONMENT

> *"Boy, this is so neat."*
> – Charlie Duke, Apollo 16

Working on the Moon is fun. The 1/6 gravity gives our Earth-trained brains and muscles a long time to react. Drop a golf ball from shoulder height – say, 1.4 meters – on Earth and it will hit the ground in just over 0.5 seconds; on the Moon, 1.3 seconds. Drop something on the Moon, and you have time to grab it before it falls very far. And, if *you* start to fall, you have plenty of time to react and, usually, either break the fall or get yourself into position for a safe landing on hands and knees. In 1/6 gravity you can throw things a long way and, when you run, you float along above the ground for what seems like a very long time. It's fun to imagine that in a century or two, after we have large indoor facilities, the action in ballet, basketball, and new, uniquely lunar sports, will be spectacular.

Adapting to One-Sixth Gravity

All of the moonwalkers adapted to lunar gravity quickly and easily. About 1 minute after stepping onto the lunar surface for the first time, Neil Armstrong said, "There seems to be no difficulty in moving around – as we suspected. It's even perhaps easier than the simulations of 1/6 g that we performed in the various simulations on the ground. It's absolutely no trouble to walk around." He and Buzz Aldrin were fairly cautious about moving quickly but, as Pete Conrad said about his experience,

Figure 9.4. "Never go exploring without a roll of duct tape." In the Backroom during Apollo 17, (left to right) John Young, Charlie Duke, Deke Slayton, Rocco Petrone, and Ron Blevins discuss a replacement fender devised for the Rover. Gene Cernan had accidentally caught the handle of his geology hammer under the fender and torn off the rear extension. This jury-rigged replacement would use spare maps and duct tape the crew had on-board. After Young did a trial installation in a pressure suit to test the method, Houston passed on the instructions to Cernan and Schmitt for making their own replacement fender. During the EVA-1 closeout, CapCom Bob Parker told the crew, "We'll take a look at it here while you're sleeping." Incidents of this sort illustrate the importance of the support teams as well as ingenuity in using materials at hand. The Apollo crews didn't have the time or tools to do any but the simplest of repairs. Such capabilities will be important on the longer duration missions of the future.

"Of course, we knew that it was going to work, because Neil and Buzz got around okay." He said that adaptation took no more than 10 minutes. By the time the last two crews flew, they weren't even taking that long. Charlie Duke tells us, "We were the fifth landing and we had all the experience of the other guys and watching them on the tapes and seeing that they didn't have any problem. And we practiced in the 1/6-g airplane, which was real [good training]. And after practicing there and watching what the other guys had done, we just knew we were going to be okay. That the training was good." And Gene Cernan adds, "You adapt very, very quickly. You very quickly realize, probably in the first couple of minutes, that you don't need to take baby steps or regular steps to get anywhere. Somehow your brain and your body coordinate your movements and, if you're going to go any distance – 10 or 12 or 15 feet or further away – you start skipping or hop-skipping to get where you're going. And it's not like you start running. It's just that you move with such ease. Later on, when you start moving at faster paces than we were doing here [early in the first EVA], if you decide to turn or change directions you have to think about your high center-of-mass and plan how you're going to handle that – or you're going to go tail-over-teakettle. But you adapt very readily, very quickly, physiologically and psychologically. You're conscious, as soon as you're on the surface, that you're in this 1/6-g environment and that you can move around so much more easily. I don't think we ever really said anything about it, but in 5 or 10 minutes you knew in a general way what you could do and what you couldn't do in terms of using 1/6 gravity to your advantage. Within the first few minutes, after we got the Rover down, we just picked it [the Rover] up – one of us on each end – and turned it around. The human being is a very unique, very adaptable creature."

On the longer missions, the crews had the advantage of the waist bellows, which made torso movement far easier, and enough time on the surface to become more efficient. Neither Scott nor Irwin ran any appreciable distance, but according to the Apollo 15 Mission Report, the walking speed for both astronauts increased from 1.0 feet/second (1.1 kilometers/hour) to 1.5 feet/second during the first EVA and from 1.5 feet/second to 2.0 feet/second during the second EVA; having achieved that rate, there was no further increase on the third EVA. Running speeds were much higher. On Apollo 11, Armstrong ran about 60 meters out to East Crater to take a partial pan and averaged about 3.6 kilometers/hour each way but, by the time of Apollos 16 and 17, Duke and Schmitt were reaching speed of 5 kilometers/hour over comparable distances.

Hard Work

> "Going to stop (drilling) for a second, Bob."
> – Gene Cernan, Apollo 17

Working on the Moon can sometimes be difficult and tiring. We've seen some examples in each of the missions: gripping objects for extended periods of time, repetitive motions such as using the hammer or operating the jack to extract the deep core, carrying the ALSEP from the LM to the deployment site, climbing slopes,

Figure 9.5. This TV screenshot shows Charlie Duke running back to the LM from Station 10. With both feet off the ground, he appears to "float" between steps in a very unearthly stride, and we can see the spray of dust he kicks up with his left boot as he launches himself forward. Each of the astronauts quickly learned to move efficiently on the lunar surface, but their choices were quite individual: Charlie, Ed Mitchell, and Gene Cernan tended to use a skipping stride; Jack Schmitt and the others preferred a loping, foot-to-foot stride. *(screen shot courtesy of Ken Glover)*

and – for the crews who did not have the waist bellows that allowed later crews to sit on the Rover – running more than a few meters, even on a reasonably level surface. For many of these activities, the effort required to bend various parts of the suit made the work far more complicated and arduous than it would have been in shirtsleeves.

Jack Schmitt, Charlie Duke, and others commented on how tired their forearms became from gripping things. Dean Eppler, who evaluates advanced suit designs and EVA equipment under field conditions on Earth for NASA, comments, "The times I have deployed our mocked up experiments package – in both the Mark III suit and wearing what my Edwards friends refer to as 'ramp pajamas' – the fatigue levels are higher in the suit, both from working things in gloves (even though our gloves are immensely more flexible than they were on Apollo) and just generally carrying things. The other thing that I think is worth mentioning, but somewhat hard to quantify, is just the physical and psychological effort associated with working in a pressure suit. Most suit/PLSS combinations turn out to be around 200 pounds or 95 kg, and while the suit does carry a bit of it's own weight (and weighs a lot less on the Moon), you still have to get it moving, change directions, or stop; so you are generally working against the inertia of the suit."

The ALSEP packages weighed up to 240 pounds (109 kilograms) on Earth and 60 pounds (27 kilograms) on the Moon. Even in shirtsleeves on Earth, carrying that

Figure 9.6. Al Shepard pulled the MET during the climb up Cone Ridge in this frame from an up-Sun pan Ed Mitchell took during a 2-minute rest stop. At this point, they were climbing an approximately 15-degree slope and stopped for another rest several minutes later. With the possible exception of the Apollo 15 core extraction, the climb to Cone may have been the hardest work done by any of the Apollo crews. The J-mission astronauts had a decided advantage in the Rover: although their traverses were far longer, they could rest between geology stops and carry far more samples and gear.

weight over a distance of 100 meters would be tiring. Add the stiffness of the suit and the problem of controlling 109 kilograms of inertia, and it is easy to understand why the LMPs had to rest frequently while making the trip to the deployment site. During Apollo 14, the hardest work was climbing toward Cone Crater; in that case, the soft surface and the lack of waist bellows may also have been important factors.

Dust

> *"Boy, everything is stiff. Everything is just full of dust. There's got to be a point where the dust just overtakes you, and everything mechanical quits moving."*
> – Gene Cernan, Apollo 17

The Moon is dusty. The lack of an atmosphere means that the surface has been subject to a steady rain of impactors, large and small – a process that over the last few billion

Figure 9.7. Dave Scott's shot of the Apollo 15 Rover from the right rear at the end of EVA-3 shows a lot of dust: caked on the back of the Rover next to the wheel, on Jim's glove and sleeve, and even on Dave's Hasselblad, in the form of a smudge at the center of the lens. Dust remains one of the chief problems of lunar exploration: fouling equipment, abrading gloves and tool grips, scratching visors, obscuring readouts on instruments, and increasing heat absorption.

years has created a layer of fine particles a few meters thick. Similarly, the lack of an atmosphere means that other forms of erosion so important on Earth simply don't happen on the Moon. Consequently, many of the dust particles are angular and sharp. Although the astronauts describe lunar dust as feeling "greasy," it is actually very abrasive. It gets into everything, clogging Velcro, tools, and latches. Woven materials become impregnated with dust and, as the cloth stretches, compresses, and twists during use; the tiny rock fragments nick the fibers and weaken the fabric. Enough wear was seen in the Apollo 15 gloves that later crews wore fingerless cover gloves over their regular gloves to reduce wear.

Dust on a reflective surface can increase the amount of heat absorbed from sunlight and can lead to equipment overheating and failing. In the case of the Apollo 17 SEP experiment, the failure of adhesive backing on Velcro strips that secured protective flaps over the thermal radiator allowed dust to accumulate, which degraded radiator performance during the rest periods. The crews used large dust brushes to keep the suits and equipment clean; and, before they climbed back in the cabin, they stomped their feet on the LM ladder, which did seem to remove a lot of dust from the suit legs and boots. Finally, they resorted to putting the suit legs in bags once they were inside in an effort to get some control of the dust. Future landers and habitats will probably require some kind of a "dustlock," where suits, equipment, and containers can be cleaned and where some items can be stowed instead of being taken inside. After Apollo 17, Jack Schmitt remarked, "One of the big issues for any kind of permanent lunar habitat is going to be dust control. And it just occurred to me that the design of the air/dust lock ought to be such that, when you depress to go out on an EVA, the air currents remove the dust you brought in the last time. Never thought of that before, and I've been dealing with this – along with a lot of other people – for a long time. I ought to patent that idea!"

Cables Underfoot

"Boy, you really have to be careful of these cables, don't you?"

– Al Bean, Apollo 12

Cables were a snaky problem on all the missions. Because of the bulkiness of the suit and, particularly, the chest-mounted RCU and cameras that protruded in front, the astronauts couldn't see their own feet. On the early missions, there were cables connecting the TV or the big S-Band antenna to the MESA, and, from Apollo 12 onward, there were cables connecting the various experiment packages to the ALSEP Central Station. These cables were stowed in the LM several weeks before mission launch and, like a garden hose left coiled in the garage through the winter, acquired some "set." As the astronauts were deploying the TV and experiments on the Moon, the cables still retained some memory of that "set." During the Apollo 16 debriefing, John Young said, "If the cables are way up off the ground, you never knew whether you were stepping on them or not. When you are standing in 1/6 gravity with (an RCU) on, you're looking about 3 to 4 inches in front of your toes – unless you make a positive effort to look over at them. Every one of those cables had a memory and were all at some distance off the surface." Couple that problem with the soft, uneven lunar surface, and there were plenty of opportunities to get a boot under a cable. If your partner didn't happen to notice, it could mean a fall and/or damaged equipment.

Aldrin: Watch it, Neil! Neil, you're on the (TV) cable.

Armstrong: Okay.

Aldrin: Yeah. Lift up your right foot. Right foot. It's still … Your toe is still hooked in it.

Armstrong: That one?

Aldrin: Yeah. It's still hooked in it. Wait a minute. *(helping out)* Okay. You're clear now.

Armstrong: Thank you.

Miraculously, there was only one significant cable accident. During Apollo 16, John Young got his foot caught in the cable connecting the heat flow experiment to the Central Station and, before he knew what was happening, tore the cable loose and, with it, any chances of getting heat flow data.

A PREDICTABLE ENVIRONMENT

"Can you imagine that, Joe? Here sits this rock, and it's been here since before creatures roamed the sea in our little Earth."

– Dave Scott, Apollo 15

Although the lunar environment is very different from Earth's, it is far more predictable. No wind. No weather. No rain. No storms – except for the solar variety, which is a separate issue. Only the thermal environment varies, but in a very predictable way, as the Sun moves across the black sky.

Judging Distance and Size

On Earth, distant objects appear more obscured, or at least, less distinct (fuzzier) than nearby objects. Dust, water vapor, and other components in the air create a haze, which obviously increases with distance. In addition, small-scale turbulence in the atmosphere – the same process that makes the stars appear to twinkle – gives us a relatively fuzzy view of far away objects. Subconsciously, from the time we are children, we learn to use these subtle clues to estimate distance.

On the Moon, there is no atmosphere, no floating dust, no smog, and no air turbulence. Distant objects look just as clear and sharp as nearby objects. Shortly after landing on the Ocean of Storms during Apollo 12 – as he was trying to figure out exactly where he'd landed – Pete Conrad described a crater "about 5 degrees off to the left of the Sun line of my [LM] shadow. And it is a very blocky rim. Big blocks. Depending on how far away it is, there's some blocks over there that may be 8 feet (2.4 meters)." Pete was thinking that this might be Bench Crater, which is about 300 meters southwest of where they landed and has a diameter of about 35 meters. The crater he was describing was actually 4500 meters west of them, has a diameter of 500 meters and blocks up to 20 meters across. Without any obscuration or fuzziness to suggest distance, the blocks seemed much, much closer than they really were. In addition, there were no familiar objects of any kind in the scene to give a sense of scale. Some of the astronauts seemed to get better at estimating distances after they been on the Moon for a while, perhaps learning to use other clues such as the apparent position of nearby objects relative to distant ones (parallax) as an indicator.

Geologist/astronaut Jack Schmitt made good distance estimates almost immediately during the Apollo 17 EVAs because he had arranged to have Houston let him know how long his shadow was at the start of each EVA. As he said later, his shadow was the only ruler he had with him all the time.

Color of the Moon

Shortly after landing at Fra Mauro, Ed Mitchell commented that "the regolith is mostly a mouse brown – or sometimes looking gray – a powdery material." The Apollo crews described some variation in color relative to Sun direction but, essentially, agreed that – with rare exceptions like the orange soil at Shorty – the Moon is all the same color: gray, with a hint of brown. The brownish tint is due to tiny droplets of metallic iron formed within small blobs of glass that are created when a very small, very fast, impactor hits the regolith. The glass tends to bind to neighboring dust particles, and the resulting collection of material, called an *agglutinate*, is quite common in the regolith. Whatever the slight difference in color, the fact that there isn't much real variation tends to increase to problem of judging distances and picking features out of the general scene.

Visibility – Up-Sun, Down-Sun, and Cross-Sun

> *"We probably ran smack into a small crater – or maybe a boulder – that we couldn't see facing east and looking right into the Sun."*
>
> – Gene Cernan, Apollo 17

Another factor affecting visibility is the bright Sun, especially as it is coupled with the lack of an atmosphere. On Earth, with the Sun directly overhead, the Sun appears roughly 30 percent dimmer – and somewhat redder – than it does on the Moon. This 30 percent figure can vary considerably, depending on the water vapor and particulate content of the atmosphere above a particular location on the Earth, but even with variations, the result is that the Sun is brighter and the glare and reflections can be more of a problem on the Moon than on Earth.

During Apollo, the crews landed early in the local morning with the Sun roughly 13 degrees above the eastern horizon. Looking in the direction of the Sun (up-Sun) was difficult, even with visors in place. The Apollo 15 Mission Report mentioned, "Looking up-Sun, the surface features are obscured when direct sunlight is on the visor, although the sunshades on the lunar extravehicular visor assembly helped in reducing the Sun glare." Scratches on the visor surface further reduced visibility and, when driving the Rover up-Sun, the astronauts tended to tack, following a slightly zigzag course to avoid facing directly into the Sun. Of course, they frequently needed to maneuver around craters and – less frequently – around rocks big enough to damage the Rover's undercarriage, so they rarely drove in a straight line anyway. The problem was worst on the first EVA of any mission but, because the Sun rose about 12 degrees in 24 hours, by the end of the third EVA on the last three missions, the Sun was halfway to the zenith and direct sunlight was much less of a problem.

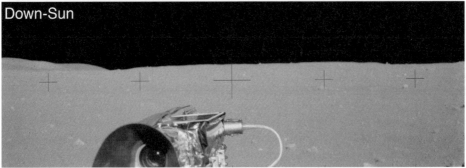

Figure 9.8. These three views, taken by Charlie Duke during the EVA-1 traverse, illustrate the difficulty of driving or walking toward the Sun. Together, they also demonstrate the need to take lunar photographs "cross-Sun." The top picture shows a cross-Sun view taken during the drive to Station 1 from the right-hand seat of the Apollo 16 Rover; it has good definition of rocks and craters. The picture in the middle is a down-Sun view taken just several minutes later and is almost completely washed out. The bottom picture, an up-Sun view Charlie took more than 1 hour later during a stop at Spook Crater, has a lot of glare and contrast that obscures features.

"Man, I'll tell you, (driving) down-Sun is really grim. I was scared to go more than 4 or 5 kilometers an hour. Going out there (to Station 1), looking dead ahead, I couldn't see the craters. I could see the blocks all right and avoid them. But I couldn't see craters."

– John Young, Apollo 16

Driving away from the Sun (down-Sun), particularly on a first EVA, presented a different set of challenges than driving up-Sun. Because the scene is essentially monochromatic, it can be difficult to see the surface rise and fall ahead. In addition, the angular dust grains tend to reflect sunlight most strongly directly back toward the Sun. And, finally, in the down-Sun direction, rocks and craters hide their own shadows. The result is that, directly down-Sun, the surface appears very bright and has a washed-out appearance.

Visibility is best cross-Sun; there is no glare and variations of shadows and brightness reveal the rocks and craters.

CREATURE COMFORTS

In-Suit Dining

"Tell you one thing I'd go for, is a good drink of ice water."
– Pete Conrad, Apollo 12

Working outside on the lunar surface was the hardest work of the missions. As is the case with any hard work, dehydration is always a concern and, except for the first two missions with their relatively short EVAs, all the crews had drink bags inside their suits so they could take occasional sips through a "drink valve" that protruded slightly above the neck ring. The drink bag held approximately 0.95 liters of water. Beginning with Apollo 15, the crews also had a "food stick/fruit bar" attached inside the suit that allowed them to have a bite to eat. Gene Cernan said, "We were going to be locked in the suit for maybe 9 hours and we had a little water bag that we suspended from the inside rim of the suit. The bag hung on your chest and had a little one-way valve on the top of it so that you could turn your head and take a drink. It was like sucking on a nipple. And then we also had one of these high-protein or high-calorie sticks shaped like a ruler. It was a soft stick and you could chew it. We had it inside a little bag, and it was probably about 8 inches long. It, too, was Velcroed just inside the helmet ring and we could put our chin down and pull a little bit out with our teeth, take a bite, chew it for some energy. It was typical candy-tasting stuff. It was nice to be able to suck up a few ounces of water now and then and have something to chew on." In fact, at one place in the mission audio, Dave Scott can be heard chewing.

Before heading out on EVAs, the astronauts drank their fill from the LM water gun – "saturated ourselves" as Gene Cernan described it – and once they were back in and the helmets were off, had another big drink. Although there doesn't seem to be any firm information about how much Shepard and Mitchell drank from the bags

Figure 9.9. Apollo dietician Rita Rapp poses with some of the Apollo 16 food items, including, presumably, the makings of the sticky, potassium-fortified "orange juice" that Charlie Duke managed get in his hair while in lunar orbit. The variety and quality of food available has greatly improved since Apollo.

during each of their 4.5-hour EVAs, Mitchell later said, "I don't remember that I used it all that much but I did take a sip from time to time. It was welcome relief."

With one notable exception, the Apollo 15, 16, and 17 moonwalkers had adequate water during the EVAs. Dave Scott's drink valve slipped out of reach during EVA-2, but he didn't report any problems. Because Apollo 15 EVA-3 was only going to be 5 hours, Scott did not take the drink bag out on that EVA. On Apollo 16, before their 7-hour first EVA, John Young forgot to install his drink bag until it was too late to do so conveniently. After the EVA, he told Charlie that the fact that he didn't have a drink bag was the reason "you saw me dogging it [meaning, going slow] out there." Like Dave Scott, John suffered no apparent ill effects. The dehydration Jim Irwin suffered because his drink mechanism wasn't working and he was unable to get anything to drink during Apollo 15's three EVAs had serious medical consequences, as is discussed below.

Food – and a Difference of Opinion

"Did we mention that we ran out of food there in the LM? We could have used a little more food."

– Jim Irwin, Apollo 15

"We had too much food. Dave [Scott] said they ate everything; but we couldn't have possibly eaten everything."

– Charlie Duke, Apollo 16

Except for the fact that the LM astronauts had no way to heat water for food preparation and had to eat everything cold, they generally ate well. Before each mission, the crews sampled and selected individual menus from the available items: about 100 choices at the time of Apollo 14. Sufficient supplies to provide each crewmember approximately 2100 kilocalories/day were stowed in the spacecrafts on the missions up to Apollo 14. This was increased to 2400 kilocalories per day for Apollo 15 and to 2800 kilocalories for Apollo 16 after Scott and Irwin reported that they would have liked more. After Young and Duke complained that they had enough food "to feed the Trojan Army," the ration was decreased to 2600 kilocalories for Apollo 17. Cernan and Schmitt ate most of what they had with them, but not everything.

The LM crews had two meals per day: one before each EVA and one after. Average consumption – based on logs and verbal reports by the crews – was 2156 kilocalories/day for the 12 moonwalkers, compared to 1868 kilocalories for the 6 CMPs. Not surprisingly, there was considerable variation within these groups. Among the moonwalkers, Pete Conrad and Al Bean on Apollo 12 only averaged about 1700 kilocalories/day; Dave Scott averaged 2800. Among the CMPs, Al Worden on Apollo 15 and Ron Evans on Apollo 17 each averaged about 2400 kilocalories. Data from long-duration stays aboard the ISS show an average intake of 2284 ± 627 kilocalories. Although it might have been expected that working in the Moon's 1/6 gravity field would have required more food energy than working in the Space Station's zero gravity, this doesn't seem to have been the case.

The Apollo astronauts generally lost a few pounds of body weight during their missions. Except for Al Shepard and Ed Mitchell, who each returned to Earth 1 pound heavier than when they left, the moonwalkers lost an average of 6 pounds each and the CMPs lost an average of 7 pounds. Al Bean lost the most: 11.5 pounds. All of the astronauts were conscious of the need to eat enough food and, more importantly, to drink enough water. With occasional encouragement from the ground, they made sure they did.

On a purely practical note, NASA's nutritionists ensured that all of the food items were low-residue so that the LM crews wouldn't have to take the time and trouble to defecate in the 3 days or less they were away from the CM. In his public talks, Charlie Duke frankly – and humorously – describes how difficult, messy, and unpleasant this normal bodily function became in the tight quarters of the CM in zero gravity. Even with the help of 1/6 gravity on the Moon, this was not something the astronauts

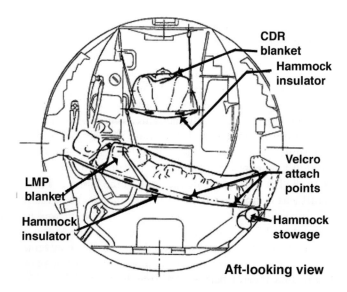

CDR
blanket
Hammock
insulator

Velcro
attach
points

LMP
blanket

Hammock
insulator

Hammock
stowage

Aft-looking view

Figure 9.10. The astronauts rested and slept in hammocks on all the flights after Apollo 11. This planning-document drawing shows the view aft from the front of the spacecraft. In his book about designing and building the LM, Project Manager Tom Kelly at Grumman described the cabin as being "the size of a modest walk-in closet." The hammocks were an effective use of the small space and, especially for the crews who were able to get out of their suits for the rest periods, provided a good night's sleep.

wanted to do in the even smaller LM cabin. Not surprisingly, although they wore diaper-like undergarments while in their suits, the prospect of defecating while outside was even less appealing. The low-residue diet – along with the precaution of defecating before they left the CM – seems to have solved a potential problem. No one has yet defecated while on the Moon. Urinating was less of a problem. While out on the surface in their suits, they each wore a condom-like fitting connected to a collection bag that they emptied after each EVA.

Sleep

With a great deal of activity crammed into a short period of time, the early crews did not sleep well.

Armstrong and Aldrin did not have hammocks, so Buzz made himself as comfortable as possible on the floor and Neil described being "on the engine cover with a loop that I'd rigged up of some kind to hold my legs up." Buzz reported getting about 2 hours of fitful sleep; but, Neil, who was the one being monitored by the flight surgeons, only occasionally had a heart rate low enough to indicate sleep, and never for very long. They were disturbed by equipment noises, by light leaking around the window shades and shining through the alignment telescope eyepiece, and by the cold in the cabin. All these problems were addressed as NASA planned for the later flights.

The Apollo 12 and 14 crews had hammocks but did not get out of their suits. The Commander's hammock was rigged fore and aft just below the overhead, and the LMP's was rigged laterally, just above the floor at the front. After his flight, Shepard commented, "I don't know what you can do to make that rest period more comfortable. There needs to be some place to rest your head. I didn't have a feeling I could put my head on anything." Later, Ed Mitchell added, "The neck rings were part of the problem as they were big, bulky, and clumsy. But a pillow or something to support the head in a more natural alignment with the spine was needed and [to] fill the space between the neck ring and the head." Shepard and Mitchell slept only fitfully, not only because they couldn't get comfortable but also because – having landed with the LM tilted 7 degrees to the left – when they were lying in the hammocks, they had a feeling that the spacecraft was beginning to tip over. They opened the window shades several times during the rest period to assure themselves that there was no problem. It is interesting to note that although the Apollo 15 LM sat on the Moon with a greater tilt than the Apollo 14 LM, Scott and Irwin did not report any adverse sensations. On Apollo 12, Conrad and Bean only got about 3 hours of sleep because, as Conrad said later, "I made a technical error before I left [Earth] when the suits were sent back to ILC [the suit manufacturer] and the boots were put on. We knew that we had to refit the suits, and I let myself get conned into refitting my suit in [while wearing] long underwear and not with the (bulkier) LCG because the flight LCG was [already accepted for flight]. That was a mistake. I wound up with the legs being too tight [that is, too short]. I realized this prior to lift-off [from Earth] while staying in the suit for a long time. I had spent only about an hour or so [in it prior to getting into the Saturn V rocket], getting fitted in my long underwear. [And during the rest period on the Moon], it became unbearable. It spoiled my rest period. I did not want to take the suit off, so I stayed that way all night. I slept only about 4 hours, and it was mainly because of suit discomfort on my shoulders. The next morning, Al did an outstanding job on letting my legs out for me [by readjusting a network of laces], which took him about an hour."

Beginning with Apollo 15, the crews removed the suits before each of the rest periods and that let them get far more comfortable in the hammocks. All six Apollo J-mission moonwalkers had three good sleep periods on the Moon – some had the help of mild medication, but it is clear that doffing the suits made all the difference.

AND SOME DEFINITE RISKS

Radiation Exposure

On Earth, a person's average radiation dosage from all sources is about 0.3 rem/year, although there can be great differences from location to location, depending on elevation, local geology, and other factors. On the lunar surface, the normal background is approximately 5 to 10 rem/year , which compares with the U.S. Department of Energy(DOE)'s limit of 5 rem/year for employees who regularly work with radiation sources. During the lunar missions, each crew member wore a personal radiation dosimeter. The crews reported the readings to Houston once a day. Crew-average exposures ranged from 0.18 rad on Apollo 11 to 1.14 rad on Apollo 14.

Dosage (measured in rem) is not the same as exposure (measured in rad) and the relationship between the two depends on the character of the radiation because some types penetrate deeper into the body than others. Using NASA's recent determination of the relationship between exposure and dosage for conditions aboard the ISS, the Apollo 14 dosage is calculated at about 2.85 rem – well under the U.S. DOE-regulated limit. None of the lunar astronauts has shown any convincing signs of ill effects from these exposures, nor were any expected. The one serious radiation risk that had to be accepted during Apollo was the small chance of solar flare large enough to cause serious medical problems. These occur no more than a few times in an 11-year solar cycle, so the risk was probably smaller than the other risks inherent in these pioneering missions. Only one major solar event occurred during Apollo, during August 1972, which was 4 months after Apollo 16 and 4 months before Apollo 17. Recent estimates by experts indicate that in the relatively shielded environment of the CM, crew exposures would have been roughly 360 rads, and much higher for a crew in the lightly shielded LM. Exposures at these levels would have been very serious and, for unshielded astronauts out on the lunar surface, even fatal.

Future long-duration operations on the Moon will require close attention to solar flare activities and ensuring a well-shielded environment to protect personnel.

Minor Physical Effects

During Apollo, there were a few instances of minor injuries. While pulling the Apollo 15 deep core out of the ground by brute force, Dave Scott badly strained his right shoulder. Characteristically, he didn't bother telling the ground about what had happened, but over the rest of the mission took a total of 14 aspirin tablets to relieve the pain. Perhaps more important than Scott's injury in the overall context of Apollo, he and Irwin spent far more time removing the core than had been planned; for Apollo 16 and 17, a treadle-and-jack were used to make core removal quicker and easier.

Several of the astronauts suffered some damage to their hands, especially their fingers and fingernails, because of abrasion and pressure inside the gloves. Irwin commented, "The wrist ring became a sore point, just because of pressure at that point. And, then, because of the heat and the perspiration, your fingers became immersed in sweat and the tissue receded from the nail and, of course, the nails grew from being immersed in sweat most of the day. They really did grow. So, at one point, I cut my nails back, 'cause I could see that was a problem. I tried to get Dave to cut his nails, but he wouldn't do it. And, as a result, all of his nails turned black. I think he was a little reluctant to cut his nails because he thought he might lose a little dexterity that he might have needed." Charlie Duke added, "You were jammed into the gloves so that you'd have dexterity at the end of your fingers. And, if your fingernails were long, it would tend to hook up and it would tear 'em loose and you'd get blood blisters underneath your fingernails. So, by cutting 'em back, it would eliminate that tendency to hang up and snag. But it didn't seem to have any effect, because we ended up black and blue at the end of the finger tips; and I think it was just pressing hard against the

Figure 9.11. Dave Scott makes some pre-dinner remarks aboard the Apollo 15 recovery carrier U.S.S Okinawa. An inset shows the damage to his fingernails caused by working in the EVA gloves.

suit and working against the suit that caused it." On Apollo 17, Cernan and Schmitt had the same problem.

Medication

The astronauts used a surprisingly small variety of medication during their missions. *Apollo 11*: Armstrong and Aldrin each took 1 Lomotil before the lunar landing to retard bowel movements; Aldrin took 2 aspirins before every rest period to help him sleep. *Apollo 12*: all three crewmen took Actifed tablets to relieve nasal congestion, and Al Bean took a total of 11 during the mission; all three took occasional aspirin tablets; and Bean had 1 Seconal tablet on six occasions to help him sleep. *Apollo 14*: the crew did not report use of any medication other than nose drops on the third day of the flight for stuffiness. *Apollo 15*: aspirin and nose drops. *Apollo 16*: Charlie Duke took 1 Seconal for the rest period before the landing and again before each of the first two rest periods on the Moon. *Apollo 17*: all three crewmen took Seconal for sleep on various occasions, Cernan (3), Evans (7), Schmitt (6); Cernan took a total of 29 tablets of Simethicone, an antiflatulent, and 1 Dexidrine for motion sickness; Schmitt took 4 aspirin and 1 Lomotil, the latter on the trip home; and Evans took 1 Simethicone, 4 Lomotil, 1 Actifed, and also used nose drops.

And One Potentially Serious Situation

Jim Irwin had a fatal heart attack on August 8, 1991, the last in a series of heart problems that started even before the crew got back to Earth. In an interview, Jim indicated that his inability to get a drink during any of the Apollo 15 EVAs contributed to the problems he had during the mission. He said, "I was really dehydrated after a day's work. That's why we [Irwin and Scott] both ended up with heart problems, irregular heart beats, just after we left the Moon."

Post-mission analysis indicated that a contributing factor to the occasional irregular heartbeat experienced by both Scott and Irwin was an unexpected potassium deficiency that had built up during training and the early stages of the flight. Two countermeasures were taken that eliminated the problem on the later flights. First, during outdoor training in the hot Florida sun, the Apollo 16 and 17 crews used water cooling in the suits – instead of air cooling – to prevent overheating and dehydration. And, second, during training and the flights, provisions were made to ensure that deficiencies would not develop. Today's sports drinks contain additives that would

Figure 9.12. Jim Irwin believed that the series of heart problems that he experienced, which began even before he got back to Earth, was caused by the dehydration he endured during all three of the Apollo 15 EVAs when his in-suit drink bag wouldn't work. The later crews received better training on installation of the drink bags.

undoubtedly have improved the wellbeing of many explorers, both Earth and Lunar, but in the early 1970s they had yet to be developed.

However, Irwin's problems were more severe and more complicated than Scott's and the dehydration of three waterless EVAs may well have contributed to his condition. In his excellent book, "A Man on the Moon," Andrew Chaikin writes that, following the flight, doctors told Irwin that "after he and Scott had returned to the command module, he'd had a problem with his heart. At the time, Irwin had no idea anything was wrong; he only knew that he was suddenly very tired." This was the first of the series of heart problems he experienced up until his death in 1991.

During a partial review of Apollo 15 conducted in December 1989, Irwin said, "You know, just going to the Moon gave me a new appreciation for the Earth, an appreciation for life, and new purpose for my life, which I never dreamed would come about as a result of going to the Moon. So my life was greatly changed. Physically, because of all the heart problems that developed; and then psychologically because I realize that people put you in a little different category; and spiritually. So my life has changed greatly. And I don't regret that."

"Even knowing the heart problems, I'd gladly do it again. In fact, I wish I had a chance to do it again, because I think I'm a little cooler than I was then and I think I'm really in better shape now than I was when I made the flight. (chuckling) I was surprised that it looked as if my resting heart rate was up in the 60s and 70s and, now, my resting heart rate is down in the 40s. And I'm 30 pounds lighter than I was when I made the flight. And, back then, I thought I was in good shape. Hopefully, we learn more as we progress through life; but I realize that, man, I could have brought back 30 pounds more of lunar material, if I'd only known that. I'd have been smaller. I would have had a little more room to move around, (chuckling) and I probably could have gotten out the hatch a little easier, too. I would like to go back again, to complete the exploration that we had planned, get up to the Northern Complex, to bring back some of the things that I forgot and left there, and then to see if the experience has as much meaning today as it did 18 years ago."

– Jim Irwin, Apollo 15

Afterword – The Spirit of Exploration

About six decades after Amundsen's intense preparations for his remarkable journey to the South Pole set the standard for all who followed him into the unknown, mankind made its first steps on a world other than Earth.

From a strictly practical point of view, the Apollo crews demonstrated that safe and fairly accurate landings could be accomplished with what we now regard as the frighteningly limited technologies of the 1960s. The astronauts triumphantly proved that human beings could adapt easily to the lunar environment and could do useful work on the Moon. Apollo also confirmed the premise that manned missions could best conduct the systematic investigations of this new environment – returning significant data, meaningful photography, and carefully selected and documented samples that would trigger great leaps of understanding in the emerging disciplines of planetary science.

Important as these achievements were, there was far more to Apollo than these practical aspects. Historians will debate for a long time to come about the role Kennedy's decision and the success of Apollo played in Cold War politics and the eventual dissolution of the Soviet Union. In addition to fueling our country's patriotism, Apollo fired our collective imagination and sense of man's destiny. In July 1969, commentators compared our first steps on the Moon with the first, hesitant forays into Earth's unknown territories by our ancestors. Historically, human cultures have been concerned about humanity's place in the Universe, seeking answers in theology, philosophy, and science. Although it may be centuries before the extent of humanity's future in space becomes clear, the success of our first voyages to the Moon and the ease with which we seem to be able to function in that new environment opens a new chapter in our quest for answers.

In this new century, we will have the opportunity to answer the question Jim Irwin asked: Will our next experience on the Moon have as much meaning as Apollo did nearly four decades ago? Perhaps so. Dave Scott's first words when stepping onto the Moon spoke truth to us all:

"Okay, Houston. As I stand out here in the wonders of the unknown at Hadley, I sort of realize there's a fundamental truth to our nature. Man must explore. (pause) And this is exploration at its greatest."

Man Must Explore

Acronyms

AGC	automatic gain control
ALSCC	Apollo lunar surface close-up camera
ALSEP	Apollo lunar surface experiment package
ALSJ	Apollo Lunar Surface Journal
c.g.	center of gravity
CCIG	cold cathode ion gauge (earlier CCG)
CMP	command module pilot
CDR	Commander
CVSC	core sample vacuum container
DAC	data acquisition camera
DOE	(U.S.) Department of Energy
DSKY	display panel and keyboard
EASEP	early Apollo scientific experiments package (later ALSEP)
EMU	extravehicular mobility unit
ETB	equipment transfer bag
EVA	extra-vehicular activity
FCSD	Flight Control Support Division
FOD	Flight Operations Directorate
GCA	Ground-controlled approach
GM	General Motors, Inc.
HFE	heat-flow experiment
HTC	hand-tool carrier
IAU	International Astronomical Union
ISS	International Space Station
LCG	liquid-cooled garment
LCRU	lunar communications relay unit
L&A	landing and ascent
LEC	lunar equipment conveyor

LEVA	Lunar extravehicular visor assembly
LM	lunar module
LMP	lunar module pilot
LOS	loss of signal
LPM	lunar portable magnetometer
LRL	Lunar Receiving Laboratory, Manned Spacecraft Center, Houston, Texas
LRRR	laser ranging retroreflector (also known as *LR-cubed*)
LRV	lunar roving vehicle (popularly: *Rover*)
LSM	lunar surface magnetometer
MESA	modularized equipment stowage assembly
MET	modular equipment transporter
MOLAB	mobile laboratory
MSC	Manned Spacecraft Center, Houston, Texas (renamed Johnson Spacecraft Center, JSC, in 1973)
NASA	National Aeronautics and Space Administration
O&C	Operations & Control
PBI	push-button indicator
PLSS	portable life support system (backpack)
PRCB	Program Review Control Board
PSE	passive seismic experiment
R&D	research and development
RCU	remote control unit
RFP	request for proposal
RTG	radioisotope thermoelectric generator
SCB	sample collection bag
SEP	surface electrical properties experiment
SEQ	scientific equipment bay
SESC	special environmental sample container
SEVA	standup extravehicular activity
SIDE	suprathermal ion detector experiment
SRC	sample return container
SWC	solar wind collector
SWS	solar wind spectrometer
TCU	television control unit
TELMU	telemetry, electrical, and EVA mobility unit
TGE	traverse gravimeter
USGS	U.S. Geological Survey
UHT	universal handling tool

Suggested Reading

BOOKS

Aldrin, B. and McConnell, M., 1989. *Men from Earth*. Bantam, 239 pp.

Baker, D., 1981. *The History of Manned Spaceflight*. New Cavendish Books, 544 pp.

Beaglehole, J. C., 1964. *Journals of Captain Cook*. Cambridge University Press, 4134 pp.

Cernan, E. and Davis, D., 1999. *The Last Man on the Moon*. St. Martin's, 356 pp.

Chaikin, A., 1994. *Man on the Moon: The Voyages of the Apollo Astronauts*. Viking, 670 pp.

Collins, M., 1974. *Carrying the Fire*. Farrar Straus Giroux, 478 pp.

Compton, W. D., 1989. *Where No Man Has Gone Before – A History of Apollo Lunar Exploration Missions*. NASA History Series, NASA SP-4214, 415 pp.

Cortright, E. M. (ed.), 1975. *Apollo Expeditions to the Moon*. NASA Publication SP-350, 313 pp.

Duke, C. and Duke, D., 1990. *Moonwalker*. Oliver Nelson, 284 pp.

Hansen, J. R., 2005. *First Man*. Simon and Schuster, 768 pp.

Harland, D. M., 1999. *Exploring the Moon*. Springer-Praxis, 411 pp.

Harland, D. M., 2006. *First Men on the Moon: The Story of Apollo 11*. Springer-Praxis, 416 pp.

Heiken, G. H., Vaniman, D. T., and French, B. M., 1991. *Lunar Sourcebook: A User's Guide to the Moon*. Cambridge University Press, 736 pp.

Kelly, T. J., 2001, *Moon Lander: How We Developed the Apollo Lunar Module*. Smithsonian, 283 pp.

Light, M., 1999. *Full Moon*. Alfred Knopf, 232 pp.

Lindsay, H., 2001. *Tracking Apollo to the Moon*, Springer-Verlag, 426 pp.

Mellberg, W. F., 1997. *Moon Missions – Mankind's First Voyages to Another World*. Plymouth Press, Ltd., 195 pp.

Miller, R. and Durant, F. C. III (eds.), 2001. *The Art of Chesley Bonestell*. Paper Tiger, 256 pp.

Murray, C. and Cox, C. B., 1989. *Apollo: The Race to the Moon*. Simon and Schuster, 512 pp.

Orloff, R. W. and Harland, D. M., 2006. *Apollo: The Definitive Sourcebook*, Springer-Praxis, 633 pp.

Schmitt, H., 2006. *Return to the Moon – Exploration, Enterprise, and Energy in the Human Settlement of Space*. Copernicus-Praxis, 335 pp.

Scott, D. R. and Leonov, A., 2004. *Two Sides of the Moon*. Simon and Schuster, 416 pp.

Spudis, P. D., 1996. *The Once and Future Moon*. Smithsonian, 308 pp.

Sullivan, S., 2004. *Virtual LM*. Apogee, 256 pp.

Wilhelms, D. E., 1987. *The Geologic History of the Moon*. U. S. Geological Survey Professional Paper 1348, large format; 302 pp.

Wilhelms, D. E., 1993. *To A Rocky Moon*. University of Arizona Press, 477 pp.

Young, A., 2006. *Lunar and Planetary Rovers: The Wheels of Apollo and the Quest for Mars*. Springer-Praxis, 370 pp.

WEBSITES

[Accurate at the time of this printing: January 2007]

Alan Bean Online Gallery, *http://www.alanbeangallery.com*
Apollo Archives, *http://www.retroweb.com/apollo.html*
Apollo Flight Journal, *http://www.hq.nasa.gov/office/pao/History/ap15fj/index.htm*
Apollo Image Atlas, *http://www.lpi.usra.edu/resources/apollo/*
Apollo Lunar Surface Journal, *http://www.hq.nasa.gov/alsj/*
Eagle Lander, 3D LM Simulator, *http://www.eaglelander3d.com/*
Honeysuckle Creek Tracking Station, *http://www.honeysucklecreek.net/*
Lunar Map Catalog, *http://www.lpi.usra.edu/research/mapcatalog/*
My Little Space Museum, *http://www.myspacemuseum.com/*
National Aeronautics and Space Administration, *http://www.nasa.gov/*
Saturn V Reference Page, *http://www.apollosaturn.com*
Virtual Rover, *http://www.batsinthebelfry.com/rover/index.php*

Sources for Figures

Cover photo   AS-17-142-21811

Chapter 1
Frontispiece   Telescopic photo (Ulli Lotzmann)
Site Map   Apollo 11 EVA map (adapted from *Apollo 11 Preliminary Science Report, Thomas Schwagmeier*)
Fig. 1.1   AS11-40-5864-69
Fig. 1.2   AS11-40-5878
Fig. 1.3   S69-45002
Fig. 1.4   AS11-37-5505
Cameos   Aldrin: AS-11-37-5534; Armstrong: AS-11-37-5528

Chapter 2
Frontispiece   AS12-49-7317
Site Map   (from Lunar Orbiter image)
Fig. 2.1   AS12-46-6726
Fig. 2.2a,b,c   original art (Ulli Lotzmann)
Fig. 2.3   A12.cuffchecklist.tif
Fig. 2.4   AS12-46-6789
Fig. 2.5   AS12-47-6921
Fig. 2.6   69-H-1637fa
Fig. 2.7   AS12-49-7242
Fig. 2.8   AS12-49-7278

Chapter 3
Frontispiece   AS12-52-7596
Fig. 3.1   detail of AS12-52-7506
Fig. 3.2   AS14-68-9486

Sidebar Map A (from Lunar Orbiter photos)
Fig. 3.3   S70-56721
Fig. 3.4   AS14-68-9405
Sidebar Map B (from Lunar Orbiter photos)
Fig. 3.5   AS14-68-9407-8
Fig. 3.6   AS1421s1-eva2
Fig. 3.7   AS14-64-9089
Fig. 3.8   AS14-64-9121
Sidebar Map C (from USGS map)
Sidebar Map D (from Lunar Orbiter photos)
Fig. 3.9   AS14-68-9448

Chapter 4
Frontispiece   Boeing-LRV2A297777
Site Map   Apollo 15 mapping camera M-1537
Fig. 4.1   Original artwork (Jerszy Zulawski)
Fig. 4.2   Original artwork (©Mark Wade)
Fig. 4.3   Life Magazine, June 11, 1971; inset: ap15-S71-30463
Fig. 4.4   71-H-646
Fig. 4.5   AS15-85-11470
Fig. 4.6   AS15-11449
Fig. 4.7   AS15-8-11901

Chapter 5
Frontispiece   AS15-84-11324
Site Map   as15Site.tif
Fig. 5.1.A   lunar drill (from J. Allton, 1989. Catalog of Apollo Lunar Surface
                Geological Sampling Tools and Containers)
Fig. 5.1.B   heat-flow diagram (from Apollo 15 Preliminary Science Report,
                1972, NASA SP-289
Fig. 5.2   S71-37218
Fig. 5.3   AS15-87-1184
Fig. 5.4   AS15-82-11057
Fig. 5.5   AS15-85-11429
Fig. 5.6   15001core

Chapter 6
Frontispiece   AS16-116-18607
Site Map   AS16-4618N1b
Fig. 6.1   AP16-S71-35588
Fig. 6.2   pan: a16.1665537; inset: AS16-106-17255
Fig. 6.3   AS16-117-18736
Fig. 6.4   AS16-106-17336
Fig. 6.5   AS16-116-18649

Chapter 7

Frontispiece	AS17-147-22465
Site Map	Detail from AS17-2309
Fig. 7.1	AS17-137-21010
Fig. 7.2	AS17-137-20990
Fig. 7.3	A17pan1454903
Fig. 7.4	(Scanning electron micrograph images, NASA Johnson Space Center)

Chapter 8

Frontispiece	AS17-140-21368
Fig. 8.1	AS17-141-21550
Fig. 8.2	AS17-144-21991
Fig. 8.3	AS17-141-21568
Fig. 8.4	AS17-141-21598
Fig. 8.5	AS17-140-21493
Fig. 8.6	AS17-146-22294
Cameos	Schmitt: AS17-145-2222; Cernan: AS17-145-22225

Chapter 9

Frontispiece	Lick Observatory photo L18
Fig. 9.1	S64-23847
Fig. 9.2	AS15-85-11435
Fig. 9.3	S74-41836
Fig. 9.4	S72-55170
Fig. 9.5	screen grab from Apollo 16 television of EVA
Fig. 9.6	AS14-68-9422
Fig. 9.7	AS15-82-11200
Fig. 9.8A	AS16-109-17758
Fig. 9.8B	AS16-109-17761
Fig. 9.8C	AS16-109-17820
Fig. 9.9	S72-19887
Fig. 9.10	a12/a12.hammocks
Fig. 9.11	S71-42195
Fig. 9.12	S71-42258

Afterword   Original art (Ulli Lotzmann)

Index

Printing: Mercedes-Druck, Berlin
Binding: Stein+Lehmann, Berlin